OXFORD LOGIC GUIDES

GENERAL EDITOR : DANA SCOTT

ELEMENTS OF INTUITIONISM

BY

MICHAEL DUMMETT

WITH THE ASSISTANCE OF ROBERTO MINIO

CLARENDON PRESS · OXFORD
1977

Oxford University Press, Walton Street, Oxford OX2 6DP

OXFORD LONDON GLASGOW NEW YORK
TORONTO MELBOURNE WELLINGTON CAPE TOWN
IBADAN NAIROBI DAR ES SALAAM LUSAKA ADDIS ABABA
KUALA LUMPUR SINGAPORE JAKARTA HONG KONG TOKYO
DELHI BOMBAY CALCUTTA MADRAS KARACHI

ISBN 0 19 853158 3

Printed in Great Britain
by Thomson Litho Ltd.,
East Kilbride

PREFACE

The purpose of this book is to provide, in a form readily intelligible to one having no prior knowledge of the subject, basic information about the fundamental ideas of intuitionistic mathematics. It is particularly aimed at those who are concerned to acquire an explicit knowledge of intuitionistic logic and the established results concerning it. It does not attempt to give a survey of the intuitionistic reconstructions so far effected of actual mathematical theories, because, of the several excellent books on intuitionism now available, the only one easily readable by someone without prior knowledge is Heyting's *Intuitionism*, which does exactly this. In Chapter 2, therefore, only a very preliminary sketch is given of the way in which the theory of real numbers is developed intuitionistically: just enough to give a sample taste of what an intuitionistic theory is like, so as to motivate the discussion of foundational matters which follows. That is not to say that the rest of the book is concerned solely with logic. There are two basic ideas underlying the whole intuitionistic reconstruction of mathematics. One is the general theory of meaning for a mathematical language, according to which the understanding of a mathematical statement is to be thought of as given in terms of the mental constructions which may serve to prove the statement, rather than in terms of the state of affairs within objective mathematical reality which renders it true or false: it is this which causes the principles of intuitionistic reasoning to deviate from those of classical logic. The second is the characteristic conception of infinite sequences, namely as generated by successive free choices but as never completed. In studying, rather than just using, the notion of a choice sequence, we are concerned with the foundations of mathematics, but not with anything that can be said to belong to logic properly so called.

In keeping with the prime objective of the book, the

tone has been kept informal and explanations have been made as explicit as possible. The objective will have been attained if readers wishing to study the subject further find that the introductory account given here enables them to tackle without difficulty other books with which, without any introduction, they would, at best, have wrestled. Heyting's book, as already remarked, needs no introduction, but contains only a cursory and inadequate treatment of logic and of the notion of choice sequences (it does not even mention Bar Induction): it is highly to be recommended as an introduction to the actual development of intuitionistic mathematics. Bishop's *Foundations of Constructive Analysis*, while written from the standpoint of a constructivism more rigorous than that of the intuitionists (in particular, one that repudiates appeal to the Fan Theorem), is also very useful for this purpose. Troelstra's *Principles of Intuitionism*, while a bit difficult for a beginner, although expository in tone, gives a much more thorough survey of the mathematical foundations than is attempted here, but pays scant attention to logic; while the volume *Metamathematical Investigation of Intuitionistic Arithmetic and Analysis*, edited and in large part written by him, is an encyclopedia of existing results concerning intuitionistic formal systems. His book on *Choice Sequences* in the present series provides an extended philosophical treatment of the notion, which is briefly expounded and commented on in the last section of this book. Kleene and Vesley's *The Foundations of Intuitionistic Mathematics* is a valuable exploration of the fundamental principles of intuitionistic analysis, rendered virtually inaccessible for all but the most resolute by the authors' insistence on setting out every proof in, and only in, an almost completely formalized style. With the partial exception of the volume edited by Troelstra, none of these books attempts a thoroughgoing treatment of intuitionistic logic: almost the only books to do so are Fitting's *Intuitionistic Logic, Model Theory and Forcing*, written from a quite special viewpoint, which deals with Kripke trees, Beth's *The Foundations of Mathematics*, which of course treats of Beth trees, and Rasiowa and Sikorski's

The Mathematics of Metamathematics, which discusses the topological interpretation, without regard to genuinely semantic considerations. It is in large part in the hope of making the reading of such books as these, and of articles that will be found to be referred to in them, a much less formidable task than otherwise that the present introductory book has been written.

In the last three sections of Chapter 5, there has been some divergence from the main plan. Though it is still intended that they will be no harder going, for anyone who has read to that point, than the preceding sections, the policy of giving only the salient ideas relating to the topic under discussion has been abandoned in favour of a more comprehensive treatment. Section 5.6 aims to set out fully all results relating to the completeness of intuitionistic first-order logic that had been established before the recent work of the Nijmegen school. This was thought worth doing, because these results have not previously been collected together and are scattered and, in many cases, rather inaccessible. Section 5.7 deals with the Nijmegen results, in a critical but not destructive spirit, and I hope will be found useful as an assessment of what they in fact achieve. The brief section 5.8 sets out an adaptation of a theorem of Gödel and Kreisel to the compactness of sentential logic, which, while it presents no difficulty, has not, so far as I know, actually been stated previously.

Two topics have been left untouched. The mathematical theory of constructions is of the greatest importance for the foundations of intuitionistic logic, and it was with great regret that I omitted all but a mention of its existence; but it is as yet in an imperfect state, and its formulation is far too complicated to permit of a brief summary. I have said nothing, either, about the Gödel *Dialectica* translation of statements of intuitionistic arithmetic into $\exists\forall$-formulas of the theory of effectively calculable functionals of finite type. This is a powerful tool for metamathematical investigation, but does not purport fully to preserve the intended meanings. The same could, admittedly, be said of realizability;

but there just was not room to include everything.

Chapter 7 is of a purely philosophical character; but brief philosophical discussions are scattered throughout the book. This is not because, being a philosopher, I have been unable to resist the temptation to pursue my own *métier;* it is because intuitionistic mathematics is pointless without the philosphical motivation underlying it. Mathematical logicians may respond to the challenge to establish metamathematical results concerning a whole separate range of formal systems; but mathematical logic is not pure mathematics but applied mathematics, and has only as much interest as the formal theories which it studies. Intuitionistic mathematics cannot be justified by its purely 'mathematical interest': one subject-matter may differ from another according to the degree of mathematical interest which they have; but a set of principles of mathematical reasoning, diverging in both directions from those usually accepted, is devoid of interest unless there is some way of understanding mathematical statements in accordance with which those principles are justified and other principles are not. Intuitionism is a scandal to those who think that philosophy is of no importance, or that it cannot affect anything outside itself, or at least that there are some things which are sacrosanct and beyond the reach of philosophy to meddle with, and that among them are the accepted practices of mathematicians. Intuitionists are engaged in the wholesale reconstruction of mathematics, not to accord with empirical discoveries, nor to obtain more fruitful applications, but solely on the basis of philosophical views concerning what mathematical statements are about and what they mean. An individual may be converted to the intuitionistic viewpoint, without wishing thereafter to scrutinize more closely the philosophical arguments for and against it, just as he may be converted to a religious faith without wishing to become a theologian: but intuitionism will never succeed in its struggle against rival, and more widely accepted, forms of mathematics unless it can win the philosophical battle. If it ever loses that battle, the practice of intuitionistic mathematics itself and the metamathematical study of intuitionistic systems will alike become a waste of time.

It is hoped that this book will be found of interest by philosophers not specifically concerned with mathematics or even with logic, and that, because it presupposes so little and has been made as easy reading as possible, the subject-matter will not be an obstacle for those whose concern is purely philosophical. Nowhere in the whole field of mathematical logic and of the foundations of mathematics are such deep philosophical issues involved as in the study of intuitionism; and these are not restricted to the philosophy of mathematics. The dispute between intuitionists and platonists relates to the acceptability of a realistic interpretation of mathematical statements as referring to an independently existing and objective reality. This dispute bears a strong resemblance to other disputes over realism of one kind or another, that is, concerning various kinds of subject-matter (or types of statement), including that over realism about the physical universe: but intuitionism represents the only sustained attempt by the opponents of a realist view to work out a coherent embodiment of their philosophical beliefs. Phenomenalists might have attained a greater success if they had made a remotely comparable effort to show in detail what consequences their interpretation of material-object statements would have for our employment of our language. What is at the root of the dispute between intuitionists and adherents of all other philosophies of mathematics is what is at the root of all disputes over realism of any sort: a disagreement about the form which should be taken by a theory of meaning -- in the present case, for the language of mathematics. This disagreement is far from being irrelevant to theories of meaning for other areas of our language: on the contrary, it seems highly likely that the contentions both of the intuitionists and of their various opponents can be generalized so as to bear on the form that a theory of meaning should take for any part of language.

In preparing this book, I have had help from many people. Above all, I owe thanks to my former research assistant, Mr. Roberto Minio. To him the actual composition of large parts of the book, on the basis of a lecture course by me, is due, though all has been subject to revision by me, and I take full responsibility for the contents. He also undertook a great deal of the labour of typing, duplicating, etc., and was unfailingly cheerful and helpful. I have also had great help in the drudgery of completing the later part of the book from Dr. Dan Isaacson, currently Lecturer in the Philosophy of Mathematics at Oxford, who wrote up the proof of one of the theorems and was of much assistance to me during a long period when I was hampered by illness. I am also extremely grateful to Dr. van Dalen for having taken the trouble to go through the manuscript and make very helpful comments and suggestions, and also for interesting and stimulating general discussions about intuitionism. Dr. Troelstra also kindly read the manuscript of the first half, and likewise made useful comments. To Professor Scott I owe the suggestion that I should write this book at all, and constant encouragement and help in the process of writing it. To my College of All Souls I owe the fact that I had time enough to get it written.

CONTENTS

INTRODUCTORY REMARKS

It was Frege who first forced both philosophers and mathematicians to acknowledge the lack of any satisfactory philosophical account of the nature and epistemological basis of mathematics. He himself constructed a complete system of philosophy of mathematics; and, in the early part of the present century, others, most notably Hilbert and Brouwer, constructed alternative systems from quite different philosophical standpoints. Of the various attempts made in that period to create over-all philosophies of mathematics providing, simultaneously, solutions to all the fundamental philosophical problems concerning mathematics, only the intuitionist system originated by Brouwer survives today as a viable theory to which, as a whole, anyone now could declare himself an adherent. The two main rival systems, those of Frege and of Hilbert, contributed much of lasting value to the foundations of mathematics; but neither remains as an integral structure of doctrines to which allegiance could still be given. In more recent times, others have defended general positions in the philosophy of mathematics widely divergent from that of the intuitionists, for instance Gödel's version of platonism in which the objects of mathematics appear no longer as logical objects, as with Frege, but, in effect, as posits. The philosophy of mathematics is not, at present, a field in which the same intense activity takes place as in the earlier part of the century, and it has yet to be seen whether any of these later views can survive as a tenable position. Intuitionism is, however, the only unified system which survives intact from the earlier period when several rival philosophies of mathematics were in conflict, each claiming to hold the key which would unlock all doors.

This is in part due to the fact that, while both Frege and Hilbert thought that a mathematical foundation was needed for mathematics, intuitionism repudiated any such requirement.

For both Frege and Hilbert, classical mathematics stood in need of a justification: for Frege, the justification was to be a direct one, by means of which the basic principles of the various branches of mathematics (other than geometry) would be demonstrated by appeal to the yet more fundamental discipline of logic; for Hilbert, the justification was to be indirect, by showing that higher mathematics, though not to be taken at face value, could be shown to be a reliable means of deriving very elementary mathematical results whose meaningfulness was unproblematic. But both embraced a philosophy of mathematics whose acceptability depended upon the success of a specific mathematical programme: for Frege, the derivation of other mathematical theories from logic; for Hilbert, the execution of finitist consistency proofs for mathematical theories. In both cases, therefore, the philosophical system, considered as a unitary theory, collapsed when the respective mathematical programmes were shown to be incapable of fulfilment: in Frege's case, by Russell's discovery of the set-theoretic paradoxes; in Hilbert's, by Gödel's second incompleteness theorem. Of course, since the mathematical programmes were formulated in vague terms, such as 'logic' and 'finitistic', the fatal character of these discoveries was not inescapably apparent straight away; but in both cases, it eventually became apparent, so that, much as we now owe both to Frege and to Hilbert, it would now be impossible for anyone to declare himself a whole-hearted follower of either.

Intuitionism took the fact that classical mathematics appeared to stand in need of justification, not as a challenge to construct such a justification, direct or indirect, but as a sign that something was amiss with classical mathematics. From an intuitionistic standpoint, mathematics, when correctly carried on, would not need any justification from without, a buttress from the side or a foundation from below: it would wear its own justification on its face. Since classical mathematics patently did not have this character, what was needed was not to prop it up, but to reconstruct it. When a correct philosophical understanding of the nature of mathematical activity was arrived at, it would be seen that the reconstruction of mathematics had to penetrate even to the most fundamental level, that of sentential logic; even the understanding of the sentential operators had been distorted by the philosophical misconceptions of mathematicians concerning what it was that they were about.

Intuitionism has this in common with Frege's philosophy of mathematics (and with other varieties of platonism), that it takes the sentences of a mathematical theory to be meaningful statements, to which the notions of truth and falsity may appropriately be applied; intuitionism therefore diverges, in just the same way that platonism does, from formalism, according to which mathematical sentences have only the outward form of declarative statements, but lack any genuine content that could be truly or falsely asserted. Furthermore, the intuitionists also agree with Frege in regarding each mathematical sentence as having an individual content, determined by the way it is constructed out of its constituent expressions. They thus implicitly repudiate, just as Frege did, a holistic view of the language of mathematics. On a holistic view, no mathematical sentence, nor even an entire mathematical theory, has any significance on its own: it has a significance only as forming part of other theories, particularly scientific theories, which can be judged correct or incorrect on the basis of experience, but, again, only as a whole. There can, therefore, on such a holistic view, be no possibility of isolating the contribution made to a physical theory by the mathematics that is used in it; still less, therefore, of judging the mathematical part of the theory to be correct or incorrect on its own. On such a view, a mathematical theory is in itself incomplete: its value, if any, will lie in the possibility of incorporating it into some empirical theory; and to the extent that classical mathematical theories have been incorporated into successful scientific theories, no critique of them can be in place.

Intuitionism rejects such a holistic view of mathematics: for it, just as for Frege, each mathematical statement has a completely specific meaning of its own, a meaning which renders it capable of those applications which are made of it, but which is independent of any supplementary empirical hypotheses upon which such applications may hang. It is for this reason that, from an intuitionistic standpoint, existing mathematical practice is subject to criticism: forms of reasoning employed within mathematics are required to be valid, that, is truth-preserving, relative to the appropriate notion of truth for mathematical statements; and it is the meaning of a statement which determines what is the appropriate notion of truth for it, in what we may take its being true to consist. We have, therefore, to look at the way in which meaning is in fact conferred by us on our mathematical statements, and then to inquire whether the notion of truth for those statements that is yielded by that meaning does in fact guarantee the validity of the forms of reasoning that we are accustomed to employ.

The assumption underlying classical mathematics is that we bestow upon our mathematical statements a meaning which renders them determinately true or false, independently of the means available to us of recognizing their truth-value. This can be seen in even the simplest type of statement for which we possess no effective means of obtaining a proof or disproof, for instance a number-theoretic statement of the form $\exists x\ A(x)$, where $A(x)$ is a decidable predicate. There are several unsolved problems concerning statements of this form, for instance whether there exists an odd perfect number. If the statement is true, then we are, on any view, capable in principle of recognizing the fact, namely by hitting on an instance. But as soon as we consider what, in general, is required if we were to be able to disprove such a statement, i.e. to prove $\forall x\ \neg A(x)$, we recognize that no a priori ground exists for supposing that we must be capable either of proving the statement or of refuting it. In order to give a proof of a universal statement such as $\forall x\ \neg A(x)$, we have to be able either to give a single uniform reason why each natural number must satisfy the predicate $\neg A(x)$, or, at least, to find a partition of the natural numbers into finitely many classes such that, for each class, we can give a uniform reason why the numbers in that class satisfy $\neg A(x)$. (Here, of course, the 'uniform reason' might be of an inductive character: we might be able to prove that $\neg A(x)$ held for any number in a certain class, provided that it held for all smaller numbers in that class, or, again, provided that it held for all numbers in that class preceding the given one in a certain well-ordering.) On the classical or platonistic understanding of a universal statement like $\forall x \neg A(x)$, however, the truth of the statement in no way depends upon the existence of a uniform reason, or finitely many uniform reasons, why it should hold. If, on this interpretation, the statement is true, then, for any one particular natural number, it is not, of course, accidental that $\neg A(x)$ is true of it: but it may simply so happen that $\neg A(x)$ is true of every natural number in turn, without there being any finite set of reasons which should explain why this was so. In such a case, the statement $\forall x \neg A(x)$ would be true but yet would lie beyond our capacity to find a proof of it: what enables us, on a platonistic view, to conceive of such a possibility is that our understanding of the universal quantifier consists in our awareness of what it is for a universally quantified statement to be true, rather than being directly related to the means by which we can establish such a statement as true.

From an intuitionistic standpoint, such a conception of truth for mathematical statements is an illusion. One way of looking at the matter is as follows. When we first acquire the practice of using statements involving quantification over infinite totalities of mathematical objects, what we actually learn is to recognize what counts as justifying the assertion of such a statement, that is, what constitutes a proof of it, together with what can be inferred from a statement of this kind, and what counts as a refutation of it. In learning these principles, much carries over from the case of quantification over finite totalities, for instance the laws of universal instantiation and existential generalization; if this were not so, the use of the same form of expression would be unreasonable. In fact, all that we actually need to learn is what kind of proof is needed for a universally quantified statement: for existential generalization gives the general form for the proof of an existential statment (viz. by deriving it from a specific instance), a refutation of a statement $\exists x\ A(x)$ consists in a proof of $\forall x\ \neg A(x)$, a refutation of a statement $\forall x\ A(x)$ consists, in general, in deriving a contradiction from assuming it, the inferential powers of a universally quantified statement are given by the rule of universal instantiation, and those of an existential statement are governed by the principle that B can be inferred from $\exists x\ A(x)$ just in case we can prove $\forall x\ (A(x) \rightarrow B)$. The fact that quantification over an infinite totality shares so much in common with quantification over a finite one tempts us to overlook the crucial difference from the case in which we are quantifying over a finite totality which we can survey, namely that we do not have, even in principle, a method of determining the truth-value of a quantified statement by carrying out a complete inspection of the elements of the domain and checking, for each one, whether the predicate applies. The use which we learn to acquire of statements involving quantification over an infinite domain does not provide any understanding of what it is for such a statement to be true independently of our ability to prove it: all that we learn is how to recognize a proof or a refutation of such a statement. In the case of a statement involving quantification over a finite, surveyable, domain, our knowledge of what it is for the statement to be true consists in our knowledge of how we might, at least in principle, set about to determine whether or not it is true: but in that of a statement involving quantification over an infinite domain, we have no such capacity, and hence to conceive of the statement as possessing a determinate, objective truth-value independently of our being able to prove or disprove it is to make a fallacious assimilation of the infinite to the finite case; our grasp of the use of mathematical statements cannot supply us with any such conception of truth for them.

From an intuitionistic standpoint, therefore, an understanding of a mathematical statement consists in the capacity to recognize a proof of it when presented with one; and the truth of such a statement can consist only in the existence of such a proof. From a classical or platonistic standpoint, the understanding of a mathematical statement consists in a grasp of what it is for that statement to be true, where truth may attach to it even when we have no means of recognizing the fact; such an understanding therefore transcends anything which we actually learn to do when we learn the use of mathematical statements. Hence the platonistic picture is of a realm of mathematical reality, existing objectively and independently of our knowledge, which renders our statements true or false. On an intuitionistic view, on the other hand, the only thing which can make a mathematical statement true is a proof of the kind we can give: not, indeed, a proof in a formal system, but an intuitively acceptable proof, that is, a certain kind of *mental* construction. Thus, while, to a platonist, a mathematical theory relates to some external realm of abstract objects, to an intuitionist it relates to our own mental operations: mathematical objects themselves are mental constructions, that is, objects of thought not merely in the sense that they are thought about, but in the sense that, for them, *esse est concipi*. They exist only in virtue of our mathematical activity, which consists in mental operations, and have only those properties which they can be recognized by us as having.

It is for this reason that the intuitionistic reconstruction of mathematics has to question even the sentential logic employed in classical reasoning. The most celebrated principle underlying this revision is the rejection of the law of excluded middle: since we cannot, save for the most elementary statements, guarantee that we can find either a proof or a disproof of a given statement, we have no right to assume, of each statement, that it is either true or false; nor, therefore, to offer as a proof of a theorem a demonstration that it is derivable from the assumption either of the truth or of the falsity of some as yet undecided proposition, for instance the Riemann hypothesis (a well-known proof of Littlewood's proceeds in exactly this way). The intuitionistic reconstruction does not consist only in a revision of the underlying logic: of almost equal importance is its treatment of the notion of an infinite sequence.

However, it is from the reassessment of the basic forms of logical argument that the reconstruction starts; and we must start from there also.

1
PRELIMINARIES

1.1. CONSTRUCTIVE PROOF

What everyone who has heard of intuitionism knows is that intuitionists want their proofs to be constructive. The notion of a constructive proof is, however, by no means restricted solely to intuitionistic or other forms of 'constructivist' mathematics: the distinction between constructive and non-constructive proofs arises within classical mathematics, and is perfectly intelligible from a completely platonistic standpoint. From such a standpoint, the distinction arises for proofs of existential and disjunctive statements: any proof proves more than just the theorem which is its conclusion, and to call a proof of an existential or disjunctive statement (with or without initial universal quantifiers) 'constructive' is to say something quite specific about the additional information which the proof provides. A proof of a closed statement of the form $\exists x\, A(x)$, say one in which the variable ranges over the natural numbers, is constructive just in case it either itself proves a specific instance $A(\bar{n})$ or yields an effective means, at least in principle, for finding a proof of such an $A(\bar{n})$. Likewise, a proof of a closed statement of the form $A \vee B$ is constructive if it either is in fact a proof either of A or of B, or yields an effective means, at least in principle, for obtaining a proof of one or other disjunct. If an existential theorem contains a parameter, i.e. if it is of the form $\forall x \exists y\, A'(x,y)$ (with no free variables), then its proof is constructive if it yields an effectively calculable function f such that $A'(\bar{m},f(\bar{m}))$ holds for each m; if a disjunctive theorem contains a parameter, i.e. is of the form $\forall x\, (A'(x) \vee B'(x))$ (with no free variables), then its proof is constructive if it yields an effective means for finding, for each m, a proof either of $A'(\bar{m})$ or of $B'(\bar{m})$.

Here is a very straightforward example, due to Benenson,

of a non-constructive proof:

Theorem There are solutions of $x^y = z$ with x and y irrational and z rational.

Proof. $\sqrt{2}$ is irrational, and $\sqrt{2}^{\sqrt{2}}$ is either rational or irrational. If it is rational, put $x = \sqrt{2}$, $y = \sqrt{2}$ so that $z = \sqrt{2}^{\sqrt{2}}$, which, by hypothesis, is rational. If, on the other hand, $\sqrt{2}^{\sqrt{2}}$ is irrational, put $x = \sqrt{2}^{\sqrt{2}}$ and $y = \sqrt{2}$, so that $z = (\sqrt{2}^{\sqrt{2}})^{\sqrt{2}} = (\sqrt{2})^2 = 2$, which is certainly rational.

Thus in either case a solution exists.

Another example is the standard proof of the Bolzano-Weierstrass Theorem:

Theorem If $\underline{S}$ is an infinite subset of the closed interval $[a,b]$, then $[a,b]$ contains at least one point of accumulation of $\underline{S}$.

Proof. We construct an infinite nested sequence of intervals $[a_i,b_i]$ as follows. Put $a_o = a$, $b_o = b$. For each i, consider two cases:

(i) if $[a_i, \frac{a_i + b_i}{2}]$ contains infinitely many points of $\underline{S}$, put $a_{i+1} = a_i$, $b_{i+1} = \frac{a_i + b_i}{2}$;

(ii) if $[a_i, \frac{a_i + b_i}{2}]$ contains only finitely many points of $\underline{S}$, put $a_{i+1} = \frac{a_i + b_i}{2}$, $b_{i+1} = b_i$.

Then it is plain by induction that each $[a_i,b_i]$ contains infinitely many points of $\underline{S}$: for $i = 0$ this is a hypothesis of the theorem, in case (i) it holds by definition, and in case (ii) it holds by the induction hypothesis that $[a_i,b_i]$ contains infinitely many points of $\underline{S}$.

The sequence of nested intervals must converge to a point of $[a,b]$ every neighbourhood of which contains infinitely many, and hence at least one, point of *S*.

Each of these proofs establishes the existence of something without providing an effective means of finding it. In the first case, the proof shows that one or other of two specific solutions of the equation satisfies the conditions of the theorem, without giving us a means of determining which. In the second case, the proof specifies a 'construction' which we cannot, in general, carry out, because we may be unable to decide whether case (i) or case (ii) applies; we could say that it shows that at least one of non-denumerably many points is a point of accumulation of *S*, without giving us a means of finding a particular one. In both cases this arises because of an appeal to the law of excluded middle in a case where we are not given a way of deciding which alternative holds.

The point is not just that the intuitionist prefers constructive proofs to a greater degree than other mathematicians. A classical mathematician may spend a considerable amount of time looking for a constructive proof of a result for which he already has a non-constructive one. The intuitionist is not in this position; he must have a constructive proof because the intuitionistic interpretation of the conclusion is always such that no non-constructive proof could count as a proof of it. The classical meaning of the logical constants is given by truth tables, which in turn depend on every statement having a determinate truth-value. Instead of appending to a proof a note to the effect that the proof was constructive, the classical mathematician might use new logical constants alongside the usual ones; e.g.

'$A \cup B$' for 'we have a constructive proof of $A \vee B$' and '3x A(x)' for 'we have a constructive proof of $\exists x\, A(x)$'. $\cup$ and 3 clearly could not be truth-functional, but such new connectives still do not satisfy the intuitionist's requirements. 'We have a constructive proof of $A \vee B$' and 'we have a constructive proof of $\exists x\, A(x)$' are unintelligible to the intuitionist because the classical meanings of '$\vee$' and '$\exists$' have no clear sense. The classical constants with their truth-functional meanings are rejected together with the assumption that every statement has a determinate truth-value whether we know anything about it or not. So, for example, the only possible interpretation of '∃' is one under which to prove ∃x A(x) is to prove, or at least provide an effective means of proving, of a specific element of the domain, that it satisfies A(x). Thinking of a statement as true or false independently of our knowledge involves a supposition of some external mathematical reality, whereas thinking of it as being rendered true, if at all, only by a mathematical construction does not.

1.2. THE MEANINGS OF THE LOGICAL CONSTANTS

The meaning of each constant is to be given by specifying, for any sentence in which that constant is the main operator, what is to count as a proof of that sentence, it being assumed that we already know what is to count as a proof of any of the constituents. The explanation of each constant must be faithful to the principle that, for any construction that is presented to us, we shall always be able to recognize effectively whether or not it is a proof of any given statement. For simplicity of exposition, we shall, in the preliminary part of this account, assume that we are dealing with arithmetical statements: an atomic sentence will then be a numerical equation, and a proof of it will consist in a computation. Every instance $A(\bar{n})$ will count as being a constituent of a quantified sentence ∀x A(x) or ∃x A(x).

The logical constants fall into two groups. First are &, ∨, and ∃. A proof of A & B is anything that is a proof of A and of B. A proof of A ∨ B is anything that is a proof either of A or of B. A proof of ∃x A(x) is anything that is a proof, for some n, of the statement $A(\bar{n})$. Note that a proof of any sentence containing only the constants &, ∨, and ∃ is a computation or finite set of computations.

The second group is composed of ∀, →, and ¬. A proof of ∀x A(x) is a construction of which we can recognize that, when applied to any number n, it yields a proof of $A(\bar{n})$. Such a proof is therefore an *operation* that carries natural numbers into proofs. A proof of A → B is a construction of which we can recognize that, applied to any proof of A, it yields a proof of B. Such a proof is therefore an operation carrying proofs into proofs. Note that it would be incorrect to characterize a proof of $\forall x\ A(x)$ merely as 'a construction which, when applied to any number n, yields a proof of $A(\bar{n})$', or a proof of $A \rightarrow B$ as 'a construction which transforms every proof of A into a proof of B', since we should then have no right to suppose that we could effectively recognize a proof whenever we were presented with one. We therefore have to require explicitly that a construction is to count as a proof of $\forall x\ A(x)$ only if we can recognize it as yielding, for each n, a proof of $A(\bar{n})$, or as a proof of $A \rightarrow B$ only if we can recognize it as effecting the required transformation of proofs of A into proofs of B.

A proof of $\neg A$ is usually characterized as a construction of which we can recognize that, applied to any proof of A, it will yield a proof of a contradiction. This is unsatisfactory because a 'contradiction' is naturally understood to be a statement $B \ \& \ \neg B$, so that it seems we are defining $\neg$ in terms of itself. We can avoid this in either of two ways. We can choose some one absurd statement, say $0 = 1$, and say that a proof of $\neg A$ is a proof of $A \rightarrow 0 = 1$. In this case, in order to validate the laws of intuitionistic logic, we must allow that, given a proof of $0 = 1$, we can find a proof of any other statement whatever. This is, however, entirely plausible: it is obvious that we have a systematic method of deriving, from $0 = 1$, a proof of any numerical equation; and from this it is easily seen that we can prove every arithmetical statement. Admittedly, if we were considering statements other than arithmetical ones, it might not be so obvious that every statement was derivable from $0 = 1$ by standard reasoning; but, if there were any doubt, we could take it as a stipulation: we shall *count* any proof of $0 = 1$ as being, simultaneously, a proof of any other statement. Alternatively, we may regard the sense of $\neg$, when applied to atomic statements, as being given by the computational procedure which decides those statements as true or false, and then define a proof of $\neg A$, for any non-atomic statement A, as being a proof of

$A \to B \& \neg B$, where B is an atomic statement. Again, we are required to acknowledge that, given a proof of $B \& \neg B$, for an atomic statement B, we can find a proof of any other statement.

Suppose we have a proof of the free-variable statement A(x). Such a proof is a proof-skeleton, and provides a proof of $\forall x\, A(x)$; for we have a simple method of finding, for any n, a proof of $A(\bar{n})$, namely by replacing the free variable x, in all occurrences in the proof-skeleton, by the numeral $\bar{n}$. This is a *uniform* operation on natural numbers to obtain proofs. However, a proof of $\forall x\, A(x)$ does not have to take this simple form. We can have an operation which, applied to any number n, yields a proof of $A(\bar{n})$, even though the structure of the proof depends on the value of n. An easy example of this is given by the intuitionistic justification of induction. Suppose that we have a proof of A(0) and a proof of $\forall x(A(x) \to A(x+1))$, which we may suppose for simplicity to have been obtained by means of a free-variable proof of $A(x) \to A(x+1)$. Then, for each n, we can find a proof of $A(\bar{n})$. When n = 1, we apply modus ponens to A(0) and $A(0) \to A(1)$; when n = 2, we first obtain A(1) by the preceding modus ponens step, and then apply modus ponens again to A(1) and $A(1) \to A(2)$; and so on. There is no uniform proof-skeleton (except one which allows explicit appeal to induction); the length of the proof (number of applications of modus ponens) depends on n: but we have an operation which we can recognize as yielding a proof of $A(\bar{n})$ for each n.

Similar remarks apply to $\to$. Suppose we have a proof of B from the hypothesis A: i.e. something that is like a proof of B save that A is cited as a premiss without justification. Then we have a method of transforming any proof of A into a proof of B: namely, by appending the proof of B from A to the proof of A. Such an operation on proofs of A to obtain a proof of B is a *uniform* operation: it does not depend upon the structure of the proof of A. Again, a proof of $A \to B$ does not have to take this simple form; it may be that we can recognize some operation which involves internal transformation of any given proof of A as nevertheless always yielding a proof of B. If this were not so, then we could not admit an inference from $\forall x\ (A(x) \rightarrow B(x))$ to $\exists x\ A(x) \rightarrow \exists x\ B(x)$ as intuitionistically valid, since it would be impossible to derive a constructive proof of $\exists x\ B(x)$ by merely appending something to a proof of $\exists x\ A(x)$; we should need to know for which particular natural number n the proof of $\exists x\ A(x)$ yielded a proof of $A(\bar{n})$. Intuitionistic mathematicians have seldom, however, found a way of fully exploiting the intuitionistic meaning of $\rightarrow$: most proofs actually given of conditional statements appeal only to very obvious properties of a possible proof of the antecedent, the only notable exception being Brouwer's proof of the Bar Theorem, which will be discussed later.

It has been emphasized that, while an intuitionistic proof of $\forall x\ A(x)$ is an operation upon natural numbers yielding proofs, and an intuitionistic proof of $A \rightarrow B$ is an operation upon proofs yielding proofs, such operations do not have to be uniform. It may be thought that this point can be reinforced by claiming that such a notion of a uniform operation of these kinds - a free-variable proof of A(x), or a proof of B from A as hypothesis - depends essentially upon the context of a particular formal system in which the proofs are carried out, and so would be inappropriate where we are concerned with intuitive proofs, not restricted to any formal system. Such a claim may be correct; but it is not evidently so, and reliance on it would therefore be imprudent. Only much closer analysis of that notion of (intuitive) proof in terms of which the explanations of the intuitionistic logical constants are given would reveal whether or not, relative to it, there are determinate notions of free-variable proofs and of proofs from hypotheses.

The explanation of $\rightarrow$ must be understood *extensionally* in the sense that the so-called paradoxes of material implication hold for intuitionistic $\rightarrow$ also. If we already have a proof of B, then there is a very simple operation which we can recognize as yielding a proof of B, given a proof of A: namely, throw away the proof of A, and replace it by the already known proof of B. Thus we may always infer $A \rightarrow B$

from B. Likewise, suppose that we have a proof of $\neg A$, i.e. an operation which will take any proof of A into a proof of $0 = 1$. We assumed that, for any statement B, we have an operation which will transform a proof of $0 = 1$ into a proof of B. Hence, by combining the two operations, we obtain one that will carry a proof of A into a proof of B, and so have a proof of $A \to B$. We may thus always infer $A \to B$ from $\neg A$.

In general, the truth-tables for the connectives are correct intuitionistically in the following sense. Construe each entry as a rule of inference with two premisses: one premiss is A if A is assigned the value True, and $\neg A$ if A is assigned the value False, and the other premiss is, likewise, either B or $\neg B$; the conclusion is either the complex statement or its negation according as it receives the value True or False. Then all such inferences are intuitionistically correct. (For example, the truth-table for & embodies, in this sense, the four rules of inference:

$$\frac{A \quad B}{A \,\&\, B} \qquad \frac{A \quad \neg B}{\neg(A \,\&\, B)} \qquad \frac{\neg A \quad B}{\neg(A \,\&\, B)} \qquad \frac{\neg A \quad \neg B}{\neg(A \,\&\, B)}$$

and that for $\to$ embodies:

$$\frac{A \quad B}{A \to B} \qquad \frac{A \quad \neg B}{\neg(A \to B)} \qquad \frac{\neg A \quad B}{A \to B} \qquad \frac{\neg A \quad \neg B}{A \to B}$$

and similarly for v.) Of course, the classical assumption that the various assignments exhaust all possibilities is not intuitionistically correct.

In some very vague intuitive sense one might say that the intuitionistic connective $\to$ was stronger than the classical $\to$. This does not mean that the intuitionistic statement $A \to B$ is stronger than the classical $A \to B$, for intuitively, the antecedent of the intuitionistic conditional is also stronger. The classical antecedent is that A is *true*, irrespective of whether we can recognize it as such or not. Intuitionistically, this is unintelligible: the intuitionistic antecedent is that A is (intuitionistically) *provable*, and this is a stronger assumption. We have to show that we could prove B on the supposition, not merely that A happens to be the case (an intuitionistically meaningless supposition), but that we have been given a *proof* of A. Hence intuitionistic $A \to B$ and classical $A \to B$ are in principle *incomparable* in respect of strength. We may sometimes have a classical proof of $A \to B$ where we lack an intuitionistic one; but there is no reason why the converse should not sometimes hold too. (This observation applies to intuitionistic mathematical theories generally: intuitionistic first-order *logic* is in fact a subsystem of classical logic.)

Since $\neg$ is really a case of $\to$, the same applies to intuitionistic negation. Classically, what we have to show to be absurd is the supposition that A should be *true*, irrespective of our knowledge; but, intuitionistically, all that we have to show absurd is the supposition that we should have a proof of A. It is impossible, therefore, that we should ever be in a position to assert, of any statement A, that A is (absolutely) neither provable nor refutable. For a demonstration that A is not provable serves as a proof of $\neg A$, i.e. as a refutation of A; for by the above remark about the paradoxes of material implication, if we have established that A can never be proved, we have established that $A \to B$ holds for any B, and hence in particular that $A \to 0 = 1$, i.e. $\neg A$, holds. If we make the assumption that we can never get a proof of $0 = 1$, then a proof of $\neg A$ can be identified with a demonstration of the unprovability of A; for if, conversely, in advance of knowing whether A can be proved or not, we find a means of transforming any proof of A into a proof of $0 = 1$, then we can infer that A will never be proved. Since a proof of $\neg A$ is thus tantamount to a proof that A will never be proved, it would be a complete mistake to try to replace the classical dichotomy true/false by a trichotomy provable/refutable/undecidable.

The Law of Excluded Middle. Evidently the statement $A \vee \neg A$ is not intuitionistically valid. This means, in particular, that we are not entitled to infer B from $A \to B$ and $\neg A \to B$.

The failure of the law of excluded middle is often explained by the different meaning of intuitionistic disjunction: a proof of $A \vee B$ is a proof either of A or of B, and hence a claim to have proved $A \vee \neg A$ amounts to a claim either to have proved A or to have proved $\neg A$. Such an explanation of the matter is correct as far as it goes, but it will naturally leave a platonist with the feeling that the meaning imposed upon $\vee$ is arbitrary: on any view on which either A or $\neg A$ must be true, irrespective of whether we can prove it, to repudiate that sense of $\vee$ in which we can assert $A \vee \neg A$ a priori is to deny ourselves the means of expressing what we are able to apprehend. No account of the intuitionistic rejection of the law of excluded middle is adequate, therefore, unless it is based on the intuitionistic rejection of the platonistic notion of mathematical truth as obtaining independently of our capacity to give a proof. When this is taken into account, the intuitionistic interpretation of disjunction no longer appears arbitrary, but as the only possible one, and the failure of the law of excluded middle no longer appears as depending on any peculiarity in the interpretation of $\vee$.

Nothing that has so far been said about the intuitionistic attitude to the notion of truth has determined whether the predicate 'is true' is to be regarded as tensed or tenseless: whether, as on a platonistic view, it attaches timelessly to any mathematical statement to which it attaches at all, or whether it comes to attach to such a statement only at the time when a proof of it is given. On the latter interpretation, 'is true' would have to be equated with 'has been proved' and 'is false' with 'has been refuted'. On this use, any statement A that has not yet been decided is neither true nor false; but this does not preclude its later becoming true or becoming false. Naturally, on a view according to which these tensed notions of mathematical truth are the only admissible ones, the law of excluded middle will, strictly speaking, be true only for statements that have already been decided.

It is not necessary, however, to adopt such a radical attitude to the notion of mathematical truth in order to reject the law of excluded middle on intuitionistic grounds.

It would be possible for a constructivist to agree with a platonist that a mathematical statement, if true, is timelessly true: when a statement is proved, then it is shown thereby to have been true all along. To say this is, in effect, to equate 'A is true' with 'We can prove A' rather than with 'A has been proved', and 'A is false' with 'We cannot prove A'. Such an interpretation of 'true' and 'false' remains faithful to the basic principles of intuitionism only if 'We can prove A' ('A is provable') is not interpreted to mean either, at one extreme, that, independently of our knowledge, there exists something which, if we became aware of it, we should recognize as a proof of A, nor, at the other, that as a matter of fact we either have proved A or shall at some time prove it. In the former case, we should be appealing to a platonistically conceived objective realm of proofs; in the latter, we should be entitled to deny that A was provable on non-mathematical grounds (e.g. if the obliteration of the human race were imminent). 'We can prove A' must be understood as being rendered true only by our actually proving A, but as being rendered false only by our finding a purely mathematical obstacle to proving it. From any standpoint, therefore, there can, again, be no guarantee that every mathematical statement is either true or false.

On the first interpretation of 'A is true', as significantly tensed, i.e. as meaning 'A has been proved', the statement 'A is false', that is '$\neg$A is true', is much stronger than 'A is not true'. But when 'A is true' is interpreted as tenseless, i.e. as meaning 'We can prove A', then 'A is not true' and 'A is false' can be equated, since the only thing in virtue of which the former can hold is a demonstration of the unprovability of A, that is, a proof of $\neg$A. Hence, on this interpretation, the truth of A $\vee \neg$A depends on our being able either to prove A or to show that we can never do so, which we cannot in general claim to be able to do.

The Assertion of a Mathematical Statement. We have explained disjunction by saying that a proof of A $\vee$ B is a proof either of A or of B, and existential quantification by saying that a proof of $\exists x\, A(x)$ is a proof of a statement $A(\bar{n})$ for some n. In practice, however, intuitionistic mathematicians do not confine their assertions of disjunctive and existential statements, even in the course of giving a demonstration of the truth of some other theorem, to those for which they actually have a proof, as thus specified; it is considered sufficient that we have a means, at least in principle, for obtaining a proof. The most striking case of this is an instance $A \vee \neg A$ of the law of excluded middle, when A is an effectively decidable statement, e.g. a statement that some very large number is prime. It is perfectly in order intuitionistically to demonstrate a theorem by showing that it follows equally from the supposition that some large number N is prime and from the supposition that N is composite: it is not required that we should actually decide the matter before regarding the theorem as established. In general, it is licit to assert $A \vee B$ provided that we have an effective means of which we can recognize that it would yield a proof either of A or of B, and to assert $\exists x\, A(x)$ if we have an effective means of which we can recognize that it would yield a particular number n and a proof of $A(\bar{n})$. We might seek to take account of this practice by revising our specification of what is to count as a proof of a disjunctive or existential statement. It is better, however, to leave that specification unchanged, but to stipulate that the assertion of a mathematical statement is to be construed, not as a claim to have a proof of it, but only as a claim to have an effective means, in principle, for obtaining a proof. Note that, in the case of a conditional, a negation, or a universally quantified statement, this stipulation makes no difference: an effective means for obtaining a proof of a statement of any of these forms is already a proof of that statement. Note also that, even from a classical standpoint, the significance of an act of assertion is a matter for conventional agreement: one can imagine people who assert a mathematical statement outright even on the basis of mere plausible reasoning (in the sense of Polya), but who have as much interest in the notion of conclusive proof as we do, and whose mathematics is in every other respect the same as ours.

The assertion of $A \vee \neg A$ is therefore a claim to have, or to be able to find, a proof or disproof of A. Likewise we may use $\forall x\ (A(x) \vee \neg A(x))$ to express that $A(x)$ is an effectively decidable predicate.

<u>More on Constructive Proof</u>. The meanings of the classical constructive operators $\cup$ and $\boldsymbol{\exists}$ envisaged in section 1.1 differ from those of the intuitionistic $\vee$ and $\exists$ just in virtue of the fact that the former are embedded in a language in which the classical logical constants are taken as meaningful. This is not surprising, since, for an operation to be effective, it is required, not only that each step be rigidly determined, but that the operation should at some stage terminate, and hence an existential quantifier is built into the very meaning of 'constructive'. Classically, a statement of the form

$$\neg \forall x\ \neg A(x) \rightarrow \exists x\ A(x)$$

is logically true; but we cannot assert this intuitionistically, since, from the fact that $\forall x\ \neg A(x)$ leads to a contradiction, it does not in the least follow that (taking x to range over the natural numbers) we can find any n such that $A(\bar{n})$. By the same token, we should not be classically justified in asserting

$$\neg \forall x\ \neg A(x) \rightarrow \boldsymbol{\exists} x\ A(x)$$

either; but we *can* assert classically

$$\forall x\ (A(x) \cup \neg A(x)) \quad \& \quad \neg \forall x\ \neg A(x) \rightarrow \boldsymbol{\exists} x\ A(x):$$

if we have an effective means of deciding, for every n, whether $A(\bar{n})$ or $\neg A(\bar{n})$, and we know that not every number falsifies

A(x), then we can actually find a number that satisfies it. The ground for this is simply the familiar observation that, if $\neg\forall x \neg A(x)$ holds, then the procedure of testing each number 0, 1, 2, ... to see whether it satisfies A(x), breaking off when we find one that does, must terminate. But the corresponding principle (known as Markov's Principle)

$$\forall x\ (A(x) \vee \neg A(x)) \quad \& \quad \neg\forall x \neg A(x) \rightarrow \exists x\ A(x)$$

does not hold intuitionistically. The classical version involved taking $\neg\forall x \neg A(x)$ to mean 'A(x) will not as a matter of fact be found to fail for every number', a statement which is not intuitionistically intelligible: intuitionistically, we can only take it to mean 'We can derive a contradiction from supposing that we could prove that A(x) failed for every number', a proposition from which no guarantee can be extracted that, by testing each number in turn, we shall eventually find one which satisfies A(x). An acceptance of the classical meanings of the logical constants enables us to recognize, as constructive, proofs which otherwise we should not so view.

Domains of Quantification. So far, in this preliminary account of the logical constants, we have assumed that the statements to which they are applied are arithmetical ones. Only a small part of the advantage of this lay in the clarity, for this case, of the notion of a proof of an atomic statement; the chief motivation was to simplify the discussion of the quantifiers.

It is essential to the classical account of quantification that a uniform account is possible for all non-empty domains; the crucial assumption of classical logic is that the interpretation of the quantifiers remains the same, whether their domain be finite or infinite, denumerable or non-denumerable. Intuitionistically, we must demand of each domain $\underline{D}$, not only that it be non-empty ($\neg\forall x\ x \notin \underline{D}$), but that it be *inhabited* ($\exists x\ x \in \underline{D}$) -- that we can instance at least one element which we can positively assert to belong to

D. But, although there is indeed a sense in which the intuitionistic quantifiers have determinate meanings, constant from one domain to another, we cannot assume without more ado that they can be explained in the same way for all inhabited domains.

Classically, the only significant distinction between domains is in respect of their cardinality. Intuitionistically, we may, in the first place, distinguish between domains according to whether they are decidable: a domain D is *decidable* if we can decide, for any object, whether or not it belongs to D ($\forall x\ (x \in D \vee x \notin D)$). The simplest case is that of a finite domain. Just as in a classical context, D is *finite* if there exists a one-one map of D on to some initial segment of the natural numbers. Because of the constructive meanings of the existential quantifiers involved in this definition, however, to say that D is finite is stronger than to say that it can be mapped *into* some initial segment (i.e. that there is a finite upper bound to its size). A language in which the primitive predicates are themselves decidable and in which the variables range over a finite domain will satisfy the laws of classical logic: every statement will be decidable. There is no general requirement that a domain of quantification be decidable, even when we can give a finite upper bound for its size: a domain may be specified by any meaningful predicate. An artificial example would be the domain consisting of the extension of the predicate '$x = 0 \vee (F \ \& \ x = 1)$', where 'F' abbreviates the statement of Fermat's Last Theorem. Here we know that the domain has at least one and at most two elements, and that 0 belongs to it; if Fermat's Last Theorem is proved, we shall know that 1 belongs to it, and, if it is refuted, we shall know that only 0 belongs to it; but, until Fermat's Last Theorem is decided, we do not know whether 1 belongs to it or not. Another such case would be the domain given by the predicate '$x = 0 \vee [(F \vee \neg F) \ \& \ x = 1]$', where 'F' has the same meaning. In this case, we shall know that 1 belongs to the domain as soon as Fermat's Last Theorem is decided, if it ever is; but we can never be in a position to say that 1 does not belong to the domain.

In intuitionistic mathematics, any mathematical object must always be considered as identifiable via some finite description, that determines *which* particular object we are referring to. (It does not hold in every kind of case that the identifying description will of itself enable us to derive all that we can know of the object.) A decidable domain must be characterized in such a way that the identifying description of any element always enables us to recognize it as belonging to the domain. This cannot be required, however, for undecidable domains such as those cited above, for otherwise a different description of the number 1 would be required when it was regarded as belonging to the first of those two domains, or to the second, or simply to the domain of natural numbers, and we should hardly want to say that we were dealing with three intensionally distinct objects in these three cases. It is therefore necessary to incorporate this requirement into the explanations of the quantifiers whenever the domain is undecidable: a proof of $\exists x\, A(x)$ is a proof of some statement of the form $A(t)$, together with a proof that the object denoted by the term t belongs to the domain; and a proof of $\forall x\, A(x)$ is a construction of which we can recognize that, when it is applied to any term t and to a proof that the object denoted by t belongs to the domain, it yields a proof of $A(t)$. Evidently, when the domain of quantification is undecidable, then, even when we know a finite upper bound for its size, and all the atomic statements are decidable, the law of excluded middle will not hold for all statements.

The domain of natural numbers is the simplest instance of an infinite domain: not merely is it decidable, but we can effectively enumerate its elements. Of course, the decidability of a domain is relative to the way in which its elements are thought of as being given; we can perfectly well describe a natural number so that a proof is required that it is a natural number, e.g. by describing it as an ordinal of a certain kind; to recognize the ordinal as a natural number, we shall need a proof that it is finite. However, since the natural numbers can be enumerated, we could, in the explanations of the quantifiers, assume that they were presented (e.g. in terms of 0 and the successor operation) in such a way that no question arose as to their being natural numbers. As soon as we consider a domain which cannot be effectively enumerated, the quantifiers need to be explained in the above manner, namely by reference to a proof that a given object belongs to the domain. A simple case would be the domain of all effectively calculable functions from natural numbers to natural numbers. We have a perfectly clear intuitive idea of what such a function is, and any rule of computation giving the values of a function of this kind may be finitely stated; but we have no effective means of enumerating the totality of such functions, since we cannot circumscribe the means of stating rules of computation so as both to include a rule governing every effectively calculable function and to make it immediately recognizable that the rule in question is effective. We therefore have to think of each such function as given by a finite statement of the rule for computing its values, but explain existential and universal quantification over these functions in the way stated above, where now the terms considered will be ones embodying rules of computation, and a proof that the function belongs to the domain will be a proof that the rule is effective and everywhere applicable.

The description by means of which a mathematical object is given must always be such as to enable it to be distinguished from other objects of the same kind. However, since mathematical objects are mental constructions, and the mental construction is expressed by means of the description in terms of which the object is given, the objects of intuitionistic mathematics must, in general, be considered as intensional objects; that is to say, that criterion of identity which is given together with the manner in which the object is presented relates to the identity of the description. Thus, for example, if an effectively calculable function is thought of as given by means of a rule of computation, different rules will determine intensionally distinct functions, even if these functions are extensionally equivalent, i.e. have the same values for the same arguments. Usually, intensional identity is symbolized by the sign $\equiv$, and is assumed to be decidable, that is, we have $\forall x\ \forall y\ (x \equiv y \vee \neg x \equiv y)$; for we can always effectively recognize whether or not two descriptions are the same. For most purposes, however, even although the objects of our theory are intensional ones, we are interested in their extensional properties, and will be able to introduce a notion of extensional equality, symbolized by the ordinary equality sign =; for instance, in the case of functions, '$f = g$' may be taken as defined by '$\forall x\ (f(x) = g(x))$'. Extensional equality will not, in general, be decidable; in the cases in which it is, it must be proved to be such.

As well as domains such as that of all effectively calculable functions on natural numbers, we may also consider domains in which not every element can be completely characterized by a finite description; since such an object must in any particular context be given by means of a finite description, it can never be completely given. For a domain of this kind, a further stipulation will be needed of the intended meanings of the quantifiers; this matter will be taken up again later.

1.3. EXAMPLES OF LOGICAL PRINCIPLES

All the usual introduction and elimination rules for a classical system of natural deduction, in which &, $\vee$, $\rightarrow$, $\neg$, $\forall$, and $\exists$ are all taken as primitive, are intuitionistically valid save the double negation rule $\frac{\neg\neg A}{A}$. We have a proof of $\neg\neg A$ when we can show that we shall never have a proof of $\neg A$, that is, when we can show that we shall never have a proof that A will never be proved; clearly, however, this does not amount to a proof of A itself, and hence $\frac{\neg\neg A}{A}$ is not a valid form of inference. On the other hand, a proof of A does count as a proof that A will never be disproved, for otherwise the possibility of deriving a contradiction would remain open; hence $\frac{A}{\neg\neg A}$ is valid. Contraposition holds in general except, of course, in those cases where use of the invalid double negation rule is suppressed.

Thus, given $A \rightarrow B$, if we have a proof that B can never be proved then clearly A can never be proved either, since we could transform any proof of A into a proof of B. So $\frac{A \rightarrow B}{\neg B \rightarrow \neg A}$ is valid, and likewise $\frac{A \rightarrow \neg B}{B \rightarrow \neg A}$. However, $\frac{\neg B \rightarrow \neg A}{A \rightarrow B}$ and $\frac{\neg A \rightarrow B}{\neg B \rightarrow A}$ do not hold. By contraposition, $\neg\neg\neg A$ is equivalent to $\neg A$, since $\frac{A}{\neg\neg A}$ holds. Since v elimination is valid and we can derive $\neg\neg A \rightarrow A$ from A and from $\neg A$, the inference $\frac{A \vee \neg A}{\neg\neg A \rightarrow A}$ is allowed. So, if the law of excluded middle holds for some particular A, then $\neg\neg A \rightarrow A$ also holds. The converse is not true since $\frac{\neg\neg A \rightarrow A}{A \vee \neg A}$ is invalid. Notice that the consequent of any conclusion established by contraposition must be a negated statement. For the same reason (the failure of $\frac{\neg\neg A}{A}$), although appeal to reductio ad absurdum is plainly acceptable, it can only be used to obtain negated conclusions; in other words, while the inferences $\frac{A \rightarrow B \quad A \rightarrow \neg B}{\neg A}$ and $\frac{\neg A \rightarrow B \quad \neg A \rightarrow \neg B}{\neg\neg A}$ are valid, $\frac{\neg A \rightarrow B \quad \neg A \rightarrow \neg B}{A}$ is not.

Whereas the classical connectives can all be defined in terms of $\neg$ and any one other, all three connectives listed above are intuitionistically independent ($A \longleftrightarrow B$ can be understood as defined to mean $(A \rightarrow B)$ & $(B \rightarrow A)$). Although the paradoxes of material implication hold, and hence $\frac{\neg A \vee B}{A \rightarrow B}$ is valid, $A \rightarrow B$ is not equivalent to $\neg A \vee B$: it is clear that the fact that we can transform any proof of A into a proof of B in no way implies that we can get either a proof of $\neg A$ or a proof of B. Likewise, $\frac{A \rightarrow B}{\neg(A \;\&\; \neg B)}$ is valid, but the converse is not. From $\neg(A \;\&\; \neg B)$ we are only entitled to infer the weaker conditional $A \rightarrow \neg\neg B$. In fact, these two formulas are equivalent to one another and to several others, as is shown by the following cycle of inferences. From $A \rightarrow \neg\neg B$, by $\frac{A \rightarrow \neg B}{B \rightarrow \neg A}$, we have $\neg B \rightarrow \neg A$; by contraposing this we get $\neg\neg A \to \neg\neg B$. Now, using $\dfrac{A \to B}{\neg(A \,\&\, \neg B)}$ and $\dfrac{\neg\neg\neg A}{\neg A}$, we have $\neg(\neg\neg A \,\&\, \neg B)$. Suppose we have a proof of $\neg(A \to B)$, then, since the paradoxes of material implication hold, we evidently can never have a proof of B and can never have a proof of $\neg A$: so $\dfrac{\neg(A \to B)}{\neg\neg A \,\&\, \neg B}$ is a valid inference, and by contraposition, so is $\dfrac{\neg(\neg\neg A \,\&\, \neg B)}{\neg\neg(A \to B)}$. Finally, from $\neg\neg(A \to B)$ we can infer $\neg(A \,\&\, \neg B)$ by contraposing $\dfrac{A \,\&\, \neg B}{\neg(A \to B)}$, which is plainly valid. We have now shown that the following are intuitionistically equivalent:

$\neg(A \,\&\, \neg B)$, $A \to \neg\neg B$, $\neg B \to \neg A$, $\neg\neg A \to \neg\neg B$, $\neg(\neg\neg A \,\&\, \neg B)$, and $\neg\neg(A \to B)$.

The classical interdefinability of $\vee$ and & fails intuitionistically because of the invalidity of the inferences $\dfrac{\neg(\neg A \,\&\, \neg B)}{A \vee B}$ and $\dfrac{\neg(\neg A \vee \neg B)}{A \,\&\, B}$. However, De Morgan's laws still hold in the form $\dfrac{\neg(A \vee B)}{\neg A \,\&\, \neg B}$ (double line). (A double line indicates derivability in both directions.)

A final important example to note is the failure of the inference $\dfrac{A \to (B \vee C)}{(A \to B) \vee (A \to C)}$. Suppose we can transform any proof of A into a proof of $B \vee C$. Then, given any proof of A, we can convert that proof either into a proof of B or into a proof of C, but it might very well depend on the particular proof of A whether it can be transformed into a proof of one rather than the other; it may be that some proofs of A can be converted into proofs of B and others into proofs of C. It therefore does not follow that we can either transform every proof of A into a proof of B or transform every proof of A into a proof of C.

We here remark (without proof) that whenever $\neg A$ is a theorem of classical sentential logic, it is also a theorem of intuitionistic logic. So if A and B are inconsistent (i.e. $\neg(A \,\&\, B)$ is provable) by classical sentential logic, then they are inconsistent intuitionistically, and if A is provable in classical sentential logic, then $\neg\neg A$ is provable intuitionistically. Hence, in particular, $\neg\neg(A \vee \neg A)$ is provable. If we add the schema $A \vee \neg A$ to any axiom system for intuitionistic sentential logic, we obtain the classical system. It is a consequence of the fact that $\neg\neg(A \vee \neg A)$ is provable, that the same is true if we add $\neg\neg A \rightarrow A$.

It is clear directly from the meanings of the quantifiers that $\dfrac{\forall x \ \neg Fx}{\neg \exists x \ Fx}$ and $\dfrac{\exists x \ \neg Fx}{\neg \forall x \ Fx}$ hold. $\dfrac{\neg \forall x \ Fx}{\exists x \ \neg Fx}$ fails because, taking the quantification to be over the natural numbers, we might know that we can never prove $\forall x \ Fx$ without being able to produce a particular n for which we have a proof of $\neg F\bar{n}$. From these rules together with sentential logic, the positive relationships shown in the following diagram between formulas constructed by means of the quantifiers and negation can easily be established.

$$\begin{array}{c} \forall x \ Fx \\ \hline \neg\neg\forall x \ Fx \\ \hline \left\{ \begin{array}{c} \forall x \ \neg\neg Fx \\ \neg\neg\forall x \ \neg\neg Fx \\ \neg\exists x \ \neg Fx \end{array} \right\} \end{array} \qquad \left\{ \begin{array}{c} \forall x \ \neg Fx \\ \neg\neg\forall x \ \neg Fx \\ \neg\exists x \ \neg\neg Fx \\ \neg\exists x \ Fx \end{array} \right\}$$

$$\begin{array}{c} \exists x \ Fx \\ \hline \exists x \ \neg\neg Fx \\ \hline \left\{ \begin{array}{c} \neg\neg\exists x \ Fx \\ \neg\neg\exists x \ \neg\neg Fx \\ \neg\forall x \ \neg Fx \end{array} \right\} \end{array} \qquad \begin{array}{c} \exists x \ \neg Fx \\ \hline \left\{ \begin{array}{c} \neg\neg\exists x \ \neg Fx \\ \neg\forall x \ \neg\neg Fx \end{array} \right\} \\ \hline \neg\forall x \ Fx \end{array}$$

Here bracketed expressions are all equivalent. We also of course have $\dfrac{\forall x \ Fx}{\exists x \ Fx}$, and therefore $\dfrac{\forall x \ \neg Fx}{\exists x \ \neg Fx}$ and $\dfrac{\forall x \ \neg\neg Fx}{\exists x \ \neg\neg Fx}$ provided, of course, that we require that the domain of quantification be inhabited, i.e. that we know at least one element of it; but, within each of the four groups, no implications hold other than those shown. For instance, we may be able to show that we can never prove that every natural number lacks a certain property, without being able to find a specific number n of which we can show that we can never prove that it lacks the property, and hence the inference $\frac{\neg\forall x\ \neg Fx}{\exists x \neg\neg Fx}$ fails. In particular, since $\neg\forall x\ Fx$ does not imply even the double negation of $\exists x\ \neg Fx$ (nor $\neg\exists x\ \neg Fx$ the double negation of $\forall x\ Fx$), the conjunction $\neg\forall x\ Fx\ \&\ \neg\exists x\ \neg Fx$ is consistent: we cannot rule out the possibility that we may be able simultaneously to show that we can never prove that every object has a certain property and that we can never find a specific object which lacks it. This conjunction provides an example to show that the concluding remark on sentential logic does not apply to predicate logic: its negation is unprovable intuitionistically, although it is provable classically; the two conjuncts constitute formulas which are classically, but not intuitionistically, inconsistent with one another. The possibility is thus opened of there being theorems of intuitionistic mathematics which are contradictory by classical predicate logic, though not by sentential logic.

Intuitionistic sentences, unlike classical ones, do not in general have equivalents in prenex normal form. The simplest counter-example is $\neg\forall x\ Fx$, which is not only not equivalent to $\exists x\ \neg Fx$ but, in fact, to no other, more complex, prenex formula. Some of the classical equivalences used to obtain normal forms hold intuitionistically: for example, $\frac{\exists x(Fx\ \&\ A)}{\exists x\ Fx\ \&\ A}$, $\frac{\forall x(Fx\ \&\ A)}{\forall x\ Fx\ \&\ A}$, $\frac{\exists x(Fx \vee A)}{\exists x\ Fx \vee A}$, $\frac{\forall x(A \rightarrow Fx)}{A \rightarrow \forall x\ Fx}$, and $\frac{\forall x(Fx \rightarrow A)}{\exists x\ Fx \rightarrow A}$. However, while $\frac{\exists x(Fx \rightarrow A)}{\forall x\ Fx \rightarrow A}$, $\frac{\exists x(A \rightarrow Fx)}{A \rightarrow \exists x\ Fx}$, and $\frac{\forall x\ Fx \vee A}{\forall x\ (Fx \vee A)}$ all hold, none of their converses does. The failure of the first of these three classically valid inferences is the most obvious: the fact that we can transform any proof that everything has the property F into a proof of the proposition A in no way entails tht we can find a particular object such that we can transform a proof that it has the property F into a proof of A. The failure of the second inference parallels the invalidity of $\frac{A \rightarrow (B \vee C)}{(A \rightarrow B) \vee (A \rightarrow C)}$: even though, from each proof of A we can find an object having the property F, there is no reason to suppose that the object so found is independent of the particular proof of A; we may not be able to find any object such that every proof of A can be transformed into a proof that it has the property F. For the last case, which is the most interesting, suppose once more that the quantification is over the natural numbers. Then to have a proof of $\forall x\ (Fx \vee A)$ is to have an effective operation which we can recognize as associating to each number n a proof either of $F\bar{n}$ or of A. However, since there are infinitely many cases to consider, we cannot in general tell whether the operation will ever actually provide a proof of A, or will provide a proof of $F\bar{n}$ for every n; we are therefore not, in general, in a position to assert either A or $\forall x\ Fx$, and have no guarantee that we shall be in such a position after any finite number of applications of the operation which constituted the proof of $\forall x\ (Fx \vee A)$.

2
ELEMENTARY INTUITIONISTIC MATHEMATICS

2.1. INTUITIONISTIC ARITHMETIC

The name 'intuitionism' is due to Brouwer's acceptance of the Kantian thesis that our concept of the natural number series is derived from temporal intuition, our apprehension of the passage of time; not, indeed, from any particular details of our experience, but from the a priori form of that experience as involving temporal succession. (Brouwer rejected Kant's complementary thesis that geometry is based upon our a priori intuition of space; in this he was a mirror image of Frege, who accepted Kant's thesis about spatial intuition, but rejected that about temporal intuition.) Important as this conception appeared to Brouwer, it is by no means essential for the acceptance of an intuitionistic conception of arithmetic. What is essential is to regard the natural numbers as mental constructions, generated in a determinate manner by the repeated application of the successor operation to 0. Considered as an infinite structure, the totality $\underline{N}$ of natural numbers is uniquely determined: there cannot be non-isomorphic structures each with an equally good claim to represent $\underline{N}$. But an infinite structure is always to be thought of as something in process of generation, not as something the construction of which can be completed; hence we cannot interpret quantification over the elements of such a structure in the platonistic manner, as yielding a statement with a determinate truth-value given as the logical product or sum of the truth-values of the infinitely many instances. Rather, we must understand it in the way already explained, as yielding a statement for which we are provided with a definite criterion for what constitutes a proof of it, though not one that can be regarded as determinately true or false in advance of finding a proof or disproof: a proof of $\exists x\, A(x)$ will involve generating (or showing how to generate) an actual number n for which we can prove $A(\bar{n})$; a proof of $\forall x\, A(x)$ will be an operation recognizable as yielding a proof of $A(\bar{n})$ for any number n which we generate. To say that $\underline{N}$ is determinate is not to say that it forms a single, completed, structure, but only that, first, there is never any choice about how to extend any given initial segment of $\underline{N}$, and, secondly, given any mathematical object, we can always effectively recognize whether or not it can be reached by repeated application of the successor operation to 0, and hence whether or not it belongs to $\underline{N}$.

We shall, therefore, naturally take intuitionistic predicate logic as the underlying logic for first-order intuitionistic arithmetic. As in the Peano axiomatization of classical arithmetic (due in fact to Dedekind), we may take the number 0 and the successor operation ' as basic notions. If, for natural numbers given in terms of 0 and ', we conceive of equality as determined by comparison of how often ' has been applied to yield each number, and express it by =, then we may evidently assume the standard principles governing equality, namely its reflexivity and substitutivity:

$$\forall x\ x = x$$

$$\forall x \forall y\ (x = y \ \&\ A(x) \rightarrow A(y))$$

$$\forall x \forall y\ (x = y \rightarrow t(x) = t(y))\ .$$

Here A(x) represents any open sentence containing one or more occurrences of x, and A(y) the result of replacing some or all of them by y; hence the schema yields the symmetry and transitivity of equality. t(x) represents any term containing one or more occurrences of x, and t(y) the result of replacing some or all of them by y. (We could, if we wished, restrict A(x) to be an atomic open sentence, and t(x) to be a primitive term.)

Assuming that our variables are restricted to range only over natural numbers, the Peano axioms reduce to three, all of which hold intuitionistically. The third and fourth Peano axioms

$$\forall x\ x' \neq 0$$

$$\forall x \forall y\ (x' = y' \rightarrow x = y)$$

are evident from the manner in which natural numbers are generated from 0 by the successor operation ' ($x \neq y$ of course abbreviates $\neg x = y$). The fifth Peano axiom is the principle of induction, which, in the absence of second-order variables, may be represented by the schema

$$A(0) \;\&\; \forall x\, (A(x) \rightarrow A(x')) \rightarrow \forall x\, A(x),$$

where $A(x)$ is any open sentence containing the variable x (and possibly other free variables), and $A(0)$ and $A(x')$ result from replacing every occurrence of x by 0 or by x' respectively. It was already explained in section 1.2 why the principle of induction is evident intuitionistically.

We should note, however, that the least number principle

$$\exists x\, A(x) \rightarrow \exists x\, (A(x) \;\&\; \forall y_{y < x}\, \neg A(y))$$

is *not* intuitionistically valid: unless $A(x)$ happens to be decidable, the fact that we can find a definite number n of which we can prove that it satisfies $A(x)$ is no guarantee that we can find any number m satisfying $A(x)$ of which we can show that no smaller number satisfies it.

The manner in which the natural numbers are generated makes it legitimate to introduce any primitive recursive functions we please by means of recursion equations of the standard type. Thus we may in particular introduce + and ., considered as governed by:

$$\forall x\; x + 0 = x$$

$$\forall x \forall y\; x + y' = (x + y)'$$

$$\forall x\; x.0 = 0$$

$$\forall x \forall y\; x.y' = x.y + x.$$

$x < y$ can then, of course, be defined as $\exists z\, (z \neq 0 \;\&\; x + z = y)$.

In most intuitionistic theories there is, as remarked in section 1.2, a sharp distinction between intensional identity ≡ and extensional equality =. In arithmetic, however, there is no place for such a distinction: there is no stricter notion of the identity of natural numbers than the ordinary relation of equality. Hence we should be quite entitled to argue that, since equality coincides with intensional identity, and intensional identity is always decidable, equality over this domain must be decidable also. However, it makes no difference whether we assume

$$\forall x \forall y \ (x = y \lor x \neq y)$$

on these grounds or not, since it is in any case provable from the third and fourth Peano axioms cited above by induction. (It is not, of course, a law of logic.)

Note that a statement, predicate, or relation A is said to be *decidable* if we can prove (the universal closure of) $A \lor \neg A$. If we can prove (the universal closure of) $\neg\neg A \rightarrow A$, then A is said to be *stable*. We saw in section 1.3 that, for statements, decidability implies stability, but not vice versa; the reasons given there clearly extend to predicates and relations. The above formula thus expresses that equality between natural numbers is a decidable relation.

Unsurprisingly, negation is definable in intuitionistic arithmetic by

$$\neg A \quad \leftrightarrow \quad (A \rightarrow 0 = 1);$$

a little more unexpectedly, disjunction is also definable, by

$$A \lor B \leftrightarrow \exists x \, [(x = 0 \rightarrow A) \, \& \, (x \neq 0 \rightarrow B)] \quad .$$

Intuitionistic arithmetic is not, in practice, very different from classical arithmetic: that is, there are few theorems to be found in textbooks of classical number theory which cannot be proved, sometimes after minor reformulation, in intuitionistic number theory. Since the two theories are very different in principle, this is an indication of to how small an extent classical number theorists have succeeded in exploiting their platonistic assumptions.

The first-order formal system which embodies those principles of intuitionistic arithmetic indicated above is known as HA (for 'Heyting arithmetic'), and is a subsystem of the corresponding classical system, known as PA (for 'Peano arithmetic'). Without here spelling out the details of this formalization, it is of interest to note the following mapping * of formulas of PA into formulas of HA:

$$A^* = A \text{ if } A \text{ is atomic}$$
$$(A \;\&\; B)^* = A^* \;\&\; B^*$$
$$(A \vee B)^* = \neg(\neg A^* \;\&\; \neg B^*)$$
$$(A \to B)^* = A^* \to B^*$$
$$(\neg A)^* = \neg A^*$$
$$(\forall x\, A(x))^* = \forall x\, A^*(x)$$
$$(\exists x\, A(x))^* = \neg \forall x\, \neg A^*(x).$$

<u>Theorem</u> (Gödel). $\vdash_{PA} A$ iff $\vdash_{HA} A^*$.

Proof. HA is plainly a subsystem of PA, so the implication from right to left is trivial, given the equivalence of A and A* in PA. For the implication from left to right, we first note that, for any formula B not containing $\exists$ or $\vee$, it can be established that $\vdash_{HA} \neg\neg A^* \to A^*$. It is now comparatively straightforward to establish that the proof in PA of A can be transformed into a proof in HA of $\neg\neg A^*$.

This theorem provides a proof of the consistency of PA relative to HA. The fact that no one has ever regarded this as a finitistic consistency proof for PA shows the extent to which intuitionistic arithmetic goes beyond purely finitistic arithmetic.

We note that the mapping originally used by Gödel (1932) had the clause: $(A \to B)^* = \neg(A^* \;\&\; \neg B^*)$. The version presented here was given by Gentzen in 1936.

2.2. REAL NUMBERS

Given the natural numbers, construction of the rationals poses no problem since the classical definition is perfectly acceptable intuitionistically. Moreover we can effectively correlate rationals to natural numbers so that considerations concerning sequences of natural numbers apply equally to sequences of rationals.

Fundamental to the construction of the real numbers is the notion of an *infinite sequence* (of rationals). Its treatment of this notion is one of the most remarkable aspects of intuitionism. For the present, however, we shall not need to invoke the special features of the intuitionistic notion of an infinite sequence, but may be content with assuming all such sequences to be given by effective rules for calculating each term (the so-called *lawlike* sequences). We use $\langle r_n \rangle$, $\langle s_n \rangle$, $\langle t_n \rangle$, ... to denote such lawlike sequences of rational numbers.

Real numbers are defined as equivalence classes of Cauchy sequences of rationals:

Definition. $\langle r_n \rangle$ is a *real number generator* iff

$$\forall k\, \exists n\, \forall m_{m > n} |r_m - r_n| < 2^{-k} \quad .$$

Apart from the different possible notions of sequence, there are sequences which would satisfy the Cauchy condition if the quantifiers were interpreted classically, but which cannot intuitionistically be asserted to be real number generators. For example, define $\langle r_n \rangle$ by:

$$r_n = \begin{cases} 1 & \text{if } 2n+1 \text{ is a perfect number and } \forall m_{m<n}\ 2m+1 \text{ is not} \\ 2^{-n} & \text{otherwise .} \end{cases}$$

Classically this is clearly a Cauchy sequence, but suppose the condition is satisfied with the quantifiers understood intuitionistically. Then, for any k, we can find an n such that

$|r_m - r_n| < 2^{-k}$ for all $m > n$. But, taking $k = 1$, this means that we can actually exhibit some n such that $r_m \neq 1$ for all $m > n$. So we have a proof that either there is an odd perfect number less than 2n+3 or there is no odd perfect number. Since we do not in fact know this, $\langle r_n \rangle$ cannot be claimed as a real number generator.

Definition. $\langle r_n \rangle \sim \langle s_n \rangle$ iff $\forall k \, \exists n \, \forall m_{m>n} \, |r_m - s_m| < 2^{-k}$.

Theorem. $\sim$ is an equivalence relation.

The proof of this is left as an exercise.

We wish to define real numbers as equivalence classes of real number generators under this equivalence relation. In order to do so, we must introduce the notion of a *species*, the intuitionistic analogue of the classical notion of a set, or, more exactly, that of a *property*. Given any well-defined domain for variables of quantification, for instance the natural numbers, we regard a species of elements of that domain as determined by a definite condition which any element must satisfy to be a member of that species; a condition is definite when we know what to count as a proof, for any element of the domain, that it satisfies that condition. Where 'A(a)' expresses such a condition we denote the species so determined by '{a | A(a)}' and hold 'b ∈ {a | A(a)}' as true when the element denoted by 'b' satisfies that condition, i.e. as synonymous with 'A(b)'. Since the procedure of forming species may be reiterated (so that, e.g., we may consider species of real number generators), we impose on species a hierarchy corresponding to the simple theory of types, the types being non-cumulative. It is clear, however, that we can have no constructive justification for the full impredicative comprehension axiom schema, asserting the existence of a species of objects satisfying any statable condition, including one involving quantification over species of the same or higher type; for such a condition would not be definite unless the domain of the species-variables was already determinate, and, in attempting to specify that domain in terms of definite conditions for membership of a species, we should be committing the fallacy prohibited by Russell's vicious circle principle. It is equally clear that the predicative comprehension axiom schema is acceptable: there will be a species determined by any condition involving quantification only over already specified domains. Thus, for each type, we admit, as first-level species of that type, all those determined by conditions expressible without quantification over the same or any higher type, and without free variables for species of higher type. We could then, if we wished, develop a complete ramified hierarchy by reiteration: each type would be subdivided into levels (usually called orders), an n-th level species being determined by a condition not involving quantification over species of level $\geq n$ or reference to species of level $> n$. We do not need, however, to go into such matters in any detail. Whether existence assumptions for species more generous than predicative comprehension are intuitionistically acceptable is a further question that has not been extensively investigated.

The usual set-theoretic notions of intersection, union, and inclusion can be transferred to species. A species $\underline{X}$ is *inhabited* iff $\exists a\ a \in \underline{X}$; $\underline{X}$ is a *detachable* subspecies of $\underline{Y}$ iff $\underline{X} \subseteq \underline{Y}$ & $\forall a_{a \in \underline{Y}}(a \in \underline{X} \vee a \notin \underline{X})$. A *decidable* species $\underline{X}$ is one detachable from the universe: $\forall a(a \in \underline{X} \vee a \notin \underline{X})$. The notion of species is plainly an intensional one: species are strictly identical only when they are given in the same way. However, we use the equality symbol '=' for extensional equivalence: if $\underline{X}$ and $\underline{Y}$ are species, $\underline{X} = \underline{Y}$ iff $\forall a(a \in \underline{X} \leftrightarrow a \in \underline{Y})$.

Definition. A *real number* is a species of the form

$$\{\langle s_n\rangle \mid \langle r_n\rangle \sim \langle s_n\rangle\}$$

for some real number generator $\langle r_n\rangle$.

The letters 'x', 'y', 'z', ... will denote real numbers. Instead of defining the reals as equivalence classes of real number generators, we could equally well develop the whole theory just in terms of generators. However, our definition simplifies the order relations involved, allowing us the extensional equality of species. A real number will always be taken to be given by reference to some representative real number generator belonging to it.

One relation distinguishing real numbers is $x \neq y$, but we can also define a stronger relation:

Definition. $\langle r_n \rangle \# \langle s_n \rangle$ iff $\exists k \exists n \forall m_{m>n} |r_m - s_m| > 2^{-k}$.

Definition.

$x \# y$ iff $\exists \langle r_n \rangle \exists \langle s_n \rangle (\langle r_n \rangle \in x \ \& \ \langle s_n \rangle \in y \ \& \ \langle r_n \rangle \# \langle s_n \rangle)$.

Theorem 1. $x \# y \to x \neq y$.

Proof. Suppose $x \# y$. Then, by the definition, we can find real number generators $\langle r_n \rangle \in x$ and $\langle s_n \rangle \in y$ such that $\langle r_n \rangle \# \langle s_n \rangle$, and so $\neg \langle r_n \rangle \sim \langle s_n \rangle$. So $\langle r_n \rangle \in x$ but $\langle r_n \rangle \notin y$. Therefore $x \neq y$.

Theorem 2. $x = y \ \& \ y \# z \to x \# z$.

Proof. Suppose given $\langle r_n \rangle \in x$, $\langle s_n \rangle \in y$, and $\langle t_n \rangle \in z$ such that $\langle r_n \rangle \sim \langle s_n \rangle$ and $\langle s_n \rangle \# \langle t_n \rangle$. Then for any k we can find an n_1, satisfying

$$\forall m_{m>n_1} |s_m - t_m| > 2^{-k},$$

and an n_2 such that

$$\forall m_{m>n_2} |r_m - s_m| < 2^{-k-1}.$$

Now if $n = \max(n_1, n_2)$, we have, for $m > n$,

$$|r_m - t_m| = |(s_m - t_m) - (s_m - r_m)|$$

$$\geq |s_m - t_m| - |r_m - s_m|,$$ by the triangle inequality.

So $\forall m_{m>n} |r_m - t_m| > 2^{-k} - 2^{-k-1} = 2^{-k-1}$.

$\therefore \langle r_n \rangle \# \langle t_n \rangle$

and $x \# z$.

Theorem 3. $\neg x \# y \to x = y$.

Proof. Suppose given $\langle r_n \rangle$ and $\langle s_n \rangle$ such that $\langle r_n \rangle \in x$, $\langle s_n \rangle \in y$, and $\neg \langle r_n \rangle \# \langle s_n \rangle$. Since $\langle r_n \rangle$ and $\langle s_n \rangle$ are real number generators, for any k we can compute an n sufficiently large so that

$$\forall m_{m>n} |r_m - r_n| < 2^{-k-2}$$

and $\forall m_{m>n}\ |s_m - s_n| < 2^{-k-2}$.

Since r_n and s_n are particular rationals, it is decidable whether their absolute difference is greater than or equal to 2^{-k}, or not. Suppose $|r_n - s_n| \geq 2^{-k}$.

Then $\forall m_{m>n}\ |r_m - s_m| \geq |r_n - s_n| - |r_m - r_n|$

$$- |s_m - s_n| > 2^{-k} - 2^{-k-2} - 2^{-k-2} = 2^{-k-1}.$$

But then $\langle r_n \rangle \# \langle s_n \rangle$, contradicting the hypothesis. Hence $|r_n - s_n| < 2^{-k}$.

This gives $\forall m_{m>n}\ |r_m - s_m| \leq |r_n - s_n| + |r_m - r_n|$

$$+ |s_m - s_n| < 2^{-k} + 2^{-k-2} + 2^{-k-2} < 2^{-k+1}.$$

Since for each k we can find such an n, we have

$$\langle r_n \rangle \sim \langle s_n \rangle$$

and so $x = y$.

<u>Corollary</u>. $\neg\neg x = y \rightarrow x = y$.

Proof. Contraposing Theorem 1, we obtain $\neg\neg x = y \rightarrow \neg x \# y$. The result now follows immediately from Theorem 3.

Notice that by contraposing Theorem 1 we also get the converse of Theorem 3, $x = y \rightarrow \neg x \# y$. Given an expression A, in order to prove $\neg\neg A \rightarrow A$ it is sufficient to find a B for which $A \leftrightarrow \neg B$ holds; for then, contraposing twice, we have $\neg\neg A \leftrightarrow \neg\neg\neg B$, and so $\neg\neg A \leftrightarrow \neg B$, i.e. $\neg\neg A \leftrightarrow A$. Thus, given Theorem 3 and its converse, it follows immediately that $x = y$ is stable.

<u>Theorem 4</u>. $x \# y \rightarrow x \# z \vee y \# z$.

Proof. Assume $x \# y$. As usual we can find $\langle r_n \rangle \in x$ and $\langle s_n \rangle \in y$ for which $\langle r_n \rangle \# \langle s_n \rangle$. Now we construct k and n such that

$$\forall m_{m>n}\ |r_m - s_m| > 2^{-k}$$

$$\forall m_{m>n}\ |r_{n+1} - r_m| < 2^{-k-3}$$

$$\forall m_{m>n}\ |s_{n+1} - s_m| < 2^{-k-3}$$

and $\forall m_{m>n}\ |t_{n+1} - t_m| < 2^{-k-3}$, where $<t_n>$ is any real number generator by which z is given.
So, in particular, $|r_{n+1} - s_{n+1}| > 2^{-k}$. By the triangle inequality, this gives $|r_{n+1} - t_{n+1}| + |s_{n+1} - t_{n+1}| > 2^{-k}$. Therefore, since r_{n+1}, s_{n+1}, t_{n+1} are particular rationals and order among rationals is decidable,

$$|r_{n+1} - t_{n+1}| > 2^{-k-1} \vee |s_{n+1} - t_{n+1}| > 2^{-k-1}.$$

Case 1. Suppose $|r_{n+1} - t_{n+1}| > 2^{-k-1}$. Then

$$\forall m_{m>n}\ |r_m - t_m| \geq |r_{n+1} - t_{n+1}| - |r_{n+1} - r_m| - |t_{n+1} - t_m|,$$

$$> 2^{-k-1} - 2^{-k-3} - 2^{-k-3} = 2^{-k-2}.$$

$\therefore\ <r_n> \# <t_n>$,
whence x # z.

Case 2. Suppose $|s_{n+1} - t_{n+1}| > 2^{-k-1}$, then, similarly
  y # z.
Hence x # z ∨ y # z.

Consider any binary relation, #, defined on a species S on which some extensional equality relation is defined.

Definition. # is an *apartness relation* if and only if it satisfies for all x, y, z ∈ S
(i) x # y → y # x
(ii) ¬x # y ↔ x = y
(iii) x # y → x # z ∨ y # z.

Theorem 5. The relation # defined on the reals is an apartness relation.

Proof. (i) is obviously satisfied. (ii) is proved by Theorem 3 and contraposition of Theorem 1, and (iii) is Theorem 4.

Definition. The operations $+$, $\cdot$, $-$, and inverse are defined on real number generators as follows:

(a) $\langle r_n\rangle + \langle s_n\rangle = \langle t_n\rangle$, where $t_n = r_n + s_n$;

(b) $\langle r_n\rangle \cdot \langle s_n\rangle = \langle t_n\rangle$, where $t_n = r_n \cdot s_n$;

(c) $-\langle r_n\rangle = \langle t_n\rangle$, where $t_n = -r_n$;

(d) $\langle r_n\rangle^{-1} = \langle t_n\rangle$, where $t_n = \begin{cases} r_n^{-1} & \text{if } r_n \neq 0 \\ 0 & \text{if } r_n = 0 . \end{cases}$

For these definitions to be of any use, we must check that (a)-(d) have in fact defined real number generators. The proofs are straightforward and are left as an exercise, but it must be noted that for (d) we require the additional assumption that $\langle r_n\rangle \# 0$. (Here 0 is the real number generator $\langle s_n\rangle$ such that $s_n = 0$ for all n.) So we state

Theorem 6. If $\langle r_n\rangle$ and $\langle s_n\rangle$ are real number generators then so are $\langle r_n\rangle + \langle s_n\rangle$, $\langle r_n\rangle \cdot \langle s_n\rangle$, and $-\langle r_n\rangle$. Also, if $\langle r_n\rangle \# 0$ then $\langle r_n\rangle^{-1}$ is a real number generator.

Before extending these definitions in the obvious way to the reals, we also have to check that $\sim$ is a congruence with respect to these operations.

Theorem 7. If $\langle r_n\rangle$ and $\langle s_n\rangle$ are real number generators, then:

(a) $\langle r_n\rangle \sim \langle r'_n\rangle \;\&\; \langle s_n\rangle \sim \langle s'_n\rangle \rightarrow \langle r_n\rangle + \langle s_n\rangle \sim \langle r'_n\rangle + \langle s'_n\rangle$

(b) $\langle r_n\rangle \sim \langle r'_n\rangle \;\&\; \langle s_n\rangle \sim \langle s'_n\rangle \rightarrow \langle r_n\rangle \cdot \langle s_n\rangle \sim \langle r'_n\rangle \cdot \langle s'_n\rangle$

(c) $\langle r_n\rangle \sim \langle r'_n\rangle \rightarrow -\langle r_n\rangle \sim -\langle r'_n\rangle$

(d) $\langle r_n\rangle \sim \langle r'_n\rangle \;\&\; \langle r_n\rangle \# 0 \rightarrow \langle r_n\rangle^{-1} \sim \langle r'_n\rangle^{-1}$.

The proof of this theorem involves no particular complication. We now extend the definitions of the operations to the real numbers as follows:

Definition. If x and y are real numbers, then

(a) $x + y = \{\langle r_n\rangle + \langle s_n\rangle \mid \langle r_n\rangle \in x \;\&\; \langle s_n\rangle \in y\}$;

(b) $x \cdot y = \{\langle r_n\rangle \cdot \langle s_n\rangle \mid \langle r_n\rangle \in x \;\&\; \langle s_n\rangle \in y\}$;

(c) $-x = \{-\langle r_n\rangle \mid \langle r_n\rangle \in x\}$; and

(d) If $x \# 0$, $x^{-1} = \{\langle r_n\rangle^{-1} \mid \langle r_n\rangle \in x\}$.

Concerning (d), notice that, since we cannot in general tell whether $x \# 0$ or not, we cannot always decide, given any x, whether or not x^{-1} is defined.

Theorem 8. $x \# y \rightarrow x + z \# y + z$.

The proof of this is left to the reader.

Theorem 9. $x \cdot y \# 0 \leftrightarrow x \# 0 \ \& \ y \# 0$.

Proof. From right to left the proof is simple. To prove the implication from left to right, consider $\langle r_n\rangle \in x$ and $\langle s_n\rangle \in y$, for which we can find n and k such that

$$\forall m_{m>n} \ |r_m \cdot s_m| > 2^{-k}$$

$$\text{and } \forall m_{m>n} \ |s_m - s_n| < 1.$$

But then $\forall m_{m>n} |r_m| > \dfrac{2^{-k}}{|s_n| + 1} > 2^{-k_o}$, for some

k_o which can be computed from k and s_n.

$\therefore \langle r_n\rangle \# 0$, and similarly $\langle s_n\rangle \# 0$.

Theorem 10. $x + y \# 0 \rightarrow x \# 0 \vee y \# 0$.

Proof. Assume $x + y \# 0$. Then, by Theorem 8,

$$-x + x + y \# -x.$$

$\therefore y \# -x$.

So, by Theorem 4, $y \# 0 \vee -x \# 0$.

But, if $-x \# 0$, $x - x \# x$, again using Theorem 8; so $x \# 0$.

Hence $x \# 0 \vee y \# 0$.

It is important to notice that we have not stated $x \cdot y = 0 \rightarrow x = 0 \vee y = 0$. In fact, we can produce a counter-example to this as follows: choose any decidable predicate A(x) for which $\exists x A(x)$ is an unsolved problem of arithmetic. (For example A(n) holds just in case there is a sequence 123456789 in the expansion of π such that the 9 in this sequence occurs in the n-th decimal place of the expansion.) Let B(x) also be decidable, and chosen so that we do not know whether B(k) holds, where k is the least number, if such a number can be found, for which A(k) holds. Now we define real number generators $\langle r_n \rangle$ and $\langle s_n \rangle$ by

$$r_n = \begin{cases} 2^{-n} & \text{if } \forall m_{m \le n} \neg A(m) \\ 2^{-n} & \text{if } k \le n \;\&\; A(k) \;\&\; \forall m_{m<k} \neg A(m) \;\&\; \neg B(k) \\ 2^{-k} & \text{if } k \le n \;\&\; A(k) \;\&\; \forall m_{m<k} \neg A(m) \;\&\; B(k) \end{cases}$$

$$\text{and} \quad s_n = \begin{cases} 2^{-n} & \text{if } \forall m_{m \le n} \neg A(m) \\ 2^{-n} & \text{if } k \le n \;\&\; A(k) \;\&\; \forall m_{m<k} \neg A(m) \;\&\; B(k) \\ 2^{-k} & \text{if } k \le n \;\&\; A(k) \;\&\; \forall m_{m<k} \neg A(m) \;\&\; \neg B(k). \end{cases}$$

With $\langle r_n \rangle$ and $\langle s_n \rangle$ defined like this we cannot tell, until the problem is solved, whether $\langle r_n \rangle \sim 0$ or not, or whether $\langle s_n \rangle \sim 0$ or not. So we cannot assert $\langle r_n \rangle \sim 0 \vee \langle s_n \rangle \sim 0$, but we can obviously show that $\langle r_n \cdot s_n \rangle \sim 0$. Thus, if $x = \{\langle t_n \rangle \mid \langle t_n \rangle \sim \langle r_n \rangle\}$ and $y = \{\langle t_n \rangle \mid \langle t_n \rangle \sim \langle s_n \rangle\}$, we can prove $x \cdot y = 0$ but not $x = 0 \vee y = 0$.

It should be noted that the above is not a counter-example in the standard sense; it sets against the general statement to which it is a counter-example, not an instance of it whose negation can be proved, but merely an instance which cannot be proved and will not be proved until a certain other problem is solved. It gains its force from the fact that we have a uniform way of constructing similar 'counter-examples' for each unsolved problem of the same form. Since we can be virtually certain that the supply of such unsolved problems will never dry up, we can conclude with equal certainty that the general statement will never be intuitionistically provable. Such a recognition that a universally quantified statement is unprovable does not amount to a proof of its negation, since the proposition that there will always be suitable unsolved problems is not, as it stands, a theorem or even a mathematical proposition at all. In some instances when 'counter-examples' of the present kind to some general statement can be given, appeal to more powerful principles of intuitionistic mathematics may yield an actual refutation of the general statement; but, even when this cannot be done, such 'counter-examples' are useful, as providing a practical indication of what we cannot hope to prove. Their interest indeed remains of this purely negative character, and they therefore make no contribution to the positive part of intuitionistic mathematics; but, then, the same is often true in classical mathematics of theorems which provide counter-examples of the standard kind.

2.3. ORDER RELATIONS

Definition. For real number generators $<r_n>$ and $<s_n>$,

$$<r_n> \; < \; <s_n> \text{ iff } \exists k \, \exists n \, \forall m_{m>n} \, (s_m - r_m) > 2^{-k}.$$

Theorem 11. $\sim$ is a congruence relation with respect to $<$.
We can now extend this definition to the reals in the obvious way.

Definition. If x and y are real numbers

$$x < y \text{ iff } \exists <r_n> \exists <s_n> (<r_n> \in x \; \& \; <s_n> \in y \; \& \; <r_n> \; < \; <s_n>).$$

The general classification of order relations is more complex than in a classical setting, so we consider the general case of a binary relation $<$ defined on some species $\underline{S}$ on which there is also defined some equality relation $=$ (assumed to be reflexive, symmetrical and transitive). We write '$x > y$' for '$y < x$', '$x \nless y$' for '$\neg x < y$', and '$x \ngtr y$' for '$\neg x > y$'. There is no difficulty in defining the notion of a partial order.

Definition. $<$ is a *partial order* iff, for all x, y, z $\in \underline{S}$, it satisfies

(a) $x < y \rightarrow x \ngtr y \; \& \; x \neq y;$

(b) $x = y \; \& \; y < z \rightarrow x < z;$

(c) $x < y \; \& \; y = z \rightarrow x < z;$

(d) $x < y \; \& \; y < z \rightarrow x < z.$

We can then, indeed, define:

Definition. $<$ is a *total order* (also called a *simple order*) iff it is a partial order and further satisfies

(e) $x = y \vee x < y \vee x > y$.

However, in many cases (e) is too strong a requirement to be realized; we cannot hope to define a total order upon the reals. It is therefore of interest to consider weaker requirements, classically equivalent to (e). We first define:

Definition. $<$ is a *weak order* iff it is a partial order and also satisfies

(f) $x \not< y \ \& \ x \not> y \to x = y$.

(Classically, of course, a weak order in the above sense is already a total order.) Two further requirements are also of interest.

Definition. $<$ is a *comparative order* (also called a *pseudo-order*) iff it is a weak order and also satisfies

(g) $x < y \to x < z \vee z < y$.

Definition. $<$ is a *virtual order* iff it is a weak order and also satisfies

(h) $x \not< y \ \& \ x \neq y \to x > y$.

We now prove a few theorems concerning these general notions.

Theorem 12. If $<$ is a weak order, and $\dot{<}$ is defined by

$x \dot{<} y \leftrightarrow x \not> y \ \& \ x \neq y$, then

(i) $x \dot{\not<} y \leftrightarrow x \not< y$

(ii) $x \dot{<} y \leftrightarrow \neg\neg x < y$

(iii) $\dot{<}$ is a virtual order.

Proof. (i) By (a), $x < y \to x \dot{<} y$; so, by contraposition,

$x \dot{\not<} y \to x \not< y$.

Conversely, if $x \not< y$, then, by (f), $x \not> y \to x = y$;

$\therefore \ \neg(x \dot{<} y)$,

i.e. $x \dot{\not<} y$.

(ii) $x \dot{<} y$ is equivalent to the negated expression $\neg(x > y \vee x = y)$, and is therefore stable. That is, $x \dot{<} y \leftrightarrow \neg\neg x \dot{<} y$. But contraposing (i) gives $\neg\neg x \dot{<} y \leftrightarrow \neg\neg x < y$.

$\therefore x \dot{<} y \leftrightarrow \neg\neg x < y$.

(iii) Write ($\dot{\text{a}}$) for condition (a) on $\dot{<}$, and so on. We need to show ($\dot{\text{a}}$) - ($\dot{\text{d}}$), ($\dot{\text{f}}$), and ($\dot{\text{h}}$).

$(\dot{a})$ $x \mathrel{\dot{<}} y \to x \not> y \;\&\; x \neq y$, by the definition of $\dot{<}$.
$\therefore$ $x \mathrel{\dot{<}} y \to x \mathrel{\dot{\not>}} y \;\&\; x \neq y$, by (i).
$(\dot{b})$ $x = y \;\&\; y \mathrel{\dot{<}} z \to x = y \;\&\; \neg\neg y < z$, by (ii).
So $x = y \;\&\; y \mathrel{\dot{<}} z \to \neg\neg(x = y \;\&\; y < z)$.
Now contraposing (b) twice and using (ii) again we get
$x = y \;\&\; y \mathrel{\dot{<}} z \to x \mathrel{\dot{<}} z$.
$(\dot{c})$ and $(\dot{d})$ follow exactly similarly.
$(\dot{f})$ follows directly from (f) using (i), and
$(\dot{h})$ follows in the same way from (h) and (i).

Theorem 13. Every virtual order is stable.

Proof. Let $<$ be a virtual order. Then, by (a) and (h),
$x < y \leftrightarrow x \not> y \;\&\; x \neq y$.
$\therefore$ $x < y \leftrightarrow \neg(x > y \lor x = y)$. Thus $x < y$ is equivalent to a negated statement, and is therefore stable.

Theorem 14. If a weak order can be defined on a species, then the equality relation on that species must be stable; i.e. if $<$ is a weak order,
$\neg\neg x = y \to x = y$.

Proof. By (a), if $x = y$, then $x \not< y$ and $x \not> y$.
So, by (f), $x = y \leftrightarrow x \not< y \;\&\; x \not> y$. As usual we apply De Morgan's law to see that $x = y$ is equivalent to a negated statement.
$\therefore$ $x = y$ is stable.

The next theorem is not in itself of special interest, but is mentioned because we have to appeal to it frequently.

Theorem 15. If $<$ is a comparative order, then
$x \not> y \;\&\; y < z \to x < z$.

Proof. Suppose $x \not> y \;\&\; y < z$. Then, by (g),
$x < z \lor y < x$. But $y \not< x$, by assumption.
$\therefore$ $x < z$.

From now on we return to using x, y, z for real numbers and $<$ and $=$ for the relevant relations between them.

Theorem 16. $x \# y \leftrightarrow x < y \lor x > y$.

Proof. From right to left the proof is obvious. For the converse, we find $\langle r_n\rangle \in x$, $\langle s_n\rangle \in y$, n_o, and k such that

$$\forall m_{m>n_o} |r_m - s_m| > 2^{-k}.$$

Further, we can find an n, $n > n_o$, such that

$$\forall m_{m>n} \, |r_m - r_n| < 2^{-k-2}$$

and $\forall m_{m>n} \, |s_m - s_n| < 2^{-k-2}$.

Since r_n and s_n are particular rationals, we have

$$r_n - s_n > 2^{-k} \vee s_n - r_n > 2^{-k}.$$

Case 1. If $r_n - s_n > 2^{-k}$,

$$\forall m_{m>n}(r_m - s_m) \geq (r_n - s_n) - |s_m - s_n| - |r_m - r_n| > 2^{-k} - 2^{-k-2} - 2^{-k-2} = 2^{-k-1}.$$

Hence $\langle r_n\rangle > \langle s_n\rangle$, and so $x > y$.

Case 2. If $s_n - r_n > 2^{-k}$, then by parallel reasoning, we obtain $x < y$.

Accordingly, $x < y \vee x > y$.

<u>Theorem 17</u>. $<$ is a comparative order on the reals.

Proof. It is easy to show that $<$ satisfies (a) - (d). For (f), suppose $x \nless y$ & $x \ngtr y$; i.e. $\neg(x < y \vee x > y)$.

Contraposing Theorem 16, this gives $\neg x \# y$, which by Theorem 3, implies $x = y$.

To verify (g), suppose $x < y$. Take $\langle r_n\rangle \in x$, $\langle s_n\rangle \in y$, $\langle t_n\rangle \in z$, and find n and k such that

$$\forall m_{m>n} \;\; s_m - r_m > 2^{-k}$$

$$\forall m_{m>n} \, |r_{n+1} - r_m| < 2^{-k-3}$$

$$\forall m_{m>n} \, |s_{n+1} - s_m| < 2^{-k-3}$$

and $\forall m_{m>n} \, |t_{n+1} - t_m| < 2^{-k-3}$.

Then, in particular, $s_{n+1} - r_{n+1} > 2^{-k}$, and we can derive

$$t_{n+1} - r_{n+1} > 2^{-k-1} \vee s_{n+1} - t_{n+1} > 2^{-k-1}.$$

We can now argue by cases as usual, obtaining $\langle r_n \rangle < \langle t_n \rangle$ on assumption of the first disjunct, and $\langle t_n \rangle < \langle s_n \rangle$ if we assume the second. Hence $x < y \rightarrow x < z \vee z < y$.

The corresponding classical ordering of the reals is, of course, a linear order, but intuitionistically condition (e) cannot be proved, so it may not be claimed as a simple order. This means that, in general, we cannot argue by cases as the classical mathematician can. However, a very useful alternative is to pick two separated reals, e.g. 0 and 1; then by (g), since $0 < 1$, we can argue from $0 < x \vee x < 1$.

<u>Definition</u>. $x \leqq y \leftrightarrow x < y \vee x = y$.

<u>Definition</u>. If $\dot{<}$ is defined as in Theorem 12, the corresponding weaker relation is given by $x \dot{\leqq} y \leftrightarrow x \dot{<} y \vee x = y$.

Recalling the definition of $\dot{<}$, we see that $x \dot{\leqq} y$ is equivalent to $x \not> y \;\&\; (x = y \vee x \neq y)$.

There seem to be no convincing philosophical reasons why the use of negation should be disallowed in constructions of mathematical objects, and intuitionists make free use of it. Bishop's system of constructive analysis is, by contrast, formulated entirely in positive terms. His $\leq$ relation between reals is given essentially by the following definitions.

<u>Definition</u>. For real number generators $\langle r_n \rangle$ and $\langle s_n \rangle$,

$$\langle r_n \rangle \leq \langle s_n \rangle \text{ iff } \forall k \, \exists n \, \forall m_{m>n} \; r_m - s_m < 2^{-k}.$$

It is easy to show that $\sim$ is a congruence with respect to $\leq$.

<u>Definition</u>. If x and y are real numbers,

$$x \leq y \text{ iff } \exists \langle r_n \rangle \, \exists \langle s_n \rangle (\langle r_n \rangle \in x \;\&\; \langle s_n \rangle \in y \;\&\; \langle r_n \rangle \leq \langle s_n \rangle).$$

<u>Theorem 18</u>. $x \leq y \leftrightarrow x \not> y$.

Proof. From left to right the implication is trivial.

The proof of the converse mimics the proof of Theorem 3. The details are left as an exercise.

<u>Definition</u>. $\max(\langle r_n\rangle,\langle s_n\rangle) = \langle t_n\rangle$, where $t_n = \max(r_n,s_n)$.

$\min(\langle r_n\rangle,\langle s_n\rangle) = \langle t_n\rangle$, where $t_n = \min(r_n,s_n)$.

<u>Definition</u>. $\max(x,y) = \{\max(\langle r_n\rangle,\langle s_n\rangle) \mid \langle r_n\rangle \in x \ \&\ \langle s_n\rangle \in y\}$.

$\min(x,y) = \{\min(\langle r_n\rangle,\langle s_n\rangle) \mid \langle r_n\rangle \in x \ \&\ \langle s_n\rangle \in y\}$.

<u>Theorem 19</u>. (i) $\max(x,y) = \max(y,x)$.
(ii) $\max(x,y) \not< x$ and $\max(x,y) \not< y$.
(iii) $\min(x,y) = \min(y,x)$.
(iv) $\min(x,y) \not> x$ and $\min(x,y) \not> y$.
(v) $\max(x,y) \not< \min(x,y)$.

<u>Theorem 20</u>. $z > \max(x,y) \leftrightarrow z > x \ \&\ z > y$.

Proof. Suppose $z > \max(x,y)$. By Theorem 19 (ii), $x \not> \max(x,y)$. But Theorem 15 now gives $z > x$; similarly we get $z > y$. Conversely, if x, y, z are given by $\langle r_n\rangle$, $\langle s_n\rangle$, $\langle t_n\rangle$, we can find n and k such that

$$\forall m_{m>n}\ t_m - r_m > 2^{-k}$$
$$\forall m_{m>n}\ t_m - s_m > 2^{-k}$$
$$\therefore\ \forall m_{m>n}\ t_m - \max(r_m,s_m) > 2^{-k}\ .$$

Hence $z > \max(x,y)$.

As is to be expected, we can also easily see that $\max(x,y) + \min(x,y) = x + y$.

<u>Definition</u>. $|\langle r_n\rangle| = \max(\langle r_n\rangle, -\langle r_n\rangle)$.

<u>Definition</u>. The absolute value of x is given by $|x| = \{|\langle r_n\rangle| \mid \langle r_n\rangle \in x\}$.

The familiar triangle inequality remains in the following form:

<u>Theorem 21</u>. $|x + y| \not> |x| + |y|$.

The following definition of a real interval appears long-winded; this is unavoidable since we do not, in general, know whether $x < y$ or $y < x$.

Definition.

$$[x,y] = \{z \mid \neg(z < x \ \& \ z < y) \ \& \ \neg(z > x \ \& \ z > y)\}.$$

By Theorem 20 and the corresponding result for min(x,y), this definition is clearly equivalent to $[x,y] = \{z \mid z \not> \max(x,y) \ \& \ z \not< \min(x,y)\}$.

Theorem 22. $[x,y] = [\min(x,y), \max(x,y)]$.

If we do know something about the order of x and y, we can simplify the expression for [x,y]:

Theorem 23. $x \not> y \rightarrow [x,y] = \{z \mid z \not< x \ \& \ z \not> y\}$.

Proof. Trivially, $\{z \mid z \not< x \ \& \ z \not> y\} \subseteq [x,y]$. For the converse inclusion, suppose $z < x$. Then, since $x \not> y$, $z < y$, by Theorem 15. So $z \notin [x,y]$. Therefore, $z \in [x,y] \rightarrow z \not< x$, and likewise $z \in [x,y] \rightarrow z \not> y$. So this inclusion holds. $\therefore [x,y] = \{z \mid z \not< x \ \& \ z \not> y\}$.

2.4. THE AXIOM OF CHOICE

It might at first seem surprising that in a system of constructive mathematics we should adopt as an axiom the Axiom of Choice, which has been looked at askance on constructive grounds. The fact is, however, that the axiom is only dubious under a half-hearted platonistic interpretation of the quantifiers. Consider the following two versions of the axiom

$$AC_{n,m}: \ \forall n \ \exists m \ A(n,m) \rightarrow \exists a \ \forall n \ A(n,a(n))$$

$$AC_{n,b}: \ \forall n \ \exists b \ A(n,b) \rightarrow \exists a \ \forall n \ A(n, \lambda m.a(n,m)),$$

where the variables a and b range over functions from natural numbers to natural numbers. If, for instance, in $AC_{n,m}$ we adopt a platonistic interpretation of quantification over the natural numbers, but demand that any function that is asserted to exist must be effectively calculable, then, of course, $AC_{n,m}$ is not true (at least when A(n,m) is not decidable). But on a thoroughgoing platonistic interpretation of the quantifiers (indeed, even on one which demands of a that it be describable), $AC_{n,m}$ is obviously true, for we can define a, by using the least number principle, as

$$a(n) = \min\{m \mid A(n,m)\}.$$

When we interpret all the quantifiers intuitionistically we cannot use the same justification, since the least number principle does not in general hold. However, intuitionistically, the antecedent of $AC_{n,m}$ expresses not merely that for each n we can effectively find an m for which we can prove A(n,m), but that we have a single effective procedure which we can recognize as yielding, for each n, such an m: the consequent merely makes this explicit, the constructive function a being that which, when applied to n, gives a suitable m. $AC_{n,b}$ is likewise evident when the quantifiers in the antecedent are intuitionistically understood. We therefore assume $AC_{n,m}$ and $AC_{n,b}$ as principles of intuitionistic mathematics, the variables a and b being taken as ranging over *constructive* (effectively calculable) functions of natural numbers. Unary constructive functions of natural numbers may be identified with lawlike sequences of natural numbers: it makes no difference whether we speak of the value of a for the argument n or of the n-th term of a. In $AC_{n,b}$ b is tacitly taken as ranging over unary constructive functions and a over binary constructive functions: we could, if we wished, take both as ranging over unary functions if we wrote $\lambda m.\ a(2^n.3^m)$ for $\lambda m.\ a(n,m)$. It is easy to see that $AC_{n,m}$ is derivable from $AC_{n,b}$.

The following proof of the completeness of the real number system serves as an example of the applications of the axiom. First we need two definitions.

Definition. $\langle x_n \rangle$ is a *Cauchy sequence* of reals iff

$$\forall k\ \exists n\ \forall m_{m>n}\ |x_m - x_n| < 2^{-k}.$$

Definition. $\lim_{n\to\infty}\langle x_n\rangle = y$ ($\langle x_n\rangle$ *converges to* y) iff

$$\forall k\ \exists n\ \forall m_{m>n}\ |y - x_m| < 2^{-k}.$$

Theorem 24. Every Cauchy sequence of reals converges.

Proof. Consider any Cauchy sequence $\langle x_n\rangle$, where, for each n, x_n is given by a real number generator

$<r_i^{(n)}>$. Then

$$\forall n\ \forall k\ \exists q\ \forall m_{m>q}\ |r_m^{(n)} - r_q^{(n)}| < 2^{-k}.$$

Now, by the Axiom of Choice, there is a constructive function a such that

$$(*)\quad \forall n\ \forall k\ \forall m_{m>a(n,k)}\ |r_m^{(n)} - r_{a(n,k)}^{(n)}| < 2^{-k}.$$

Define a sequence $<s_n>$ of rationals by $s_n = r_{a(n,n)}^{(n)}$. We claim that $<s_n>$ is a real number generator, and that the real number which it determines is equal to $\lim_{n\to\infty} <x_n>$. Since $<x_n>$ is a Cauchy sequence, given any k, for some $n \geq k+2$

$\forall m_{m>n}\ |x_m - x_n| < 2^{-k-1}$, and by the definition of $<s_n>$ together with (*),

$$\forall m\ \forall i_{i>a(m,m)}\ |r_i^{(m)} - s_m| < 2^{-m},$$

and so

$$\forall m\ |x_m - s_m| \not> 2^{-m}.$$

$$\therefore\ \forall m_{m>n}\ |s_m - s_n| \not> |s_m - x_m| + |x_m - x_n| + |s_n - x_n| < 2^{-m} + 2^{-k-1} + 2^{-n} < 2^{-k}.$$

So $<s_n>$ is a real number generator.

Let y be the real number determined by $<s_n>$. Then, given any k, we can certainly find an $n \geq k+1$ such that $\forall m_{m>n}\ |y - s_m| < 2^{-k-1}$.

Hence, $\forall m_{m>n}\ |y - x_m| \not> |y - s_m| + |s_m - x_m| < 2^{-k-1} + 2^{-m} < 2^{-k}$.

So $\lim_{n\to\infty} x_n = y$.

3
CHOICE SEQUENCES AND SPREADS

3.1. THE NOTION OF INFINITY

In intuitionistic mathematics, all infinity is potential infinity: there is no completed infinite.

Since the distinction between the potential and the actual infinite arises within classical mathematics in a perfectly reasonable way, this dictum may at first appear as the expression of a groundless prejudice. Characteristically, we may contrast the uses of the symbols '∞' and '$\aleph_0$' in the statements:

(i) $f(x) \to \infty$ as $x \to 0$

and

(ii) The number of finite sets of natural numbers is $\aleph_0$. In (i), the surface appearance, that reference is being made to some infinite quantity denoted by '∞', is misleading; the sentence has the same apparent structure as:

(iii) $f(x) \to 1$ as $x \to 0$,

but, in fact, when the meaning of (i) is spelt out, the apparent reference to an infinite quantity vanishes, whereas, when the meaning of (iii) is spelt out, the reference to the number 1 remains. By contrast, in (ii) '$\aleph_0$' really does denote something which is in itself infinite, a transfinite cardinal number. In (ii), therefore, we are concerned with the actual infinite, i.e. with a genuine infinite quantity; in (i) the apparent reference to an infinite quantity is merely a *façon de parler*, and we are in fact concerned only with a function which takes on an unbounded *finite* value in the neighbourhood of 0, a situation described by saying that the sentence relates to the potential infinite.

The intuitionistic rejection of the completed infinite is not intended to impugn this distinction: there is no objection to introducing into the language of intuitionistic mathematics genuinely denotative symbols, such as 'ω', for quantities which, like the denumerable ordinals, stand in no finite ratio to positive finite numbers; nor is there any ground for assailing the contrast between such symbols and the symbol '∞' as used in (i). Rather, the thesis that there is no completed infinity means, simply, that to grasp an infinite structure is to grasp the process which generates it, that to refer to such a structure is to refer to that process, and that to recognize the structure as being infinite is to recognize that the process will not terminate. In the case of a process that does terminate, we may legitimately distinguish between the process itself and its completed output: we may be presented with the structure that is generated, without knowing anything about the process of generation. But, in the case of an infinite structure, no such distinction is permissible: all that we can, at any given time, know of the output of the process of generation is some finite initial segment of the structure being generated. There is no sense in which we can have any conception of this structure as a whole save by knowing the process of generation.

This outlook is in accordance with the ordinary, common-sense notion of infinity as something which does not come to an end: it is quite literally true that we can arrive at the notion of infinity in no other way than by considering a process of generation or construction which will never be completed. It is, however, integral to classical mathematics to treat infinite structures as if they could be completed and then surveyed in their totality, in other words, as if we could be presented with the entire output of an infinite process. The basic example of this is the classical understanding of quantification over an infinite totality. Given his assumption that the application of a well-defined predicate to each element of the totality has a determinate value, true or false, the classical mathematician concludes that its universal closure has an equally determinate value, formed by taking the product of the values of its instances, and that the existential closure likewise has a determinate value, formed by taking the sum of the values of its instances. On such a conception, the truth-value of a quantified statement is the final outcome of a process which involves running through the values of all its instances; the assumption that its truth-value is well-defined and determinate is precisely the assumption that we may regard an infinite process of this kind as capable of completion.

From an intuitionistic standpoint, the platonistic conception is the result of blatantly transferring, from the finite case to the infinite one, a picture appropriate only to the former. In making this transference, the platonist destroys the whole essence of infinity, which lies in the conception of a structure which is always in growth, precisely because the process of construction is never completed. The platonistic conception of an infinite structure as something which may be regarded both extensionally, that is, as the outcome of a process, and as a whole, that is, as if the process were completed, thus rests on a straightforward contradiction: an infinite process is spoken of as if it were merely a particularly long finite one. On an intuitionistic view, neither the truth-value of a statement nor any other mathematical entity can be given as the final result of an infinite process, since an infinite process is precisely one that does not have a final result: that is why, when the domain of quantification is infinite, an existentially quantified statement cannot be regarded in advance as determinately either true or false, and a universally quantified one cannot be thought of as being true accidentally, that is independently of there being a proof of it, a proof which must depend intrinsically upon our grasp of the process whereby the domain is generated.

A possible platonist retort to the charge that his conception of mathematical infinity involves a contradiction might be that, while we can come by the notion of infinity only via that of a process which we are incapable of completing, we do not need to think of each infinite totality as in fact generated by such a process. It is perfectly intelligible, even if in fact false, to say that there are infinitely many stars, or again, that a ball bounces infinitely often before coming to rest. The meaning of saying that some totality, of stars or of bounces, is infinite relates to the incompletability of the process of counting them: but the members of the totality are not generated by that process, and so the totality can be given to us by means of a concept which does not itself determine the size of the totality; there is therefore no absurdity in thinking of an infinite totality as already formed.

From an intuitionistic standpoint, such a defence, however licit it may be when applied to empirically given objects or events, cannot be applied to mathematical totalities, whose elements are mental constructions. Naturally, a platonist would not accept this: for him, mathematical entities are eternally existing abstract objects, which are not created by our thought; mathematical thought makes us aware of them, just as aided or unaided observation makes us aware of physical objects and events, but, in either case, their existence is independent of our awareness. The question is not, however, resolved by the mere utterance of a metaphysical credo. Mathematical objects, unlike concrete ones, can be apprehended only in thought; hence if they are not regarded as themselves the products of thought, that can be only because they are viewed as the permanent possibilities of certain mental operations. In not regarding it as necessary, for the number 10^{10} to exist, that anyone should actually have counted up to that number, we do not deny that, in speaking of that number, we are envisaging the possibility of someone's doing so, a possibility that obtains whether or not it is realized. Hence, even if an infinite mathematical totality, such as the totality of natural numbers, be conceived of as a totality of actually existing objects, it must be given to us in terms of an incompletable mental process. This entails that any operation upon the abstract totality has to be explained in terms of a possible operation upon the mental constructions which are the products of the process. Even if there were infinitely many stars, it might be thought to be a theoretical possibility that we should devise an instrument which would give a certain reading just in case there were *any* star possessing a certain property. (In fact, this is almost certainly not even a theoretical possibility.) But, since mathematical objects have no effect upon us save through our thought-processes, the conception of an analogous means of determining the truth-value of a statement involving quantification over an infinite mathematical totality is an absurdity. Granted that an infinite totality of purely abstract objects can be given to us only in terms of a process for generating an infinite sequence of mental constructions, we can introduce an operation upon the totality only in terms either of an operation upon the process itself, or of an operation upon a suitable initial segment of the sequence, or of a combination of the two: the platonistic conception of an operation upon an abstract infinite totality which depends upon all the elements of the totality, but not on the way in which they were generated, belies the very way in which we apprehend such a totality, namely as a representation of the possibility of effecting an arbitrary finite number of mental constructions of a certain kind. This principle is, of course, enough to rule out the classical explanation of quantification over an infinite mathematical totality.

At this stage in the dispute, a platonist will be disposed to claim that, while it may be beyond *our* capacities to complete an infinite process, there is no infinite structure whose construction it would be contradictory to conceive of as being completed. Since our capacities are limited, we must first attain the notion of infinity by reference to the sort of process which we are in principle incapable of completing; but, having formed the conception of an infinite structure, considered as generated by such a process, we make the conceptual advance of apprehending the possibility that such a process should be completed, say by an actual or hypothetical being whose powers transcend our own; and this enables us to form a clear conception of the finished structure that would result from the completion of the process. The necessity to resort to such a defence explains why platonists are inclined to disparage the impossibility of completing an infinite process as a pretty trivial type of impossibility; not a logical impossibility at all, but one relative to the contingent capabilities of human beings. Russell goes so far as to speak of it as 'a mere medical impossibility'.

This situation is characteristic. The intuitionist holds that the expressions of our mathematical language must be given meaning by reference to operations which we can in principle carry out. The strict finitist holds that they must be given meaning by reference only to operations which we can in practice carry out. The platonist, on the other hand, believes that they can be given meaning by reference to operations which we cannot even in principle carry out, so long as we can conceive of them as being carried out by beings with powers which transcend our own. These are deep questions in the theory of meaning which we cannot pursue any further here.

Within number theory, we are concerned with an infinite domain, but the elements of that domain are themselves finite objects. Within analysis, however, the objects with which we deal are themselves infinite: real numbers must be introduced either as infinite classes of rationals, produced by Dedekind cuts, or as equivalence classes of infinite sequences - Cauchy sequences of rationals, or sequences of nested rational intervals - or by some other similar means. In practice, the notion of a real number is most characteristically introduced by appeal to the conception of an infinite sequence. For instance, the approach most usually adopted in the school classroom is via the notion of an infinite decimal expansion. (For intuitionistic purposes, it would be essential that this notion should eventually be generalized to that of a Cauchy sequence, with or without a prescribed rate of convergence, since not every real number generator may be proved to have a determinate decimal expansion; it is left as an exercise for the reader to construct a 'counter-example', of the kind already illustrated, to the proposition that every real number generator has a decimal expansion.)

Whatever the precise route taken to the introduction of the notion of a real number via that of an infinite sequence, there are two stages. The notion of an infinite sequence is introduced in the first place via that of such a sequence generated by means of an effective rule. For instance, children are first introduced to recurring decimal expansions as representing rationals, and then to irrational numbers, such as $\surd 2$, considered as represented by an infinite decimal expansion which does not recur but nevertheless proceeds in accordance with a method of computation. At this stage the conception of an irrational number has been introduced, but not yet that of the totality of all real numbers, as classically understood. In order to arrive at this, a second, more crucial, step has to be taken.

The notion of an infinite sequence was introduced in the first place in terms of the means by which such a sequence may be given to us in its entirety, viz. by a method of computing its terms. This notion could be extended by considering also infinite sequences given by some non-effective means, that is, by a general, though not effective, specification of what, for each n, the n-th term is to be; the sequence would be regarded as determinate provided that the specification, platonistically understood, conferred a definite truth-value on each statement of the form 'r is the n-th term of the sequence'. If the range of possible specifications were made precise by restricting them to any one language, however, this would still not yield the classical continuum. What, instead, is needed at this stage is to sever altogether the connection between the notion of an infinite sequence and the conception, by reference to which it was originally introduced, of a means by which we are able to grasp such a sequence as a whole. The notion thus arrived at is that of an *arbitrary* infinite sequence. It would be a mistake to take the word 'arbitrary', in this context, as a mere device for indicating the scope of an expression of generality, as it would be in the phrase 'an arbitrary natural number': when applied to infinite sequences, the word genuinely serves to characterize the kind of sequence we are thinking of.

It is strictly necessary, if we are to arrive at the classical conception of the continuum, to appeal to this notion of an arbitrary infinite sequence: such a sequence (e.g. an infinite decimal expansion) is generated, not according to any rule or other prior mathematical prescription, but by a process involving repeated arbitrary selection of one term after another. This appeal is seen very plainly when we reflect on the motivation for so defining exponentiation, as applied to cardinal numbers, that the number of infinite sequences of 0's and 1's is by definition $2^{\aleph_0}$: just as the totality of all n-tuples of 0's and 1's can be displayed by considering each as generated by a sequence of n choices between 0 and 1, so we think of each infinite sequence as generated by a denumerable sequence of such choices.

Most constructivist approaches to the theory of real numbers disallow the second of the two steps in the formation of the classical conception, namely the admission of arbitrary infinite sequences, and allow only infinite sequences generated by an effective rule, thus arriving at a restriction of the classical continuum to the recursive real numbers or the like. This method thus involves the wholesale rejection of one of the basic ingredients of the classical concept. Intuitionism aims, however, to reform mathematics, not to prune it; according to it, scarcely any of the ideas of classical mathematics is wholly spurious, but all are deformed by being systematically misconstrued. Hence intuitionism retains both fundamental ideas which go to form the classical conception of the continuum, admitting not only infinite sequences determined in advance by an effective rule for computing their terms, but also ones in whose generation free selection plays a part. Such free choices of the terms of an infinite sequence need not be absolutely unrestricted: a partial restriction may be imposed at the outset, or, at any stage in the process of generating the sequence, a further partial, or even total, restriction may be imposed upon subsequent choices of terms. An infinite sequence in the process of generating which free selection of this kind is permitted to play a part is known as a (*free*) *choice sequence*.

At first it may seem that the admission of choice sequences betrays the whole basis of constructive mathematics. This is only so, however, if such sequences are interpreted in accordance with the classical or platonistic conception of infinite structures as completed objects. On an intuitionistic conception, any infinite sequence, whether wholly determined in advance or not, must be taken as 'in process of growth'; that is, we must not regard it as something all of whose terms can be surveyed. So long as we regard an infinite sequence in the way that a constructive approach demands that we regard all infinite structures, no harm can come from admitting processes of generation which neither terminate nor are wholly determined in advance.

The fact that an infinite totality, such as that of the natural numbers, is understood as 'in process' comes out in the interpretation of quantification over such a totality. An infinite sequence being, unlike a natural number, an object itself in process of growth, its uncompleted character must come out in the way statements about any one such sequence are interpreted. Infinite sequences, whether determined by a rule (*lawlike* sequences) or not, must be regarded as intensional in character: they are given by means of a particular process of generation, and are therefore not uniquely determined by their terms, any more than a species is uniquely determined by its members. Even an extensional statement about an infinite sequence, however, i.e. one true of any sequence extensionally equivalent to a sequence of which it is true, can be recognized as true only on the basis of some finite amount of information about it which can be acquired at some time. In the case of a lawlike sequence, this will consist in the effective rule for generating its terms. In the more general case of a choice sequence, however, it will consist in some finite initial segment of the sequence, together with any restrictions which have been imposed upon subsequent choices of terms at or before the stage at which the last term in that initial segment is generated.

We shall use the letters a, b, c, ... to range over lawlike sequences of natural numbers: as already remarked, these may be identified with constructive unary functions from natural numbers to natural numbers. We shall use the Greek letters α, β, γ, ... to range, more generally, over choice sequences, the lawlike sequences being taken as included among the choice sequences. The n-th term of α (of a) will be represented by '$\alpha(n-1)$' (by '$a(n-1)$'). We shall use '=' for extensional equality: '$a = b$' is defined to mean $\forall n\ (a(n) = b(n))$, and '$\alpha = \beta$' to mean $\forall n\ (\alpha(n) = \beta(n))$. We shall also use the abbreviations '$\mathrm{Ext}_a\ A(a)$' and '$\mathrm{Ext}_\alpha\ B(\alpha)$'

to mean that 'A(a)' and 'B(α)' express, respectively, extensional properties of lawlike sequences and of choice sequences: that is, 'Ext_a A(a)' is defined to mean:

$$\forall a\ \forall b\ (a = b\ \&\ A(a) \rightarrow A(b))$$

and 'Ext_α B(α)' likewise.

What, then, will be the meaning of quantification over choice sequences, more particularly, where the quantification is into an extensional context? Suppose that A(α) is extensional. The existential quantifier must, as always, have a constructive meaning; so ∃α A(α) will normally imply that we can effectively find a lawlike sequence a such that A(a). More exactly, it will mean that we can effectively find a certain initial segment $\langle n_0,\ldots,n_{k-1}\rangle$ and a certain set of restrictions upon subsequent choices of terms such that, for every choice sequence α subject to those restrictions for which $\alpha(i) = n_i$ for $i < k$, A(α) holds. An explanation of universal quantification over choice sequences is more difficult. In general, where A(α) is extensional, '∀α A(α)' must mean that, for each α, it is possible to determine the truth of A(α) from some finite amount of information about α available at some stage; that is to say, from some initial segment $\langle\alpha(0),\ldots,\alpha(k-1)\rangle$ of α, together with any restrictions upon subsequent choices of terms of α that have been imposed by the stage at which α(k-1) was selected. In particular, we may consider a statement of the form '∀α∃n B(α,n)', where B(α, n) is extensional. As always, a form of Axiom of Choice holds good: the statement involves that there must be a uniform effective procedure for finding, for each given α, an n such that B(α, n). Where α is a choice sequence, however, it cannot be 'given' in its entirety, and hence the procedure for finding n must operate upon some finite amount of information about α that we may possess at some stage. In this particular case, the claim is made that n may be computed from some sufficiently long initial segment of α(without further regard to restrictions upon future choices of terms that may have been imposed by that stage). That is, if '∀α∃n B(α, n)' is to hold, we must have an effective rule by which we can decide, for every finite sequence, whether or not it is sufficient to determine an n such that $B(\alpha, n)$ holds for every α of which that finite sequence is an initial segment, and which enables us to compute such an n if the sequence is sufficiently long; and every choice sequence α must have some initial segment from which the rule will compute such an n. (That this is implied by a statement of the form '$\forall\alpha\exists n\, B(\alpha, n)$', where $B(\alpha, n)$ is extensional, is not an immediate consequence of the meanings of the quantifiers, but needs to be argued for on the basis of a more exact analysis of the notion of a choice sequence. It is cited here only in order to give a general indication of the complete difference in meaning between quantification over choice sequences, as intuitionistically understood, and quantification over arbitrary infinite sequences in the classical sense, where we treat the infinite sequence as a completed structure, that is, as if we were able to survey it in its entirety, prescinding from the process by which it was generated.)

3.2. THE FAN THEOREM AND BAR INDUCTION

In this section we consider infinite sequences of natural numbers generated by free choices. At this stage, we attempt no exact analysis of the process of generating such a sequence, but we do consider one type of restriction which may be placed at the outset upon further choices. Such a restriction is effected by confining the sequence to be an element of a *spread*; before defining this notion, however, we need some notation.

$\bar{\alpha}(n)$ is $\langle\alpha(0),\ldots,\alpha(n-1)\rangle$, the finite sequence consisting of the first n terms of α. Thus, for any α, $\bar{\alpha}(0)$ is the empty sequence, denoted by $\langle\ \rangle$.

$\vec{u}$, $\vec{v}$, $\vec{w}$, are variables for finite sequences of natural numbers, and u_{n-1} is the n-th term of $\vec{u}$. We define a length function on finite sequences by

$$\ell h(\vec{u}) = k \quad \text{iff } \vec{u} = \langle u_0,\ldots,u_{k-1}\rangle \ .$$

We can extend sequences by concatenation:

$$\vec{u} * \vec{v} = \langle u_0, \ldots, u_{k-1}, v_0, \ldots, v_{\ell-1} \rangle, \text{ where}$$

$\ell h(\vec{u}) = k$ and $\ell h(\vec{v}) = \ell$. We write $\vec{u}^\frown m = \vec{u} * \langle m \rangle$ for any natural number m. An ordering relation is defined on finite sequences by

$$\vec{v} \preccurlyeq \vec{u} \quad \leftrightarrow \quad \exists \vec{w}\ (\vec{v} = \vec{u} * \vec{w}).$$

Note that $\vec{v} \preccurlyeq \vec{u}$ if $\vec{v}$ is an extension of $\vec{u}$, and not vice versa.

A spread is essentially a tree, with the restriction that every path is infinite, and that we can effectively construct any subtree consisting of initial segments of finitely many paths. The paths of the tree are being identified with choice sequences, and each node with a finite sequence of natural numbers, namely the initial segment common to all choice sequences the paths corresponding to which pass through that node (with any two distinct nodes representing distinct finite sequences). Thus each node $\vec{u}$ determines a species of choice sequences, those choice sequences which have $\vec{u}$ as an initial segment. It is not necessary to introduce any special notation for this species, since we may simply define:

$$\alpha \in \vec{u} \leftrightarrow \exists n\ (\vec{u} = \bar{\alpha}(n)).$$

We can identify the restrictions on choice sequences which we are trying to represent by a constructive function on finite sequences. This function, when applied to any finite sequence, determines whether or not that sequence is admissible to the spread, i.e. whether or not it is a possible initial segment of a choice sequence satisfying the restrictions. We call this function s, the *spread-law*:

$$s(\vec{u}) = \begin{cases} 0 & \text{if } \vec{u} \text{ is admissible} \\ 1 & \text{otherwise.} \end{cases}$$

The spread-law clearly has to be effectively calculable. Further, we want to stipulate, for any spread, (i) that the empty sequence is admissible, so that the spread is not empty, (ii) that every admissible finite sequence has at least one admissible extension, and (iii) that no extension of an inadmissible finite sequence is admissible. This is effected by the following definition.

Definition. $spr(s) \leftrightarrow s(\langle\ \rangle) = 0\ \&$
$\forall\vec{u}(s(\vec{u}) = 0 \rightarrow \exists k\ s(\vec{u}^\frown k) = 0)\ \&$
$\forall\vec{u}\ \forall\vec{v}(\vec{u} \preccurlyeq \vec{v}\ \&\ s(\vec{u}) = 0 \rightarrow s(\vec{v}) = 0)\ \&$
$\forall\vec{u}(s(\vec{u}) = 0 \vee s(\vec{u}) = 1).$

Examples. (1) The *universal spread* is the spread given by s such that

$$\forall\vec{u}\ s(\vec{u}) = 0.$$

(2) For each n, the *full* n-*ary spread* is that given by s such that

$$\forall\vec{u}[(s(\vec{u}) = 0 \leftrightarrow \forall i_{i<\ell h(\vec{u})}\ u_i < n)\ \&$$
$$(s(\vec{u}) = 1 \leftrightarrow \exists i_{i<\ell h(\vec{u})}\ u_i \geq n)].$$

We can now define the notion of a choice sequence's being an *element* of a spread thus:

$$\alpha \in s \leftrightarrow \forall n\ s(\bar{\alpha}(n)) = 0.$$

The restricted quantifiers '$\forall\alpha_{\alpha\in s}$' and '$\exists\alpha_{\alpha\in s}$' are taken to range over those choice sequences subject to the restrictions represented by the particular spread s, regardless of any other conditions which some such sequences might satisfy.

So far we have spoken only of spreads of sequences of natural numbers. Such spreads are known as *naked spreads*. It is easy to see how we can extend the notion to sequences of other mathematical constructions, e.g. rationals, reals, etc. Given a spread-law s and the corresponding naked spread, we can construct an (effective) correlation law c, which associates members of some species $\underline{A}$ with admissible sequences of s. (Sometimes it is more convenient to leave c undefined on the empty sequence $\langle\ \rangle$.) Then, if $s(\vec{u}) = 0$, $c(\vec{u}) \in \underline{A}$. If $\alpha \in s$, the corresponding choice sequence of elements of $\underline{A}$ is

$$c(\alpha) = \langle c(\bar{\alpha}(0)),\ c(\bar{\alpha}(1)),\ldots\rangle\ .$$

We call the structure $\langle s,c\rangle$ a *dressed spread*, whose elements are the choice sequences $c(\alpha)$ for $\alpha \in s$, and for $\xi = c(\alpha)$ we write $\xi \in \langle s,c\rangle$.

Examples. (1) Given any naked spread s, a dressed spread $<s,c>$ can be constructed with c defined by

$$c(<u_0,\ldots,u_{n-1}>) = u_{n-1}\cdot 2^{-n+1}.$$

(2) Another example is obtained by defining the correlation law c' by

$$c'(<u_0,\ldots,u_{n-1}>) = \sum_{i=0}^{n-1} u_i\cdot 2^{-i} .$$

We now consider a classical theorem about trees. A tree is said to be *finitary* if each node has only finitely many nodes immediately below it. The theorem states that if $\underline{T}$ is a finitary tree in which every path terminates, then there is an upper bound on the lengths of the paths. Using p as a variable for paths, we can write this as

$$\text{CFT:}\quad \underline{T}\text{ is finitary}$$
$$\& \ \forall p_{p\in\underline{T}}\ \exists n\ \ell h(p) = n \rightarrow \exists m \forall p_{p\in\underline{T}} \exists n_{n\leq m} \ell h(p) = n .$$

If we hope to prove this theorem intuitionistically, we must first find a suitable way of formulating it. Obviously, we want to represent the finitary tree by a spread. A finitary spread is called a *fan*.

Definition. $\text{fan}(s) \leftrightarrow \text{spr}(s)\ \&\ \forall \vec{u}_{s(\vec{u})=0}\ \exists k\ \forall m_{m>k}\ s(\vec{u}^\frown m) = 1.$

We have so defined a spread, however, that every element of a spread is an infinite choice sequence. We can nevertheless get the effect of all paths terminating by supposing that there is some species $\underline{R}$ of finite sequences which *bars* the vertex, in the following sense:

Definition. A species $\underline{R}$ of finite sequences *bars* a node $\vec{u}$ in a spread s iff

$$\forall \alpha_{\alpha\in s,\alpha\in\vec{u}}\ \exists n\ \bar{\alpha}(n) \in \underline{R}.$$

The theorem can now be restated as

$$\text{GFT:}\ \text{fan}(s)\ \&\ \forall\alpha_{\alpha\in s}\ \exists n\ \bar{\alpha}(n) \in \underline{R} \rightarrow \exists m\ \forall \alpha_{\alpha\in s}\ \exists n_{n\leq m}\ \bar{\alpha}(n) \in \underline{R},$$

or equivalently as

$$\text{fan}(s)\ \&\ \underline{R}\text{ bars }<\ >\text{ in } s \rightarrow \exists m\ \forall\alpha_{\alpha\in s}\ \exists n_{n\leq m}\ \bar{\alpha}(n) \in \underline{R}.$$

The classical proof of this theorem (the General Fan Theorem) proceeds by proving its contraposition as a lemma.

<u>König's Lemma</u> (Unendlichkeitslemma). If there is no finite upper bound to the lengths of paths in a finitary tree, then there is at least one infinite path in the tree.

In order to make it more explicit why the proof of this lemma fails intuitionistically, we prove the version corresponding to GFT:

$$\text{fan}(s) \ \& \ \neg\exists m \ \forall\alpha_{\alpha\in s} \ \exists n_{n\leq m} \ \bar{\alpha}(n) \in \underline{R} \rightarrow \exists\alpha_{\alpha\in s} \ \forall n \ \bar{\alpha}(n) \notin \underline{R}.$$

Proof. We start by defining a relation $Q(\vec{u},m)$ which holds between a finite sequence $\vec{u}$ and a natural number m when every choice sequence of the fan with initial segment $\vec{u}$ has an initial segment of length less than or equal to m in $\underline{R}$, i.e.

$$Q(\vec{u},m) \leftrightarrow \forall\alpha_{\alpha\in s,\alpha\in\vec{u}} \ \exists n_{n\leq m} \ \bar{\alpha}(n) \in \underline{R}.$$

We then take the species $\underline{A}$ to be the domain of this relation:

$$\underline{A} = \{\vec{u} \mid \exists m \ Q(\vec{u},m)\} \quad .$$

Now suppose that, for given $\vec{u}$ such that $s(\vec{u}) = 0$, we have:

$$\forall k_{s(\vec{u}\frown k)=0} \ \vec{u}\frown k \in \underline{A},$$

$$\text{i.e. } \forall k_{s(\vec{u}\frown k)=0} \ \exists m \ Q(\vec{u}\frown k,m).$$

Since s is a fan, there exists q such that

$$\forall k_{k>q} \ s(\vec{u}\frown k) = 1.$$

Now, by the Axiom of Choice, there exists a constructive function a such that

$$\forall k_{s(\vec{u}\frown k)=0} \ Q(\vec{u}\frown k,a(k)).$$

(Since there are only finitely many k such that $s(\vec{u}\frown k) = 0$, the appeal to the Axiom of Choice is inessential, but serves merely as a convenient means of selecting a unique m for each k.) It is now evident that we have:

$$Q(\vec{u},\max\{a(k) \mid s(\vec{u}\frown k) = 0\}),$$

whence: $\vec{u} \in \underline{A}$.

We have thus shown that the species $\underline{A}$ is *hereditary upwards* in the sense that it contains any admissible finite sequence all of whose admissible immediate extensions belong to it, i.e.

$$(1) \quad s(\vec{u}) = 0 \;\&\; \forall k_{s(\vec{u}^\frown k)=0}\; \vec{u}^\frown k \in \underline{A} \rightarrow \vec{u} \in \underline{A}.$$

Classically, we can contrapose (1) to obtain:

$$(2) \quad s(\vec{u}) = 0 \;\&\; \vec{u} \notin \underline{A} \rightarrow \exists k_{s(\vec{u}^\frown k)=0}\; \vec{u}^\frown k \notin \underline{A}.$$

By the hypothesis of the lemma, we have $\langle\ \rangle \notin \underline{A}$, and so we can define an infinite path β by induction thus:

$$\beta(n) = \min\{k \mid s(\bar{\beta}(n)^\frown k) = 0 \;\&\; \bar{\beta}(n)^\frown k \notin \underline{A}\} .$$

From the construction of β we clearly have:

$$\forall n\; \bar{\beta}(n) \notin \underline{A}.$$

But plainly,

$$(3) \quad \underline{R} \subseteq \underline{A} ,$$

and so $\forall n\; \bar{\beta}(n) \notin \underline{R}$.

Since $\beta \in s$, we have proved, as required:

$$\exists \alpha_{\alpha \in s}\; \forall n\; \bar{\alpha}(n) \notin \underline{R} .$$

We can now derive the General Fan Theorem, GFT, as a corollary by contraposition and manipulation of the quantifiers.

Intuitionistically, the above proof of König's Lemma is invalid: although we can reach (1), which states that $\underline{A}$ is hereditary upwards, we cannot derive (2) from (1), since this step involves the use of the rule

$$\frac{\neg \forall x\; Fx}{\exists x\; \neg Fx}.$$

Given that we have only finitely many k to consider, this step would be legitimate if $\underline{A}$ were a decidable species, which, however, it plainly is not, even when $\underline{R}$ is assumed decidable. We therefore cannot define β, since, given $\bar{\beta}(n)$, we have no effective means of finding a k such that $\bar{\beta}(n)^\frown k \notin \underline{A}$.

Moreover, we cannot remedy the situation by modifying the proof of Konig's Lemma, since reflection on the difficulty involved shows that there is no reason to suppose Konig's Lemma to be constructively true. Intuitionistically understood, the assertion that there exists an infinite path amounts to the claim that we can effectively define such a path; but the mere fact that there is no finite upper bound on the lengths of paths does not supply us with any way of doing this, since we have no effective means of deciding, for each given node, whether or not it is the case that there is a finite upper bound on the lengths of paths going through it.

Even if we had an intuitionistic proof of Konig's Lemma, we should be unable to derive the Fan Theorem as a corollary, since this inference again involves an illicit form of contraposition, invoking the above invalid quantifier rule. However, although we have found reason for supposing Konig's Lemma not to be intuitionistically true, there is no parallel reason for supposing this of the Fan Theorem itself. On the contrary, at least as regards the first formulation, CFT, of the theorem, it is evident that, if every path in a finitary tree is finite, we have an effective method for finding a finite upper bound to the lengths of paths in the tree.

In our formulation GFT we are using the species $\underline{R}$ to represent the termination of paths in the finitary tree. The possibility of effectively finding a bound on the lengths of paths in the tree depends on our being able to recognize when a path terminates; so, in order to arrive at a formulation of the Fan Theorem in these terms under which it is intuitionistically true, we may reasonably require that R be a decidable species. (Since our notion of a spread does not in fact allow for terminating paths, this means that, if we were appealing to a liberalized notion of a spread according to which it was not required that every admissible finite sequence had an admissible proper extension, we should require that it be decidable, for every admissible finite sequence, whether or not it had such an extension.)

With this added hypothesis, the Fan Theorem takes the

form:

$$\text{FT: } \mathrm{fan}(s)\ \&\ \forall\alpha_{\alpha\in s}\ \exists n\ \bar{\alpha}(n) \in \underline{R}\ \&\ \forall\vec{u}\ (\vec{u} \in \underline{R} \vee u \notin \underline{R})$$

$$\rightarrow \exists m\ \forall\alpha_{\alpha\in s}\ \exists n_{n\leq m}\ \bar{\alpha}(n) \in \underline{R}.$$

It is to this formulation that we shall henceforth take the name 'the Fan Theorem' to refer.

Before going on to consider how we may give an intuitionistic proof of the Fan Theorem, we shall pause here to show that our suspicion that König's Lemma is not intuitionistically true is capable of more precise demonstration. The fan s that we shall consider is the full binary spread (whose elements are all infinite sequences of 0's and 1's). We shall define a decidable species $\underline{R}$ of which we can prove, not merely that there is no upper bound on the lengths of smallest initial segments belonging to $\underline{R}$, viz.

$$\text{(A)} \qquad \neg\exists m\ \forall\alpha_{\alpha\in s}\ \exists n_{n\leq m}\ \bar{\alpha}(n) \in \underline{R}\ ,$$

but the intuitionistically stronger statement that, for each m, we can find an element of s no initial segment of which of length less than or equal to m belongs to $\underline{R}$:

$$\text{(B)} \qquad \forall m\ \exists\alpha_{\alpha\in s}\ \forall n_{n\leq m}\ \bar{\alpha}(n) \notin \underline{R}.$$

This species $\underline{R}$ is defined with the help of Kleene's T-predicate. We first define:

$$W(i,r,k) \leftrightarrow [i = 0\ \&\ T_1((r)_1,r,k)\ \&\ \forall j_{j\leq k}\ \neg T_1((r)_0,r,j)]$$

$$\vee\ [i = 1\ \&\ T_1((r)_0,r,k)\ \&\ \forall j_{j\leq k}\ \neg T_1((r)_1,r,j)]$$

where $(r)_0$ and $(r)_1$ are, respectively, the exponents of 2 and 3 in the prime factorization of r, and then put:

$$\underline{R} = \{\vec{u} \mid s(\vec{u}) = 0\ \&\ \exists r_{r<\ell h(\vec{u})}\ \exists k_{k<\ell h(\vec{u})-r}\ W(u_r,r,k)\}\ .$$

Thus for $\alpha \in s$

$$\bar{\alpha}(n) \in \underline{R} \leftrightarrow \exists r_{r<n}\ \exists k_{k<n-r}\ W(\alpha(r),r,k).$$

Since W(i,r,k) is primitive recursive, $\underline{R}$ is plainly decidable. Now, in order to establish (B), we suppose m given, and define $\alpha \in s$ as follows

$$\alpha(r) = \begin{cases} 1 & \text{if } r < m \ \& \ \exists k_{k<m-r} W(0,r,k) \\ 0 & \text{otherwise} . \end{cases}$$

The constructive character of this definition is not in question, since α is again primitive recursive. It is left as an exercise for the reader to prove that $\forall n_{n \leq m}\ \bar{\alpha}(n) \notin \underline{R}$.

(B) plainly implies (A), so that, if König's Lemma were true intuitionistically for decidable $\underline{R}$, we should have:

(C) $\exists \alpha_{\alpha \in s}\ \forall n\ \bar{\alpha}(n) \notin \underline{R}$.

However, there is no hope of proving (C) constructively, since we can show that, for every general recursive $\alpha \in s$, some initial segment of α belongs to $\underline{R}$. The proof of this depends upon the fact that, where $\alpha \in s$ is general recursive, we have, for some numbers p_0 and p_1 and for every r:

$$\alpha(r) = 1 \leftrightarrow \exists k\ T_i(p_0,r,k)$$

$$\text{and } \alpha(r) = 0 \leftrightarrow \exists k\ T_1(p_1,r,k) ,$$

i.e., where $p = 2^{p_0}.3^{p_1}$,

$$\alpha(r) = 1 \leftrightarrow \exists k\ T_1((p)_0,r,k)$$

$$\text{and } \alpha(r) = 0 \leftrightarrow \exists k\ T_1((p)_1,r,k) .$$

It is left as an exercise for the reader to establish that, for some k, $W(\alpha(p),p,k)$, and that therefore, for $n = p + k + 1$, $\bar{\alpha}(n) \in \underline{R}$. We thus cannot establish (C) by means of a general recursive α. This argument would decisively show the impossibility of a constructive proof of (C) only if we accepted Church's Thesis, that every constructive function of natural numbers is general recursive. While Church's Thesis is not itself a principle of intuitionistic mathematics, it is known to be consistent with the standard intuitionistic axioms, and it is therefore highly improbable that there exists any intuitionistic proof of (C).

This example, which is due to Kleene, also shows that the Fan Theorem itself, FT, does not hold good if we restrict the elements of the fan to general recursive functions (that is, if we accept Church's Thesis, to constructive functions).

We now turn to the question how FT may be proved intuitionistically. The classical proof made use of an inductive argument. However, to use ordinary finite induction would involve us in the illicit contraposition we are seeking to avoid, because it was an induction along the paths of the tree: starting from the hypothesis that the vertex did not belong to $\underline{A}$, it 'constructed' a path which proceeded always from a node not belonging to this species to a node below it also not belonging to it. The crucial idea of the proof was the recognition that $\underline{A}$ is hereditary upwards, as expressed by line (1). So it looks as though, instead of performing the doubly illicit move of first turning (1) upside down, by means of an invalid contraposition, to obtain (2), in order to be able to use induction down the tree, and then contraposing the result (König's Lemma), again invalidly, to obtain the Fan Theorem, we should be able to use some sort of induction directly, up the tree, with (1) serving as the induction step. We shall naturally start from the finite sequences in $\underline{R}$, since $\underline{R}$ is obviously included in $\underline{A}$: this proposition (3) thus serves as the induction basis. The conclusion that we desire is that the vertex, i.e. the empty sequence $\langle\ \rangle$, is in $\underline{A}$. Of course, the induction would not be valid unless we assumed that the species $\underline{R}$ barred the vertex. We are thus led to formulate the following principle of induction, known as *Bar Induction*:

$$\mathrm{BI}_{\mathrm{DR}}\!: \quad \mathrm{spr}(s)\ \& \qquad \text{(i)}$$

$$\forall \vec{u}\ (\vec{u} \in \underline{R} \vee \vec{u} \notin \underline{R})\ \& \qquad \text{(ii)}$$

$$\forall \alpha_{\alpha \in s}\ \exists n\ \bar{\alpha}(n) \in \underline{R}\ \& \qquad \text{(iii)}$$

$$\forall \vec{u}\ (\vec{u} \in \underline{R} \to \vec{u} \in \underline{A})\ \& \qquad \text{(iv)}$$

$$\forall \vec{u}_{s(\vec{u})=0}\ (\forall k_{s(\vec{u}^\frown k)=0}\ \vec{u}^\frown k \in \underline{A} \to \vec{u} \in \underline{A}) \qquad \text{(v)}$$

$$\to \langle\ \ \rangle \in \underline{A}.$$

It is important to note that we have not required that the spread s be a fan. In the Fan Theorem, the correctness of the induction step, that $\underline{A}$ is hereditary upwards, depends on the spread's being a fan, but there is no reason to think that the general principle to which we are appealing itself depends on this.

We shall return to the subject of possible justifications of Bar Induction. For the moment let us assume that BI_{DR} has been adopted as an axiom, and use it to derive a proof of the Fan Theorem. By assumption, s is a spread, $\underline{R}$ is decidable, and $\underline{R}$ bars < > in s. Further, it follows trivially from the definition of $\underline{A}$ that $\underline{R}$ is contained in $\underline{A}$. So, in order to apply BI_{DR}, it remains only to satisfy ourselves that we can give an intuitionistic proof of hypothesis (v), i.e. of line (1) of the classical proof. Inspection of that part of the classical proof, as set out above, shows at once that it is intuitionistically valid; hence BI_{DR} yields a proof of the Fan Theorem, FT.

BI_{DR} is stated as licensing Bar Induction relative to a spread s. (In doing so, it might have seemed more natural to weaken hypotheses (ii) and (iv) by restricting the quantifiers '$\forall\vec{u}$' to admissible $\vec{u}$; but it is easily seen that the generality of the principle would not be effectively increased in this way.) A particular case of BI_{DR} occurs when s is taken as the universal spread; but it turns out that this special case, which we call BI_D, entails the general case, BI_{DR}. (The subscript D indicates that the decidability of $\underline{R}$ is among the hypotheses of the theorem; the subscript R indicates that the formulation is relativized to a spread.)

$$BI_D: \quad \forall\vec{u}\ (\vec{u} \in \underline{R} \vee \vec{u} \notin \underline{R})\ \& \qquad \text{(ii)}$$

$$\forall\alpha\ \exists n\ \bar{\alpha}(n) \in \underline{R}\ \& \qquad \text{(iii')}$$

$$\forall\vec{u}\ (\vec{u} \in \underline{R} \rightarrow \vec{u} \in \underline{A})\ \& \qquad \text{(iv)}$$

$$\forall\vec{u}\ (\forall k\ \vec{u}^\frown k \in \underline{A} \rightarrow \vec{u} \in \underline{A}) \qquad \text{(v')}$$

$$\rightarrow \langle\ \rangle \in \underline{A}\ .$$

The derivation of BI_{DR} from BI_D is left as an exercise.

Towards an intuitive justification of Bar Induction, we have remarked merely that the restriction of it to a fan seemed unnecessary. The principle, as applied to a fan, seemed reasonable, since it yielded a form of induction backwards along the fan whose effect, classically, was obtained by contraposing the induction step (v), carrying out an induction forwards through the fan, and then contraposing the conclusion. In the case of the Fan Theorem, at least, the principle of Bar Induction, as a direct means of arriving at a conclusion which, in the classical proof, was arrived at by a double indirection, seemed plausible because the conclusion itself seemed plausible. In order to convince ourselves, therefore, that we really are justified in assuming the validity of Bar Induction relative to all spreads, and not just to fans, it is reasonable to ask whether the validity of Bar Induction can be established classically by an argument which bears to it the same relation as the classical proof of the Fan Theorem has to it. This is indeed the case.

Classical Justification of Bar Induction. We shall demonstrate the validity of Bar Induction in the form BI_D. Hypothesis (ii) falls away classically as a truth of logic. We assume the truth of (iv) and (v'), together with the falsity of the conclusion, and derive the falsity of hypothesis (iii'), contraposing to obtain the last line of the proof. That is, assuming the induction basis and induction step, we show that, if the vertex $\langle\ \rangle$ is not in the species $\underline{A}$, then $\langle\ \rangle$ is not barred by $\underline{R}$. As in the classical proof of König's Lemma, on these assumptions we construct a choice sequence α which has no initial segment in $\underline{R}$, i.e. $\forall n\ \bar{\alpha}(n) \notin \underline{R}$. We again carry out this construction by means of a finite induction: suppose that we have already selected $\alpha(0),\ldots,\alpha(n-1)$ in such a way that $\bar{\alpha}(n) \notin \underline{A}$. For n = 0 this is possible because we have assumed $\langle\ \rangle \notin \underline{A}$ (and, of course, $\bar{\alpha}(0) = \langle\ \rangle$ for every α). We now set:

$$\alpha(n) = \min\{k \mid \bar{\alpha}(n)^\frown k \notin \underline{A}\} \quad .$$

That such a k always exists is shown by contraposing

hypothesis (v'). We have thus defined an α such that $\forall n\ \bar{\alpha}(n) \notin \underline{A}$. From (iv) it now follows that $\forall n\ \bar{\alpha}(n) \notin \underline{R}$, contradicting hypothesis (iii').

This proof is intuitionistically invalid in exactly the same way that the classical proof of GFT by means of König's Lemma was invalid. Namely, first, the (tacit) last step of the proof depended on an intuitionistically illicit contraposition from

$$\langle\ \rangle \notin \underline{A} \rightarrow \exists\alpha\ \forall n\ \alpha(n) \notin \underline{R} \quad ,$$

regarded as holding in the presence of (iv) and (v), to

$$\forall\alpha\ \exists n\ \bar{\alpha}(n) \in \underline{R} \rightarrow \langle\ \rangle \in \underline{A} \quad .$$

Secondly, the induction step in the classical proof depended on an intuitionistically illicit move from hypothesis (v') to

$$\forall\vec{u}\ (\vec{u} \notin \underline{A} \rightarrow \exists k\ \vec{u}^\frown k \notin \underline{A}) \quad .$$

The fact that the classical proof of Bar Induction is related to the principle of Bar Induction exactly as the classical proof of the Fan Theorem is to the Fan Theorem should encourage us to regard Bar Induction as an intuitionistically correct principle.

More positively, we may look at the matter in the following way. The statement that $\langle\ \rangle$ is barred by $\underline{R}$ in the universal spread is expressed by

$$(1) \qquad \forall\alpha\ \exists n\ \bar{\alpha}(n) \in \underline{R} \ .$$

Now obviously, the property of being barred by $\underline{R}$ is possessed by every $\vec{u} \in \underline{R}$. Equally obviously, the property is hereditary upwards. Moreover, it appears intuitively evident that the species of finite sequences $\vec{u}$ which are barred by $\underline{R}$ is the *smallest* species $\underline{A}$ which has these two features. To say that $\langle\ \rangle$ belongs to the smallest such species is to assert:

$$(2)\ \forall\underline{A}\,[\underline{R} \subseteq \underline{A}\ \&\ \forall\vec{u}\ (\forall k\ \vec{u}^\frown k \in \underline{A} \rightarrow \vec{u} \in \underline{A}) \rightarrow \langle\ \rangle \in \underline{A}] \quad .$$

The principle of Bar Induction states, in effect, that when $\underline{R}$ is a decidable species, (1) and (2) are equivalent. The implication from (2) to (1) is obvious, from the fact, already noted, that the species $\underline{A}$ of finite sequences barred by $\underline{R}$

contains $\underline{R}$ and is hereditary upwards; so by (2), $\langle\ \rangle \in \underline{A}$, i.e. $\forall\alpha\ \exists n\ \bar{\alpha}(n) \in \underline{R}$. The converse implication, from (1) to (2), is the content of the principle of Bar Induction: it amounts precisely to the assumption that, when $\underline{R}$ is assumed decidable, the species of sequences barred by $\underline{R}$ is the smallest species $\underline{A}$ satisfying:

$$\underline{R} \subseteq \underline{A} \quad \& \quad \forall\vec{u}\ (\forall k\ \vec{u}^\frown k \in \underline{A} \rightarrow \vec{u} \in \underline{A}) \ .$$

3.3. THE CONTINUITY PRINCIPLE

We have laid down two formulations of the Axiom of Choice, $AC_{n,m}$ and $AC_{n,b}$; by $AC_{n,m}$, for example, a statement of the form

$$\forall n\ \exists m\ A(n,m)$$

implies the existence of a constructive function a such that

$$\forall n\ A(n,\ a(n)) .$$

Plainly, the same considerations which led us to accept $AC_{n,m}$ and $AC_{n,b}$ will justify a suitable form of the Axiom of Choice for each hypothesis beginning with a universal quantifier followed by an existential one. For instance, we may assume the analogue $AC_{n,\beta}$ of $AC_{n,b}$ for choice sequences in place of constructive functions:

$$AC_{n,\beta} : \quad \forall n\ \exists\beta\ A(n,\beta) \rightarrow \exists\gamma\ \forall n\ A(n,\ \lambda m.\ \gamma(j(n,m))) .$$

where j is a pairing function (e.g. $j(n,m) = m + \frac{1}{2}(m + n)(m + n + 1)$). (We formulated $AC_{n,b}$ by quantifying over constructive functions of two arguments, but, when we are dealing with choice sequences, we need a pairing function to obtain the same effect.) In just the same way, a statement of the form

$$\forall\alpha\ \exists m\ B(\alpha,\ m)$$

must be understood as implying the existence of a constructive functional Φ, mapping choice sequences to natural numbers, such that

$$\forall\alpha\ B(\alpha, \Phi(\alpha)).$$

'Constructive' here means that we can effectively calculate $\Phi(\alpha)$ from whatever we may at some stage know of the choice sequence α: that is $\Phi(\alpha)$ may be calculated from the particular way in which the choices of terms of α are made, from some finite number of those terms, say those contained in some initial segment $\bar{\alpha}(n)$, and from the restriction imposed, in advance or by stage n, on subsequent choices of terms.

We shall not, however, formulate a form of the Axiom of Choice for this case, partly because we do not want to introduce quantification over functionals into our notation (though there is, of course, no objection in principle to doing so), but principally because, in the case which will almost always be that which concerns us, namely that in which the relation $B(\alpha, m)$ is extensional, we can make the requirement on Φ more precise: we can require that it be continuous in the following sense. Consider the universal spread as Baire space, that is, as a topological space whose points are the choice sequences and in which the open neighbourhoods which provide a base for the topology are the species of choice sequences sharing some initial segment: then we require that Φ be continuous with respect to this topology. Thus, where Φ is any functional from choice sequences to natural numbers, the condition that it be continuous can be written as:

$$\forall\alpha\ \exists n\ \forall\beta_{\beta\ \in\ \bar{\alpha}(n)}\ \Phi(\beta) = \Phi(\alpha).$$

Any constructive or lawlike functional which is continuous in this sense can be represented by a lawlike function on finite sequences of natural numbers. A function e from finite sequences of natural numbers to natural numbers will represent a continuous functional Φ if, for every α,

$$\text{for some } n\quad e(\bar{\alpha}(n)) = \Phi(\alpha) + 1, \text{ and, for all } m<n,$$
$$e(\bar{\alpha}(m)) = 0.$$

The fact that $e(\bar{\alpha}(m)) = 0$ for all $m \leq k$ means, in effect, that $\bar{\alpha}(k)$ is not a sufficiently long initial segment of α from which to compute $\Phi(\alpha)$: the first n for which $e(\bar{\alpha}(n))$ is positive allows us to determine $\Phi(a)$. The condition on e is thus:

$$\forall\alpha\ \exists n\ [e(\bar{\alpha}(n)) = \Phi(\alpha) + 1\ \&\ \forall m_{m<n} e(\bar{\alpha}(m)) = 0]\ .$$

This condition tells us nothing about the values of $e(\bar{\alpha}(m))$ for $m > n$. Kleene requires that, for $m > n$, $e(\bar{\alpha}(m)) = 0$, while Troelstra requires that, for $m > n$, $e(\bar{\alpha}(m)) = e(\bar{\alpha}(n))$. Both requirements are merely a matter of making some stipulation for the sake of determinateness, and, for the moment, we make no such requirement.

It is convenient to employ the notation '$e(\alpha)$', explained as follows:

$$e(\alpha) \text{ is defined} \leftrightarrow \exists n\ e(\bar{\alpha}(n)) > 0$$

$$e(\alpha) = k \leftrightarrow \exists n\ (e(\bar{\alpha}(n)) = k + 1\ \&\ \forall m_{m<n}\ e(\bar{\alpha}(m)) = 0).$$

Using this notation, we state the first, and weakest, version of the Continuity Principle, $\forall\alpha\exists!n$-continuity, as follows:

$$CP_{\exists!n} : \forall n\ Ext_{\alpha}\ C(\alpha, n)\ \&\ \forall\alpha\exists!n\ C(\alpha, n) \rightarrow$$
$$\exists e\ \forall\alpha\ [e(\alpha) \text{ is defined } \&\ C(\alpha, e(\alpha))]\ .$$

Note that, since $\exists!n\ C(\alpha, n)$ is equivalent to $\exists n\ \forall m\ (C(\alpha, m) \leftrightarrow m = n)$, $\forall\alpha\ \exists!n\ C(\alpha, n)$ implies that $C(\alpha, n)$ is decidable, since $m = n$ is decidable.

Even this, the weakest version of the Continuity Principle, is, as it stands, stronger than what we obtain by simply attending to the meanings of the quantifiers, for that tells us merely that there is some constructive functional Φ such that $\forall\alpha\ C(\alpha,\Phi(\alpha))$. The values of this functional might then depend upon intensional aspects of its arguments; if α is not lawless, $\Phi(\alpha)$ may depend, not merely on the terms of α, but on a rule for generating α, or a restriction upon the choice of its terms, imposed in advance (or at some later stage). The fact that $C(\alpha,n)$ is extensional does not immediately rule out this possibility, for to say that $C(\alpha,n)$ is extensional is not to say that we can determine its truth merely by reference to the terms of α, but only that, whenever we know that α and β coincide extensionally, then we also know that they bear the relation $C(\ ,\ )$ to the same n. Nevertheless, $CP_{\exists!n}$ asserts that, when $C(\alpha,n)$ is extensional and is satisfied, for each given α, by a unique n, the value of the functional depends only upon the terms of its argument, i.e. $\Phi(\alpha)$ is determined by the extension of α. The justification of this depends upon a more careful consideration than we have undertaken so far of the concept of a choice sequence, and this is postponed until the final chapter.

A stronger version of the Continuity Principle is $\forall\alpha\exists n$-continuity, where the requirement that n be unique is dropped, and is expressed by:

$$CP_{\exists n}: \qquad \forall n\ \mathrm{Ext}_{\alpha}\ C(\alpha,n)\ \ \&\ \ \forall\alpha\ \exists n\ C(\alpha,n)$$

$$\rightarrow \exists e\ \forall\alpha\ [e(\alpha) \text{ is defined } \ \&\ \ C(\alpha,e(\alpha))]\ .$$

Furthermore, we can extend our definition of extensionality, in an obvious way, to relations of the form $A(\alpha,\beta)$, and consider the formulation of a Continuity Principle with the hypothesis

$$\mathrm{Ext}_{\alpha,\beta}\ C(\alpha,\beta)\ \ \&\ \ \forall\alpha\ \exists\beta\ C(\alpha,\beta)\ .$$

As before, for any $A(\alpha,\beta)$, whether extensional or not, the statement

$$\forall\alpha\ \exists\beta\ A(\alpha,\beta)$$

may be taken to imply the existence of a constructive functional Ψ, mapping choice sequences to choice sequences, such that

$$\forall\alpha\ A(\alpha,\Psi(\alpha))\ .$$

Given the extensionality of $C(\alpha,\beta)$, $\forall\alpha\exists\beta$-continuity will then require that there be such a functional Ψ which is continuous in the sense that each term $[\Psi(\alpha)](n)$ of $\Psi(\alpha)$ depends on only finitely many terms $\alpha(0), \alpha(1), \ldots, \alpha(m-1)$ of α. Such a continuous functional Ψ can also be represented by a function e, namely by an e satisfying:

$$\forall\alpha\ \forall n\ \exists m[e(\langle n\rangle * \bar{\alpha}(m)) = [\Psi(\alpha)](n) + 1 \quad \& \quad \forall r_{r<m}\ e(\langle n\rangle * \bar{\alpha}(r)) = 0] \ .$$

When e represents Ψ in this way, we denote the function $\Psi(\alpha)$ by $e|\alpha$, so that:

$$e|\alpha \text{ is defined iff } \forall n\ \exists m\ e(\langle n\rangle * \bar{\alpha}(r)) > 0, \text{ and}$$

$$e|\alpha = \beta \leftrightarrow \forall n\ \exists m[e(\langle n\rangle * \bar{\alpha}(m)) = \beta(n) + 1 \quad \& \quad \forall r_{r<m}\ e(\langle n\rangle * \bar{\alpha}(r)) = 0] \ .$$

Under this representation, $\forall\alpha\ \exists\beta$-continuity may be expressed by the schema:

$$CP_{\exists\beta}: \qquad Ext_{\alpha,\beta}C(\alpha,\beta) \ \& \ \forall\alpha\ \exists\beta\ C(\alpha,\beta) \rightarrow \exists e\ \forall\alpha[(e|\alpha \text{ is defined } \& \ C(\alpha,e|\alpha)] \ .$$

Of these three principles of continuity, Brouwer certainly appealed to $\forall\alpha\ \exists!n$-continuity, and very probably to $\forall\alpha\ \exists n$-continuity as well. $\forall\alpha\ \exists\beta$-continuity was never claimed by Brouwer, and, as will be seen later, it is actually inconsistent with some of Brouwer's later, though controversial, proposals: it was formulated by Kleene as an analogue to $\forall\alpha\ \exists n$-continuity. Except in effecting a translation from statements involving choice sequences to those involving only constructive functions, there is no known use of $CP_{\exists\beta}$ which cannot be handled (sometimes with a little more trouble) by $CP_{\exists n}$. Henceforward the expression 'the Continuity Principle', if used without qualification, will be taken as referring to $CP_{\exists n}$.

Some Consequences of the Continuity Principle

In the rest of this section, the condition $\forall n\ Ext_{\alpha}\ C(\alpha, n)$ is tacitly assumed in all cases.

An obvious corollary of $CP_{\exists n}$ is the theorem:

<u>Local Continuity Principle</u> (LCP)

$$\forall\alpha\ \exists n\ C(\alpha, n) \rightarrow \forall\alpha\ \exists n\ \exists m\ \forall\beta_{\beta\,\in\,\bar{\alpha}(m)}\ C(\beta, n).$$

Proof. The proof is trivial. For assume $\forall\alpha\ \exists n\ C(\alpha, n)$; then, by $CP_{\exists n}$, there exists e such that

$$\forall\alpha\ (e(\alpha) \text{ is defined } \&\ C(\alpha, e(\alpha))).$$

For such an e and any given α, suppose

$$e(\bar{\alpha}(m)) = n + 1\ \&\ \forall k_{k<m}\ e(\bar{\alpha}(k)) = 0.$$

Then $e(\beta) = n$, and hence $C(\beta, n)$, for every $\beta \in \bar{\alpha}(m)$.

LCP is equivalent to:

$$(*)\ \forall\alpha\ \exists n\ C(\alpha, n) \rightarrow \forall\alpha\ \exists m\ \exists e\ \forall\beta_{\beta\,\in\,\bar{\alpha}(m)}\ (e(\beta) \text{ is defined } \&\ C(\beta, e(\beta))).$$

(*) says that if $\forall\alpha\ \exists n\ C(\alpha, n)$, then, for every α, the Continuity Principle holds in some neighbourhood of α. To see that LCP is equivalent to (*), assume $\forall\alpha\ \exists n\ C(\alpha, n)$. If LCP holds, then, for given α, we may suppose that

$$\forall\beta_{\beta\,\in\,\bar{\alpha}(m)}\ C(\beta, n).$$

We need then only put:

$$e(\vec{u}) = \begin{cases} n + 1 & \text{if } \vec{u} \preccurlyeq \bar{\alpha}(m) \\ 0 & \text{otherwise} \end{cases}$$

to obtain an e such that $e(\beta) = n$ for each $\beta \in \bar{\alpha}(m)$. Conversely, if (*) holds, suppose, for given α, that

$$\forall\beta_{\beta\,\in\,\bar{\alpha}(m)}\ (e(\beta) \text{ is defined } \&\ C(\beta, e(\beta))).$$

Since obviously $\alpha \in \bar{\alpha}(m)$, $e(\alpha)$ is defined, and we may suppose that

$$e(\bar{\alpha}(r)) = n + 1\ \&\ \forall k_{k<r}\ e(\bar{\alpha}(k)) = 0$$

for some r and for $n = e(\alpha)$. It follows that $C(\beta, n)$ for all $\beta \in \bar{\alpha}(r)$.

Continuity for Alternatives.

$$\mathrm{Ext}_{\alpha}A(\alpha) \ \& \ \mathrm{Ext}_{\alpha}B(\alpha) \ \& \ \forall\alpha\ (A(\alpha) \vee B(\alpha)) \rightarrow$$

$$\exists e\ \forall\alpha\ [e(\alpha) \text{ is defined } \&$$

$$((e(\alpha) = 0 \ \& \ A(\alpha)) \vee (e(\alpha) = 1 \ \& \ B(\alpha)))] \quad .$$

Proof. From $\forall\alpha\ (A(\alpha) \vee B(\alpha))$ we can easily derive

$$\forall\alpha\ \exists!n[(A(\alpha) \ \& \ n = 0) \vee (B(\alpha) \ \& \ n = 1)].$$

Now apply $CP\exists_{!n}$.

The principle of Continuity for Alternatives enables us to refute certain classically valid logical laws. That is, we can prove the negations of universal closures of certain instances of these laws involving variables for choice sequences. The basic example is the law of excluded middle: in order to refute this, we prove:

Theorem. $\neg\forall\alpha(\forall n\ \alpha(n) = 0 \vee \neg\ \forall n\ \alpha(n) = 0)$.

Proof. Suppose

$$\forall\alpha(\forall n\ \alpha(n) = 0 \vee \neg\forall n\ \alpha(n) = 0).$$

The property $\forall n\ \alpha(n) = 0 \vee \neg\forall n\ \alpha(n) = 0$ is obviously extensional. So, by Continuity for Alternatives, for some e, for each α,

$$(e(\alpha) = 0 \ \& \ \forall n\ \alpha(n) = 0) \vee (e(\alpha) = 1 \ \& \ \neg\forall n\ \alpha(n) = 0).$$

For given α, suppose

$$e(\alpha) = 0 \quad \& \quad \forall n\ \alpha(n) = 0.$$

Then, for some m,

$$e(\bar{\alpha}(m)) = 1 \quad \& \quad \forall r_{r<m}\ e(\bar{\alpha}(r)) = 0.$$

So we have

$$\forall\beta_{\beta \in \bar{\alpha}(m)}\ \forall n\ \beta(n) = 0.$$

But this is absurd, for we can define β by:

$$\beta(n) = \begin{cases} 0 \text{ for } n<m \\ 1 \text{ for } n\geq m . \end{cases}$$

Then, $\bar{\beta}(m) = \bar{\alpha}(m)$, but $\neg \forall n\ \beta(n) = 0$. So we must have:

$$\forall\alpha\ (e(\alpha) = 1\ \&\ \neg\forall n\ \alpha(n) = 0).$$

But this is equally absurd, since we may take $\alpha(n) = 0$ for all n. Hence,

$$\neg\forall\alpha\ (\forall n\ \alpha(n) = 0 \vee \neg\forall n\ \alpha(n) = 0).$$

This theorem shows that the Continuity Principle is actually inconsistent with classical logic. Many other laws of classical logic can be refuted in the same way.

We remark here that we can derive the Continuity Principle for a particular spread from the above unrestricted version, just as we can derive the restricted version of Bar Induction from Bar Induction on the universal spread.

The next theorem has often appeared in the literature, e.g. in Heyting's *Intuitionism*, under the name 'the Fan Theorem'. It is, however, important to distinguish it from our 'Fan Theorem', since, unlike ours, it depends on the Continuity Principle.

Extended Fan Theorem.

$$\text{fan}(s)\ \&\ \forall\alpha_{\alpha\in s}\ \exists n\ C(\alpha, n) \rightarrow \exists m\ \forall\alpha_{\alpha\in s}\ \exists n\ \forall\beta_{\beta\in s,\beta\in\bar{\alpha}(m)}\ C(\beta,n).$$

Proof. By $CP_{\exists n}$ restricted to s, there is an e such that

$$\forall\alpha_{\alpha\in s}\ \exists m\ \exists n\ [e(\bar{\alpha}(m)) = n+1\ \&\ \forall r_{r<m}\ e(\bar{\alpha}(r)) = 0\ \&\ C(\alpha,n)].$$

Define a species $\underline{R}$ by

$$\vec{u} \in \underline{R} \leftrightarrow e(\vec{u}) > 0\ \&\ \forall\vec{v}_{\vec{u} \prec \vec{v}}\ e(\vec{v}) = 0 .$$

Then $\underline{R}$ is decidable, and

$$\forall\alpha_{\alpha\in s}\ \exists n\ \bar{\alpha}(n) \in \underline{R} .$$

So, by the Fan Theorem, FT, there exists an m such that

$$\forall\alpha_{\alpha\in s}\ \exists n_{n\leq m}\ \bar{\alpha}(n) \in \underline{R}.$$

So, if $\alpha \in s$, then, for some $n \leq m$ and some q,

$$e(\bar{\alpha}(n)) = q+1 \ \& \ \forall r_{r<n}\ e(\bar{\alpha}(r)) = 0 \ \& \ C(\alpha,q).$$

Now, if $\beta \in s$ and $\beta \in \bar{\alpha}(m)$, then $\beta \in \bar{\alpha}(n)$, so $e(\beta) = q$ and $C(\beta,q)$.

Hence, $\exists m\ \forall\alpha_{\alpha\in s}\ \exists n\ \forall\beta_{\beta\in s,\beta\in\bar{\alpha}(m)}\ C(\beta,n)$.

The Continuity Principle also enables us to obtain stronger forms of Bar Induction. We can replace the premiss in BI_D which asserts the decidability of $\underline{R}$ by one asserting that $\underline{R}$ is monotonic: that is, if $\vec{u} \in \underline{R}$, then $\vec{v} \in \underline{R}$ whenever $\vec{v}$ is an extension of $\vec{u}$.

BI_M:

$$\forall\vec{u}\ \forall\vec{v}_{\vec{v} \succcurlyeq \vec{u}}\ (\vec{u} \in \underline{R} \rightarrow \vec{v} \in \underline{R}) \ \&$$

$$\forall\alpha\ \exists n\ \bar{\alpha}(n) \in \underline{R} \ \&$$

$$\forall\vec{u}\ (\vec{u} \in \underline{R} \rightarrow \vec{u} \in \underline{A}) \ \&$$

$$\forall\vec{u}\ (\forall k\ \vec{u}^\frown k \in \underline{A} \rightarrow \vec{u} \in \underline{A})$$

$$\rightarrow \langle\ \rangle \in \underline{A} .$$

To prove that BI_M implies BI_D we do not have to appeal to the Continuity Principle, though we do for the converse implication.

<u>Theorem</u>. $BI_M \rightarrow BI_D$.

Proof. Assume BI_M holds, and suppose $\underline{R}$ and $\underline{A}$ satisfy the premiss of BI_D. Clearly, our strategy must be to define new species to which we can apply BI_M, in such a way that the conclusion of BI_M so applied yields $\langle\ \rangle \in \underline{A}$. To this end, define species $\underline{R}^*$ and $\underline{A}^*$ by:

$$\vec{u} \in \underline{R}^* \leftrightarrow \exists\vec{v}_{\vec{u} \preccurlyeq \vec{v}}\ \vec{v} \in \underline{R}$$

and $\vec{u} \in \underline{A}^* \leftrightarrow \vec{u} \in \underline{A} \vee \vec{u} \in \underline{R}^*$.

Plainly, $\underline{R}^*$ is monotonic. Also, by assumption, $\forall\alpha\ \exists n\ \bar{\alpha}(n) \in \underline{R}$, so $\forall\alpha\ \exists n\ \bar{\alpha}(n) \in \underline{R}^*$. Trivially,

$\vec{u} \in \underline{R}^* \to \vec{u} \in \underline{A}^*$. Before we can apply BI_M we must show $\underline{A}^*$ to be hereditary upwards. Suppose $\forall k\ \vec{u}^\frown k \in \underline{A}^*$. By hypothesis $\underline{R}$ is decidable, so $\underline{R}^*$ is clearly also decidable. Thus we can argue by cases. If $\vec{u} \in \underline{R}^*$, then $\vec{u} \in \underline{A}^*$. If $\vec{u} \notin \underline{R}^*$, then we have $\vec{u}^\frown k \in \underline{R}^* \to \vec{u}^\frown k \in \underline{R}$, and so $\vec{u}^\frown k \in \underline{R}^* \to \vec{u}^\frown k \in \underline{A}$. But then, from the definition of $\underline{A}^*$, for all k, $\vec{u}^\frown k \in \underline{A}^* \to \vec{u}^\frown k \in \underline{A}$. Therefore

$$\forall k\ \vec{u}^\frown k \in \underline{A}^* \to \vec{u} \in \underline{A},$$

since $\underline{A}$ is hereditary upwards. Thus, in both cases,

$$\forall k\ \vec{u}^\frown k \in \underline{A}^* \to \vec{u} \in \underline{A}^* .$$

Now we can use BI_M to infer $< > \in \underline{A}^*$, and so $< > \in \underline{A}$.

Theorem. BI_D & $CP_{\exists n} \to BI_M$.

Proof. Assume BI_D, and take $\underline{R}$ and $\underline{A}$ satisfying the premisses of BI_M. Applying the Continuity Principle to

$$\forall \alpha\ \exists n\ \bar{\alpha}(n) \in \underline{R},$$

we get $\forall \alpha\ \bar{\alpha}(e(\alpha)) \in \underline{R}$, for some e subject to the usual conditions. Now define $\underline{R}^*$ by:

$$\vec{u} \in \underline{R}^* \leftrightarrow \exists \vec{v}_{\vec{u} \leq \vec{v}} [e(\vec{v}) > 0 \ \&\ \forall \vec{w}_{\vec{v} < \vec{w}}\ e(\vec{w}) = 0 \ \&\ \ell h(\vec{u}) \geq e(\vec{v}) - 1]$$

Plainly, $\underline{R}^*$ is decidable. Since $\underline{R}$ was assumed to be monotonic,

$$\vec{u} \in \underline{R}^* \to \vec{u} \in \underline{R}, \text{ and so } \vec{u} \in \underline{R}^* \to \vec{u} \in \underline{A}.$$

Clearly, $\forall \alpha\ \exists n\ \bar{\alpha}(n) \in \underline{R}^*$, since $e(\alpha)$ is defined for each α. Finally, $\underline{A}$ is hereditary upwards by hypothesis. Hence, by BI_D,

$$< > \in \underline{A},$$

and so BI_M holds.

The Continuity Principle permits us to drop the assumption of decidability from the Fan Theorem, FT, and prove the more general version, GFT, which thus proves after all to be intuitionistically valid.

General Fan Theorem.

$$\mathrm{fan}(s)\ \&\ \forall\alpha_{\alpha\in s}\ \exists n\ \bar{\alpha}(n) \in \underline{R} \rightarrow \exists p\ \forall\alpha_{\alpha\in s}\ \exists n_{n\leq p}\ \bar{\alpha}(n) \in \underline{R}.$$

Proof. By $CP_{\exists n}$, the second premiss gives, for some e:

$$\forall\alpha_{\alpha\in s}\ \bar{\alpha}(e(\alpha)) \in \underline{R}\ .$$

By the Extended Fan Theorem, there exists an m such that:

$$\forall\alpha_{\alpha\in s}\ \exists n\ \forall\beta_{\beta\in s, \beta\in\bar{\alpha}(m)}\ \bar{\beta}(n) \in \underline{R}.$$

Since s is a fan, there are only finitely many admissible sequences $\vec{u}$ with $\ell h(\vec{u}) \leq m$. If $\vec{u}_0, \vec{u}_1, \ldots, \vec{u}_k$ is a list of these sequences, put

$$p = \max_{i\leq k} e(\vec{u}_i) - 1\ .$$

Then $\forall\alpha_{\alpha\in s}\ \exists n_{n\leq p}\ \bar{\alpha}(n) \in \underline{R}$.

Having seen that we can alter the restriction on $\underline{R}$ in Bar Induction and omit the premiss of decidability from the Fan Theorem, it is natural to ask whether we can dispense with all restrictions on Bar Induction. The following counter-example, due to Kleene, shows that we cannot do this without contradicting the Continuity Principle.

Let B(n) be any decidable predicate of natural numbers. Define $\underline{R}$ and $\underline{A}$ by:

$$\vec{u} \in \underline{R} \leftrightarrow (\vec{u} = \langle\ \rangle\ \&\ \neg\forall n\ B(n))\ \vee\ \exists m(\vec{u} = \langle m\rangle\ \&\ B(m)),$$

$$\vec{u} \in \underline{A} \leftrightarrow (\vec{u} = \langle\ \rangle\ \&\ (\forall n B(n)\ \vee\ \neg\forall n B(n)))\ \vee\ \exists m\ (\vec{u} = \langle m\rangle\ \&\ B(m)).$$

$\underline{R}$ is obviously not monotonic, and, if it were decidable, then, from $\langle\ \rangle \in \underline{R} \vee \langle\ \rangle \notin \underline{R}$, we could infer $\neg\forall n\ B(n) \vee \neg\neg\forall n\ B(n)$.

The other hypotheses for Bar Induction do hold, however: since B(n) is decidable, we have, for each α,

$$B(\alpha(0))\ \vee\ \neg B(\alpha(0))\quad .$$

But $B(\alpha(0))$ implies $\bar{\alpha}(1) \in \underline{R}$, and $\neg B(\alpha(0))$ implies $\neg\forall n\ B(n)$, and so $\bar{\alpha}(0) \in \underline{R}$. Hence,

$$\forall\alpha\ \exists n\ \bar{\alpha}(n) \in \underline{R}.$$

Clearly $\vec{u} \in \underline{R} \rightarrow \vec{u} \in \underline{A}$, and $\forall k\ \vec{u}^\frown k \in \underline{A}$ implies $\vec{u} = \langle\ \rangle\ \&\ \forall n\ B(n)$, which yields $\langle\ \rangle \in \underline{A}$, i.e. $\vec{u} \in \underline{A}$. Now, if Bar Induction holds when $\underline{R}$ is neither decidable nor monotonic, we can apply it here to conclude

$$\langle\ \rangle \in \underline{A}\ ,$$

and so $$\forall n\ B(n) \vee \neg\forall n\ B(n)\ .$$

But if we take P(n) to be $\alpha(n) = 0$ for any α in the universal spread, we get

$$\forall\alpha\ (\forall n\ \alpha(n) = 0 \vee \neg\forall n\ \alpha(n) = 0)\ ,$$

contradicting our counter-example to the Law of Excluded Middle, derived from the $\forall\alpha\exists!n$-Continuity Principle. (The assumption that $\underline{R}$ is decidable would contradict the Continuity Principle for just the same reason.) So we have shown that Bar Induction does not hold for unrestricted $\underline{R}$.

It might momentarily appear as though we could patch up this counter-example to BI_M. We could put:

$$\vec{u} \in \underline{R}^* \leftrightarrow \neg\forall n\ P(n) \vee (\ell h(\vec{u}) > 0\ \&\ P(u_0))$$

and $$\vec{u} \in \underline{A}^* \leftrightarrow (\forall n\ P(n) \vee \neg\forall n\ P(n)) \vee (\ell h(\vec{u}) > 0\ \&\ P(u_0)).$$

Clearly there has to be something wrong with applying BI_M to this case, since the conclusion would again be inconsistent with the Continuity Principle. On examining the premisses of the induction, we find that all but one are satisfied: $\underline{A}^*$ is not hereditary upwards; for, given $\forall k\ \langle k\rangle \in \underline{A}^*$, we can only infer $\forall k\ (P(k) \vee \langle\ \rangle \in \underline{A}^*)$, but not $\langle\ \rangle \in \underline{A}^*$. To infer $\langle\ \rangle \in \underline{A}^*$, we should have to appeal to the classical law

$$(*)\quad \forall x\ (F(x) \vee B) \rightarrow (\forall x\ F(x) \vee B)\ ,$$

which is intuitionistically invalid.

The counter-example depended on the plainly correct inference from

$$\forall k\ (P(k) \vee \neg P(k))$$

to $$\forall k\ (P(k) \vee \neg\forall n\ P(n))\ .$$

An application of the law (*) would allow the further step to

$$\forall n\, P(n) \vee \neg \forall n\, P(n) \ .$$

Since, as already noted, $CP_{\exists!n}$ yields a counter-example to

$$\forall n (P(n) \vee \neg P(n)) \rightarrow (\forall n\, P(n) \vee \neg \forall n\, P(n)),$$

it thus likewise yields one to (*).

By means of BI_M and the Axiom of Choice, we can show that the Local Continuity Principle implies the full Continuity Principle ($CP_{\exists n}$). Let $AC_{n,e}$ denote the analogue of $AC_{n,b}$ for constructive functions from finite sequences to natural numbers.

Theorem. $AC_{n,e}$ & BI_M & LCP $\rightarrow$ $CP_{\exists n}$.

Proof. Assume $\forall\alpha\, \exists n\, C(\alpha, n)$.

By LCP, $\forall\alpha\, \exists m\, \exists n\, \forall\beta_{\beta\in\bar{\alpha}(m)}\, C(\beta, n)$.

In order to apply BI_M, we put:

$$\underline{R} = \{\vec{u} \mid \exists n\, \forall\alpha_{\alpha\in\vec{u}}\, C(\alpha, n)\}$$

$$\underline{A} = \{\vec{u} \mid \exists e\, \forall\alpha_{\alpha\in\vec{u}}\, (e(\alpha) \text{ is defined } \& \ C(\alpha, e(\alpha)))\}$$

The hypotheses of BI_M are satisfied for this $\underline{R}$ and $\underline{A}$.

(i) It is obvious from its definition that $\underline{R}$ is monotonic.

(ii) $\forall\alpha\exists m\, \bar{\alpha}(m) \in \underline{R}$ follows from the consequent of LCP.

(iii) In order to show that $\underline{R} \subseteq \underline{A}$, suppose that $\vec{u} \in \underline{R}$ and that

$$\forall\alpha_{\alpha\in\vec{u}}\, C(\alpha, n) \ .$$

Put:

$$e(\vec{v}) = \begin{cases} n + 1 & \text{if } \vec{v} \preccurlyeq \vec{u} \\ 0 & \text{if } \vec{u} \prec \vec{v} \\ 1 & \text{otherwise.} \end{cases}$$

Then $e(\alpha)$ is defined for every α, and for $\alpha \in \vec{u}$,

$$e(\alpha) = n.$$

Hence $\vec{u} \in \underline{A}$.

(iv) In order to show that $\underline{A}$ is hereditary upwards, suppose that

$$\forall k\ \vec{u}^\frown k \in \underline{A},$$

i.e. that

$$\forall k\ \exists e\ \forall\alpha_{\alpha\in\vec{u}^\frown k}\ (e(\alpha) \text{ is defined } \&\ C(\alpha, e(\alpha))).$$

Writing $G(k, e)$ for

$$\forall\alpha_{\alpha\in\vec{u}^\frown k}\ (e(\alpha) \text{ is defined } \&\ C(\alpha, e(\alpha))),$$

$AC_{n,e}$ implies the existence of a binary function f such that

$$\forall k\ G(k, \lambda\vec{v}.\ f(k, \vec{v})).$$

Let us write e_k for $\lambda\vec{v}.\ f(k, \vec{v})$. We now define e thus:

$$e(\vec{v}) = \begin{cases} 0 & \text{if } \vec{u} \not\leq \vec{v} \\ e_k(\vec{v}) & \text{if } \vec{v} \leq \vec{u}^\frown k \text{ and } e_k(\vec{w}) = 0 \text{ for all } \vec{w} \not\leq \vec{u} \\ r+1 & \text{if } \vec{v} \not\leq \vec{u}^\frown k,\ \vec{u} \leq \vec{w},\ e_k(\vec{w}) = r+1, \\ & \quad \text{and } e_k(\vec{w}') = 0 \text{ for all } \vec{w}' \succ \vec{w}. \\ 1 & \text{otherwise} \end{cases}$$

Then $e(\alpha)$ is defined for all α, and, for all $\alpha \in \vec{u}$, $C(\alpha, e(\alpha))$, and accordingly $\vec{u} \in \underline{A}$. Since all the hypotheses of BI_M are satisfied, we may conclude that $\langle\ \rangle \in \underline{A}$, i.e. that the consequent of $CP_{\exists n}$ holds. The proof is thus completed.

Suppose that, for some extensional predicate $C(\alpha, n)$, $\forall\alpha\ \exists n\ C(\alpha, n)$ and that e is the function whose existence is demanded by $CP_{\exists n}$. We are not entitled to assume that e satisfies:

$$\forall\alpha_{\alpha\in\vec{u}}\ C(\alpha, n) \rightarrow \exists\vec{v}_{\vec{u} \preccurlyeq \vec{v}}\ e(\vec{v}) = n + 1.$$

That is, although it might be the case that every choice sequence α with some fixed initial segment $\vec{u}$ bears the relation C(,) to n, it might be possible to discover this only by examining later terms of the choice sequences. For example where $\underline{A} = \{\vec{u} \mid [\vec{u} = \langle\ \rangle \ \&\ (\forall n\, P(n) \vee \neg\forall n\, P(n))] \vee \exists m(\vec{u} = \langle m\rangle \ \&\ P(m))\}$, as in the foregoing example of Kleene's to show that Bar Induction does not hold for unrestricted $\underline{R}$, we have

$$\forall\alpha\ \exists n\ \bar{\alpha}(n) \in \underline{A}.$$

Hence, by $CP_{\exists n}$, taking $C(\alpha, n)$ as '$\bar{\alpha}(n) \in \underline{A}$', there is an e such that

$$\forall\alpha(e(\alpha) \text{ is defined } \&\ \bar{\alpha}(e(\alpha)) \in \underline{A}).$$

Now suppose further that this e does satisfy

$$\forall\vec{u}[\forall\alpha_{\alpha\in\vec{u}}\ \bar{\alpha}(n) \in \underline{A} \rightarrow \exists\vec{v}_{\vec{u} \preccurlyeq \vec{v}}\ e(\vec{v}) = n + 1] .$$

Then, in particular,

$$\forall\alpha\ \bar{\alpha}(0) \in \underline{A} \rightarrow e(\langle\ \rangle) = 1.$$

Furthermore, since $\bar{\alpha}(0) = \langle\ \rangle$,

$$\forall\alpha\ \bar{\alpha}(0) \in \underline{A} \leftrightarrow (\forall n\ P(n) \vee \neg\forall n\ P(n)) ,$$

and so

$$(\forall n\ P(n) \vee \neg\forall\ n\ P(n)) \rightarrow e(\langle\ \rangle) = 1.$$

By contraposing twice and appealing to the stability of $e(\langle\ \rangle) = 1$ and to the valid logical law $\neg\neg(B \vee \neg B)$, we obtain from this:

$$e(\langle\ \rangle) = 1.$$

But this in turn implies $e(\alpha) = 0$ for all α, and hence $\forall\alpha\ \bar{\alpha}(0) \in \underline{A}$, and therefore

$$\forall n\ P(n) \vee \neg\forall n\ P(n).$$

Since P(n) was an arbitrary decidable predicate of natural numbers, it follows that we cannot, in general, require the existence of an e on which this further restriction is imposed.

Another counter-example of Kleene's makes it clear that such an assumption is illegitimate even when $\forall\alpha\ \exists!n\ C(\alpha, n)$. Where $T_1(m,n,k)$ is the usual Kleene T-predicate, we put:

$$C(\alpha,n) \leftrightarrow [n = 0\ \&\ T_1(\alpha(0),\alpha(0),\alpha(1))]$$
$$\vee\ [n = 1\ \&\ \neg T_1(\alpha(0),\alpha(0),\alpha(1))].$$

Obviously, $\forall\alpha\ \exists!n\ C(\alpha,n)$. Let e be the function whose existence is demanded by $CP_{\exists!n}$, so that $\forall\alpha\ C(\alpha,e(\alpha))$, and suppose also that

$$\forall\alpha_{\alpha\in\vec{u}}\ C(\alpha,n) \rightarrow \exists\vec{v}_{\vec{u}\leqslant\vec{v}}\ e(\vec{v}) = n + 1.$$

Then, if, for some m, $\forall r \neg T_1(m,m,r)$, we have

$$\forall\alpha_{\alpha\in\langle m\rangle}\ C(\alpha,1)\ ,$$

and so $$e(\langle m\rangle) = 2 \vee e(\langle\ \rangle) = 2.$$

But $e(\langle\ \rangle) = 2$ implies $\forall m\ \forall r \neg T_1(m,m,r)$, which is absurd, so $e(\langle m\rangle) = 2$. Conversely, $e(\langle m\rangle) = 2$ implies $\forall r \neg T_1(m,m,r)$.

Hence $$\forall r \neg T_1(m,m,r) \leftrightarrow e(\langle m\rangle) = 2.$$

But if this is the case, then, for familiar reasons, e cannot be general recursive, in contradiction to Church's Thesis. So if we are to maintain the consistency of our theory with Church's Thesis, we must drop the restriction on e.

Church's Thesis may be expressed by the formula

$$\forall a\ \exists m\ \forall n\ \exists k\ [T_1(m,n,k)\ \&\ a(n) = U(k)]\ ,$$

which says that every constructive function is general recursive. (A version of Church's Thesis for functions e from finite sequences to natural numbers can easily be framed by using some effective representation of finite sequences by natural numbers.) It can be shown that this formula is in fact consistent with the assumptions we have made so far.

3.4. BROUWER'S PROOF OF THE BAR THEOREM

Brouwer attempted to give a proof of the validity of Bar Induction. This proof is of great conceptual interest.

Where $\underline{R}$ is a species of finite sequences, s is a spread, and $\vec{u}$ a finite sequence admissible in s, we gave in section 3.2 the definition:

$$\underline{R} \textit{ bars } \vec{u} \textit{ in } s \text{ iff } \forall\alpha_{\alpha\in s,\alpha\in\vec{u}} \exists n\ \bar{\alpha}(n) \in \underline{R} .$$

If $\underline{R}$ bars $\vec{u}$ in the universal spread, we say simply that $\underline{R}$ bars $\vec{u}$. The premisses of Bar Induction for the universal spread can then be written:

(i) $\underline{R}$ bars $\langle\ \rangle$;

(ii) $\underline{R}$ is decidable;

(iii) $\underline{R} \subseteq \underline{A}$;

(iv) $\forall\vec{u}\ (\forall k\ \vec{u}\ k \in \underline{A} \rightarrow \vec{u} \in \underline{A})$.

From (i) to (iv), we have to prove that $\langle\ \rangle \in \underline{A}$. Brouwer's proof proceeds by assuming that we are given a proof of (i), and asking what form such a proof can take. Part of the interest of Brouwer's proof thus comes from the fact that it is an example, indeed so far the only example, of a proof exploiting the full intuitionistic interpretation of $\rightarrow$: instead of merely allowing appeal to the antecedent in deducing the consequent, it analyses the form of any possible proof of the antecedent, and attempts to show that we can subject that proof to a transformation converting it into a proof of the consequent.

The data from which we can infer that $\underline{R}$ bars $\langle\ \rangle$ consist solely in the knowledge of which finite sequences $\vec{u}$ belong to $\underline{R}$. Brouwer considers the proof that $\underline{R}$ bars $\langle\ \rangle$ as expanded into its 'fully analysed'form, that is, a form, in which every step has been broken down into a sequence of steps each of which is as short as possible, and he asserts that such a proof will employ only three distinct types of inference. These are η-inferences, of the form:

$$\frac{\vec{u} \in \underline{R}}{\underline{R} \text{ bars } \vec{u}} \; ;$$

ζ-inferences, of the form:

$$\frac{\underline{R} \text{ bars } \vec{u}}{\underline{R} \text{ bars } \vec{u}^\frown \bar{k}} \; ;$$

and $\mathcal{F}$-inferences, of the form:

$$\frac{\underline{R} \text{ bars } \vec{u}^\frown 0 \quad \underline{R} \text{ bars } \vec{u}^\frown 1 \quad \underline{R} \text{ bars } \vec{u}^\frown 2 \quad \ldots}{\underline{R} \text{ bars } \vec{u}} \; .$$

(The nomenclature comes from Brouwer and Kleene.) A $\mathcal{F}$-inference is thus an inference with denumerably many premisses.

A proof containing inferences with infinitely many premisses evidently cannot be a written proof: but Brouwer insists that proofs of the sort with which intuitionism is concerned are not written proofs, but mental constructions, and that such a construction may be an infinite structure. To be a proof, it must be well-founded: if the proof is conceived as arranged in tree form, every branch must be finite. (If it were possible to form an infinite sequence of propositions, beginning with the conclusion of the proof, each subsequent proposition being one of the premisses upon which the preceding one depended, then we should have no reason to accept as true any proposition in the sequence, including the conclusion of the 'proof'. This resembles Aquinas's denial of the possibility of an infinite regress in causes.) But it is not necessary that every proposition occurring in the fully analysed proof should depend on only finitely many premisses.

At first sight, this appeal to the conception of an infinite, though well-founded, proof is contrary to the constructivist spirit. However, we must interpret it in accordance with the general intuitionistic doctrine about the nature of infinity: an infinite structure can never be regarded as completed; rather, to grasp an infinite structure is to grasp a principle whereby any finite segment of it, however large, can be explicitly constructed. Looked at in this light, it can hardly be denied that an intuitionistic proof supplies a principle for the explicit construction of any finite segment of its fully analysed version, so long as it is understood that the direction of analysis, and thus of construction, runs counter to the direction of inference. Brouwer's notion of an analysis of a proof appears to be this: that whenever, in the course of the proof, we appeal to some operation as yielding a result of a certain kind, then, in the analysed form of the proof, that operation will actually be carried out. Thus the appearance of a universally quantified statement in a proof, for instance the statement

$$\forall k \ \underline{R} \text{ bars } \vec{u}^\frown k,$$

signifies our recognition that a certain operation will, when applied to any element of the domain (in this case, to any natural number k), yield a proof of the corresponding instance (here, of the statement '$\underline{R}$ bars $\vec{u}^\frown \bar{k}$'). In the fully analysed proof, therefore, the universal quantification does not appear: the operation is actually applied to each element of the domain, yielding a proof of the corresponding instance, and that which formerly was inferred from the universally quantified statement now appears as following from the individual instances taken together. A proof containing inferences with infinitely many premisses appears non-constructive only because we try to imagine ourselves as completing the infinite process of proving each of the premisses individually, and then drawing the conclusion, and this, of course, makes no constructive sense. But, on the intuitionistic understanding of infinity, the only way in which we can draw an inference from infinitely many premisses is by recognizing *that* each of these premisses can be proved; and that, in turn, can be accomplished only by recognizing, of some general procedure, that it will yield a proof of each of the premisses. Thus the only way of understanding the idea of an inference from denumerably many premisses A(0), A(1),... which is consistent with a constructivist outlook proves to coincide exactly with the intuitionistic interpretation of an inference from $\forall n\, A(n)$. An intuitionistic proof involving inferences from universally quantified statements really is, therefore, what Brouwer maintains, a representation of a more fully analysed proof containing inferences from infinitely many premisses.

Given a proof that $\underline{R}$ bars $\langle\ \rangle$ which uses only η-, ζ-, and $\mathcal{F}$-inferences, it may be rendered more direct by elimination of the ζ-inferences. Since the proof is well-founded, it is sufficient to show that we can eliminate any ζ-inference which has no ζ-inference occurring in any branch of the proof-tree running upwards from its premiss (the proof-tree being taken as oriented so that the vertex, i.e. the conclusion '$\underline{R}$ bars $\langle\ \rangle$', stands at the bottom). If

$$\frac{\underline{R} \text{ bars } \vec{u}}{\underline{R} \text{ bars } \vec{u}^\frown\bar{k}}$$

is such a ζ-inference, then its premiss '$\underline{R}$ bars $\vec{u}$' must have been derived either by a $\mathcal{F}$-inference or by an η-inference. If by a $\mathcal{F}$-inference, then the statement '$\underline{R}$ bars $\vec{u}^\frown\bar{k}$' must have figured as one of the denumerably many premisses of the $\mathcal{F}$-inference, which, therefore, in that occurrence, was proved without the use of any ζ-inference: we can accordingly replace the existing proof of the occurrence of that statement as the conclusion of the ζ-inference by the proof that contains no ζ-inference. If, on the other hand, the premiss '$\underline{R}$ bars $\vec{u}$' of the ζ-inference was obtained by an η-inference from the premiss '$\vec{u} \in \underline{R}$', then we consider that path in the proof-tree which leads from the conclusion '$\underline{R}$ bars $\vec{u}^\frown\bar{k}$' of our ζ-inference to the conclusion '$\underline{R}$ bars $\langle\ \rangle$' of the whole proof. Each of the statements occurring on this path is of the form '$\underline{R}$ bars $\vec{v}$' for some finite sequence $\vec{v}$ which is either an initial segment or an extension of $\vec{u}$. Moreover, in the passage from one statement to the next, the length of the finite sequence mentioned is either increased or diminished by 1, and the length of the sequence mentioned at the end of the path is 0. Hence, somewhere along the path there must occur the statement '$\underline{R}$ bars $\vec{u}$'. We accordingly replace the entire proof of that occurrence of '$\underline{R}$ bars $\vec{u}$' by a derivation of it by means of an η-inference.

Once we have a proof of '$\underline{R}$ bars $\langle\ \rangle$' in which only η- and $\mathcal{F}$-inferences occur, we proceed to transform it into a proof of '$\langle\ \rangle \in \underline{A}$' by simply replacing every statement of the form '$\underline{R}$ bars $\vec{u}$' by the statement '$\vec{u} \in \underline{A}$'. Each η-inference is transformed into a valid inference in virtue of the truth of hypothesis (iii), and each $\mathcal{F}$-inference is transformed into a valid inference in virtue of the truth hypothesis (iv).

This proof of Brouwer's is extremely interesting: we must, however, recognize that, as it stands, it is incorrect. For, if it were correct, it would establish the unrestricted validity of Bar Induction, without the requirement that $\underline{R}$ be either decidable or monotonic; for it makes no appeal to the decidability or the monotonicity of $\underline{R}$.

We can, therefore, see where Brouwer's proof fails by considering a case in which $\underline{R}$ is neither decidable nor monotonic, but in which the other hypotheses of Bar Induction are satisfied. Such a case is Kleene's counter-example to unrestricted Bar Induction. It will be recalled that, in this example, B(n) expresses some decidable property of natural numbers, and $\underline{R}$ is defined that it contains no sequences of length greater than 1, it contains $\langle\ \rangle$ just in case $\neg\forall n\, B(n)$, and it contains $\langle k\rangle$ just in case B(k). It comes as no surprise that consideration of this example shows that the flaw in Brouwer's proof lies in his unsupported assertion that any fully analysed proof that a species bars $\langle\ \rangle$ (or any other finite sequence $\vec{u}$) can contain only η-, ζ-, and $\mathcal{F}$-inferences. For the species $\underline{R}$ of the example, the proof that $\underline{R}$ bars $\langle\ \rangle$ proceeds by observing that from

$$\forall k\ (B(k) \vee \neg B(k))$$

we may infer

$$\forall k\ (B(k) \vee \neg\forall n\ B(n)),$$

which amounts to saying

$$\forall k\ (\langle k\rangle \in \underline{R} \vee \langle\ \rangle \in \underline{R}).$$

This of course yields

$$\forall k\,(\underline{R} \text{ bars } \langle k\rangle \vee \underline{R} \text{ bars } \langle\ \rangle),$$

which in turn, since the property of being barred by $\underline{R}$ is monotonic, yields

$$\forall k\ \underline{R} \text{ bars } \langle k\rangle\ ;$$

finally, since the property of being barred by $\underline{R}$ is hereditary upwards, we infer

$$\underline{R} \text{ bars } \langle\ \rangle.$$

Generalizing this, we obtain a form of inference which appears to resist reduction, under the process of complete analysis of a proof, to any one of the three forms of inference which Brouwer says will alone occur. Let us give the name 'Θ-inference' to the analysed version of an inference of the form

$$\frac{\underline{R} \text{ bars } \vec{u}^\frown\bar{k} \vee \underline{R} \text{ bars } \vec{u}}{\underline{R} \text{ bars } \vec{u}^\frown\bar{k}}\ .$$

Then, in the fully analysed version of the above proof that $\underline{R}$ bars $\langle\ \rangle$, we perform denumerably many Θ-inferences (with $\vec{u}$ as $\langle\ \rangle$), preparatory to arriving at the final conclusion by means of a F-inference. Thus a proof of a conclusion of the form '$\underline{R}$ bars $\langle\ \rangle$' may employ Θ-inferences as well as inferences of the three kinds mentioned by Brouwer.

It may well be objected that a Θ-inference is not really a distinct form of inference at all. For, since a proof of a disjunction is a proof of one or other disjunct, in a fully analysed proof the premiss '$\underline{R}$ bars $\vec{u}^\frown\bar{k} \vee \underline{R}$ bars $\vec{u}$' will reduce either to '$\underline{R}$ bars $\vec{u}^\frown\bar{k}$' or to '$\underline{R}$ bars $\vec{u}$'. In the former case, the premiss of the Θ-inference becomes the same as its conclusion, so that we do not really have an inference at all; in the latter case, the Θ-inference reduces to a straightforward ζ-inference.

A certain delicacy is required in stating why this objection fails. Earlier, it was explained why a crude objection to Brouwer's appeal to a mental proof as an infinite structure was unsound; but such a conception has to be very carefully handled if it is not to import non-constructive notions. It is true enough that, in the fully analysed proof, the premiss of a Θ-inference will not be a disjunctive statement, but a statement of the form '$\underline{R}$ bars $\vec{u}_{\bar{k}}$', where $\vec{u}_{\bar{k}}$ is, determinately, either $\vec{u}\frown\bar{k}$ or simply $\vec{u}$. What it means to say this is, however, that, when we perform an explicit construction of a finite segment of the infinite proof which contains this particular Θ-inference, we shall be able effectively to determine its premiss either as '$\underline{R}$ bars $\vec{u}\frown\bar{k}$' or as '$\underline{R}$ bars $\vec{u}$'. When we do this we shall, indeed, be able, in the former case, to eliminate the Θ-inference altogether, and, in the latter, to recognize it as reducing to a ζ-inference. But the proof, considered as a whole, is a mental construction; and this means that we cannot consider it as, so to speak, already having certain features, or displaying certain patterns, which we have failed to notice. What *we* know about the proof is that, for each k, we can prove '$\underline{R}$ bars $\vec{u}_{\bar{k}}$', and that, from this, '$\underline{R}$ bars $\vec{u}\frown\bar{k}$' follows; that is, that, for each k, we can prove either '$\underline{R}$ bars $\vec{u}\frown\bar{k}$' or '$\underline{R}$ bars $\vec{u}$'. Directly we recognize that, for some k, $\vec{u}_k$ is $\vec{u}$, we have indeed reduced that Θ-inference to a ζ-inference, which we can then eliminate; if, on the other hand, we come to recognize that, for every k, $\vec{u}_k$ is $\vec{u}\frown k$, then we thereby recognize the Θ-inferences as all of them superfluous; but we have no guarantee that we shall ever be in a position to assert either that one of the $\vec{u}_k$ is $\vec{u}$ or that none of them is; and, until we are, we cannot characterize more than finitely many of the Θ-inferences otherwise than simply as Θ-inferences.

The difference between the species $\underline{A}$ (under the hypotheses of Bar Induction) and the species of sequences barred by $\underline{R}$ is that the latter is monotonic and the former is not necessarily so: hence the necessity to eliminate ζ-inferences from the proof that $\langle\ \rangle$ is barred by $\underline{R}$ before transforming it into a proof that $\langle\ \rangle \in \underline{A}$. (If we know that $\underline{A}$ is monotonic, then we may replace the species $\underline{R}$ by the monotonic species $\underline{R}'$ of sequences with initial segments belonging to $\underline{R}$; $\underline{R}'$ will then still be included in $\underline{A}$, so that Bar Induction will be valid for this case.) If we know that a certain line

'$\underline{R}$ bars $\vec{u}^\frown\bar{k}$' of the proof was arrived at by means of a ζ-inference, we can eliminate it, since it was only a step towards proving '$\underline{R}$ bars $\vec{u}$': so something recognized as a ζ-inference indeed provides no obstacle towards proving that $\langle\ \rangle$ belongs to $\underline{A}$. Something recognized as a $\mathcal{F}$-inference is, likewise, no obstacle, since it exploits the property of being hereditary upwards which is shared by $\underline{A}$ and by the species of sequences barred by $\underline{R}$. But, if we know of a sequence $\vec{u}$ only that, for each of its immediate extensions $\vec{u}^\frown k$, either that extension or $\vec{u}$ itself belongs to $\underline{A}$, we cannot infer that $\vec{u}$ belongs to $\underline{A}$; hence, if we know of the proof that $\underline{R}$ bars $\langle\ \rangle$ that it somewhere contains denumerably many Θ-inferences which supply the premisses for a $\mathcal{F}$-inference, we cannot transform that proof into a proof that $\langle\ \rangle$ belongs to $\underline{A}$ until we obtain further information which we may never have.

If we could assert with any confidence that a fully analysed proof of a proposition of the form '$\underline{R}$ bars $\langle\ \rangle$' would contain, besides η-, ζ-, and $\mathcal{F}$-inferences, only Θ-inferences whose premisses were themselves obtained by η-inferences, then we could rectify Brouwer's proof by an appeal to the decidability or monotonicity of $\underline{R}$. Suppose that we know that, for each k, $\vec{u}_k \in \underline{R}$. If $\underline{R}$ is decidable, then we can ask whether $\vec{u} \in \underline{R}$. If $\vec{u}$ does belong to $\underline{R}$, then we can immediately infer that $\underline{R}$ bars $\vec{u}$ by an η-inference, and the Θ-inferences and subsequent $\mathcal{F}$-inference fall away as superfluous; if $\vec{u}$ does not belong to $\underline{R}$, then each $\vec{u}_k$ must be $\vec{u}^\frown k$, and we obtain the premisses for a $\mathcal{F}$-inference to '$\underline{R}$ bars $\vec{u}$' simply by η-inferences. If, on the other hand, $\underline{R}$ is monotonic, then, for each k, we can infer '$\vec{u}^\frown\bar{k} \in \underline{R}$' from '$\vec{u}_k \in \underline{R}$', and again proceed by η-inferences followed by a $\mathcal{F}$-inference. However, there does not appear to be any ground for holding that Θ-inferences of this kind form the only exception to Brouwer's claim that the only steps appearing in a fully analysed proof of '$\underline{R}$ bars $\langle\ \rangle$' are η-, ζ-, and $\mathcal{F}$-inferences.

Brouwer's idea of a fully analysed proof is certainly important, although the above discussion demonstrates that it must be handled with the greatest care. It is, however, far from being completely clear. It is evident that, in the process of analysis, any statement of the form A ∨ B occurring in the main deduction, though not one in a subordinate deduction, will be replaced either by A or by B, and, likewise, every statement ∃n A(n) occurring in the main deduction by the statement $A(\bar{k})$ for some k; further, every statement ∀n A(n) occurring in the main deduction will be replaced by the denumerably many statements A(0), A(1), ... It seems impossible, however, to subject a statement of the form A → B occurring in the main deduction to similar treatment. The underlying principle that every operation which, in the original proof, we appeal to as having a certain effect must, in the fully analysed proof, actually be applied can be complied with only if we are able systematically to generate everything to which the operation can be applied. Given a proof of A → B, this would involve generating in turn each putative proof of A, and either demonstrating it not to be a proof of A or applying to it the transformation which will convert it into a proof of B. However, it appears quite contrary to the intuitionistic insistence on the impossibility of surveying possible proofs of a given mathematical statement to suppose that we could, in any such way, systematically generate a class of constructions which should include all proofs of a given statement A. Thus the general conception of a fully analysed proof remains obscure.

Even if we could overcome these difficulties, so as to render Brouwer's proof of the Bar Theorem convincing, it is better, in any formalization of intuitionistic analysis at present possible, to assume the principle of Bar Induction axiomatically; this is quite apparent from Kleene's account of how Brouwer's proof of the Bar Theorem might be formalized. In order to formalize it, it is necessary to assume, as an axiom, a formalization of the statement 'If R̲ bars < >, then there is a proof of "R̲ bars < >" which uses only η-, ζ-, and $\mathcal{F}$-inferences'. Formalization of this statement is quite straightforward; a proof-tree can obviously be represented as a dressed spread. However, standing on its own, such an axiom appears quite ad hoc; especially when, in order to be able to exploit the well-founded character of the proof-tree, we should also have to assume axiomatically a principle of transfinite induction within such proof-trees (the general principle of transfinite induction can be proved equivalent to the principle of Bar Induction): it is therefore preferable simply to assume the principle of Bar Induction at the outset as an axiom or axiom schema.

This situation probably reflects only a transitional stage in the formalization of intuitionistic mathematics. We cannot incorporate into our first-order logic laws embodying the full intuitionistic meaning of the connective →: in that logic, the introduction rule for → entitles us to assert A → B only when B can be derived from A as hypothesis. In order to exploit the full intuitionistic meaning of →, we should need to be able, within intuitionistic analysis, to derive, for each statement A, the conditional 'A → ∃s ∃c (<s, c> is a dressed spread and <s, c> is a fully analysed proof of "A")'. Assuming that we can circumscribe the language within which a fully analysed proof of each statement A within the given system can be stated, such a conditional is intuitionistically reasonable, since it asserts that we have a method of deriving, from any proof of A, a proof that there is a fully analysed proof of A. Of course, it would be useless to take the relation expressed by ' ... is a fully analysed proof of ...' as primitive, and then simply assume such an axiom for each A, without any further assumptions about that relation: we should need, in addition, axioms which allowed us to determine, from the internal structure of any particular statement A, what form a fully analysed proof of it would have to take. If we could do this, then the assumptions necessary for carrying out (an emended version of) Brouwer's proof of the Bar Theorem would no longer appear ad hoc, but would be an application of general principles. Evidently, we cannot know how to do this until we have found some satisfactory explanation of the notion of a fully analysed proof, or of some other suitably restricted notion of proof such that any provable statement must have a proof of this restricted type. However, the exploitation of the full intuitionistic meaning of → depends on our finding such a theory of proofs; it is not merely that we do not at present know how to incorporate that meaning into our formalizations of intuitionistic mathematics, but that, within informal intuitionistic mathematics, hardly any attempt has been made to exploit it. That is why Brouwer's proof of the Bar Theorem, being virtually the sole attempt to do so, has such interest, despite its shortcomings.

3.5. THE REPRESENTATION OF CONTINUOUS FUNCTIONALS BY NEIGHBOURHOOD FUNCTIONS

Let $\underline{J}_0$ be the species of constructive functions from finite sequences to natural numbers which represent continuous functionals from choice sequences to natural numbers, i.e. the species

$$\underline{J}_0 = \{e \mid \forall\alpha\ e(\alpha) \text{ is defined}\} .$$

Under our representation of continuous functionals by such functions e, the condition for membership of $\underline{J}_0$ can evidently be given by:

$$e \in \underline{J}_0 \leftrightarrow \forall\alpha\ \exists n\ e(\bar{\alpha}(n)) > 0 .$$

This leaves the same functional represented in $\underline{J}_0$ by distinct functions which, from the point of view of the representation, differ only inessentially, i.e. on finite sequences which are proper extensions of the shortest ones for which they have a positive value. We can eliminate this redundancy by considering only a suitable subspecies of $\underline{J}_0$, for instance the species $\underline{K}_0$ of those functions e whose values, once positive, remain constant:

$$\underline{K}_0 = \{e \mid \forall\alpha\ \exists n\ \exists k\ (\forall m_{m<n}\ e(\bar{\alpha}(m)) = 0\ \&$$
$$\forall m_{m\geq n}\ e(\bar{\alpha}(m) = k + 1)\} .$$

Continuous functionals are, of course, still not represented uniquely in $\underline{K}_0$, in the sense that we can easily find extensionally distinct members e_1 and e_2 of $\underline{K}_0$ such that $\forall\alpha\ (e_1(\alpha) = e_2(\alpha))$; but every continuous functional has a representative in $\underline{K}_0$, and it may reasonably be claimed that extensionally distinct functions in $\underline{K}_0$ represent intensionally distinct continuous functionals. $\underline{K}_0$ has, moreover, the advantage of being included in the subspecies $\underline{J}'_0$ of $\underline{J}_0$ consisting of those functions e which represent continuous functionals from choice sequences to sequences, i.e. the species

$$\underline{J}'_0 = \{e \mid \forall\alpha\ e|\alpha \text{ is defined}\}$$

$$= \{e \mid \forall m\ \forall\alpha\ \exists n\ e(\langle m\rangle * \bar{\alpha}(n)) > 0\}$$

$$= \{e \mid \forall\alpha\ \exists n_{n>0}\ e(\bar{\alpha}(n)) > 0\};$$

and, again, every such continuous functional has a representative (though not a unique one) in $\underline{K}_0$. The representation of continuous functionals of either kind by functions in $\underline{K}_0$ is in no way artificial, but corresponds closely to the way in which one would naturally think of such functionals as being given.

We first prove a theorem about the composition of the species $\underline{K}_0$ as thus defined.

<u>Theorem</u>. (i) $\forall k\ \lambda\vec{u}\ .\ (k + 1) \in \underline{K}_0$

(ii) $e(\langle\ \rangle) = 0\ \&\ \forall m\ \lambda\vec{u}\ .\ e(\langle m\rangle * \vec{u}) \in \underline{K}_0 \rightarrow\ e \in \underline{K}_0$.

Proof. (i) is trivial. To prove (ii), assume the antecedent. Consider any particular α, and put

$$\beta = \lambda n.\ \alpha(n + 1).$$

By hypothesis,

$$\lambda\vec{u}\ .\ e(\langle\alpha(0)\rangle * \vec{u}) \in \underline{K}_0\ .$$

Hence, for some n and k,

$$\forall m_{m<n}\ e(\langle\alpha(0)\rangle * \bar{\beta}(m)) = 0\ \&\ \forall m_{m\geq n}\ e(\langle\alpha(0)\rangle * \bar{\beta}(m)) = k + 1.$$

Moreover, $e(\bar{\alpha}(0)) = 0$. Thus

$$\forall m_{m<n+1}\ e(\bar{\alpha}(m)) = 0\ \&\ \forall m_{m\geq n+1}\ e(\bar{\alpha}(m)) = k + 1.$$

Since, for every α, this holds for some n and k,

we have

$$e \in \underline{K}_0 \ .$$

If for any species $\underline{M}$ of functions from finite sequences to natural numbers, we define:

$$\underline{M}^* = \{e \mid \exists k\ e = \lambda\vec{u}\ .\ (k+1) \ \vee \ (e(\langle\ \rangle) = 0\ \&\ \forall m\ \lambda\vec{u}\ .\ e(\langle m\rangle * \vec{u}) \in \underline{M})\},$$

we can express the above theorem in the simple form:

$$\underline{K}_0^* \subseteq \underline{K}_0 \ .$$

The theorem says that $\underline{K}_0$ contains all constant functions with positive value, and is closed under a certain infinitary operation. We now claim that $\underline{K}_0$ is in fact the smallest species satisfying these two conditions. This claim constitutes a type of induction principle, which we call K-Induction (KI), and formulate as:

$$\text{KI:} \qquad \underline{M}^* \subseteq \underline{M} \rightarrow \underline{K}_0 \subseteq \underline{M} \ .$$

KI may be proved by appeal to Bar Induction, BI_D, and can be shown to be equivalent to it.

Theorem. The schema BI_D implies the schema KI.

Proof. We assume BI_D. Suppose $\underline{M}^* \subseteq \underline{M}$ and $e \in \underline{K}_0$; we have to show that $e \in \underline{M}$.

For an application of BI_D, we put:

$$\underline{R} = \{\vec{u} \mid e(\vec{u}) > 0\ \&\ \forall \vec{v}_{\vec{u} \prec \vec{v}}\ e(\vec{v}) = 0\} \ .$$

Plainly,

$$\forall \vec{u}\ (\vec{u} \in \underline{R} \vee \vec{u} \notin \underline{R}).$$

Also, since $e \in \underline{K}_0$,

$$\forall \alpha\ \exists n\ \bar{\alpha}(n) \in \underline{R}.$$

We now put:

$$\underline{A} = \{\vec{u} \mid \lambda\vec{v}.\ e(\vec{u} * \vec{v}) \in \underline{M}^*\} \ .$$

We claim:

(i) $\underline{R} \subseteq \underline{A}$.

For suppose $\vec{u} \in \underline{R}$. Then, for some k,

$$\forall \vec{v}\ e(\vec{u} * \vec{v}) = k + 1 ,$$

i.e. $\lambda\vec{v}.e(\vec{u} * \vec{v}) = \lambda\vec{v}.(k + 1)$,

whence $\lambda\vec{v}.e(\vec{u} * \vec{v}) \in \underline{M}^*$,

i.e. $\vec{u} \in \underline{A}$.

We also claim:

(ii) $\forall \vec{u}\ (\forall r\ \vec{u}^\frown r \in \underline{A} \rightarrow \vec{u} \in \underline{A})$.

For suppose $\forall r\ \vec{u}^\frown r \in \underline{A}$,

i.e. $\forall r\ \lambda\vec{v}.e((\vec{u}^\frown r) * \vec{v}) \in \underline{M}^*$.

Now either $e(\vec{u}) = k + 1$ for some k, or $e(\vec{u}) = 0$.

If $e(\vec{u}) = k + 1$, then, since $e \in \underline{K}_0$,

$$\lambda\vec{v}.e(\vec{u} * \vec{v}) = \lambda\vec{v}.(k + 1),$$

whence $\lambda\vec{v}.e(\vec{u} * \vec{v}) \in \underline{M}^*$,

i.e. $\vec{u} \in \underline{A}$.

If, on the other hand, $e(\vec{u}) = 0$, we have
$e(\vec{u}) = 0\ \&\ \forall r\ \lambda v.e((\vec{u}^\frown r) * \vec{v}) \in \underline{M}$,
since we assumed that $\underline{M}^* \subseteq \underline{M}$.
By the definition of $\underline{M}^*$, it follows that

$$\lambda\vec{v}.e(\vec{u} * \vec{v}) \in \underline{M}^*,$$

i.e. $\vec{u} \in \underline{A}$.

We have thus shown that all the hypothesis of BI_D are satisfied, and hence can conclude that $< \ > \in \underline{A}$, which amounts to saying that

$$e \in \underline{M}^* .$$

Since $\underline{M}^* \subseteq \underline{M}$, it follows that $e \in \underline{M}$, and the proof is concluded.

<u>Theorem</u>. The schema KI implies the schema BI_D.

Proof. We assume KI. We also suppose that, for given $\underline{R}$

and $\underline{A}$, the hypotheses of BI_D are satisfied. We want to define a species $\underline{M}$ such that

$$\text{(i)}\quad \underline{M}^* \subseteq \underline{M}$$

$$\text{and (ii)}\quad \underline{K}_0 \subseteq \underline{M} \rightarrow \langle\ \ \rangle \in \underline{A}.$$

To this end, we take $\underline{A}'$ as the species of finite sequences which either belong to $\underline{A}$ or have an initial segment belonging to $\underline{R}$,

$$\text{i.e. } \underline{A}' = \underline{A} \cup \{\vec{u} \mid \exists\vec{v}(\vec{v} \in \underline{R}\ \&\ \vec{u} \preccurlyeq \vec{v})\ ,$$

and then define $\underline{M}$ by:

$$\underline{M} = \{e \mid \forall\vec{u}[\forall\vec{v}(e(\vec{v}) > 0 \rightarrow \vec{u} * \vec{v} \in \underline{A}') \rightarrow \vec{u} \in \underline{A}']\}.$$

To see that (ii) holds, put:

$$e_0(\vec{u}) = \begin{cases} 1 & \text{if } \exists\vec{v}(\vec{v} \in \underline{R}\ \&\ \vec{u} \preccurlyeq \vec{v}) \\ 0 & \text{otherwise.} \end{cases}$$

Since $\forall\alpha\ \exists n\ \bar{\alpha}(n) \in \underline{R}$, $e_0 \in \underline{K}_0$. Hence, if $\underline{K}_0 \subseteq \underline{M}$, $e_0 \in \underline{M}$,

i.e. $$\forall\vec{u}[\forall\vec{v}\ (e_0(\vec{v}) > 0 \rightarrow \vec{u} * \vec{v} \in \underline{A}') \rightarrow \vec{u} \in \underline{A}']\ .$$

In particular, putting $\vec{u} = \langle\ \ \rangle$,

$$\forall\vec{v}\ (e_0(\vec{v}) > 0 \rightarrow \vec{v} \in \underline{A}') \rightarrow \langle\ \ \rangle \in \underline{A}'\ .$$

Since $\langle\ \ \rangle$ has only itself as an initial segment, and $\underline{R} \subseteq \underline{A}$, if $\langle\ \ \rangle \in \underline{A}'$, then $\langle\ \ \rangle \in \underline{A}$. Hence, in order to obtain the required conclusion, that $\langle\ \ \rangle \in \underline{A}$, we have only to prove

$$\forall\vec{v}\ (e_0(\vec{v}) > 0 \rightarrow \vec{v} \in \underline{A}')\ .$$

But this is obvious, since if $e_0(\vec{v}) > 0$, then, by definition of e_0, $\exists\vec{w}\ (\vec{w} \in \underline{R}\ \&\ \vec{v} \preccurlyeq \vec{w})$, whence, by the definition of $\underline{A}'$, $\vec{v} \in \underline{A}'$.

If, now, we can prove (i), then by appeal to KI, we can derive

$$\underline{K}_0 \subseteq \underline{M}\ .$$

The proof is therefore completed as soon as we have established that (i) holds.
To prove this, suppose that $e \in \underline{M}^*$,

i.e. $\exists k\ e = \lambda\vec{u}.(k + 1) \vee (e(<\ >) = 0\ \&\ \forall m\ \lambda\vec{u}.e(<m>* u) \in \underline{M})$.

We have to show that $e \in \underline{M}$.

Case 1: $e = \lambda\vec{u}.(k + 1)$.
Then $\forall\vec{v}\ e(\vec{v}) > 0$,
so that $\forall\vec{v}\ (e(\vec{v}) > 0 \to \vec{u} * \vec{v} \in \underline{A}')$
reduces to $\forall\vec{v}\ \vec{u} * \vec{v} \in \underline{A}'$.
But evidently, by putting $v = <\ >$, we have

$$\forall\vec{v}\ \vec{u} * \vec{v} \in \underline{A}' \to \vec{u} \in \underline{A}'.$$

Thus in this case $e \in \underline{M}$.

Case 2: $e(<\ >) = 0\ \&\ \forall m\ \lambda\vec{u}.e(<m> * \vec{u}) \in \underline{M}$.
We wish to show that $e \in \underline{M}$, i.e. that

$$\forall\vec{u}[\forall\vec{w}\ (e(\vec{w}) > 0 \to \vec{u} * \vec{w} \in \underline{A}') \to \vec{u} \in \underline{A}'].$$

We therefore assume:

(a) $\forall\vec{w}\ (e(\vec{w}) > 0 \to \vec{u} * \vec{w} \in \underline{A}')$,

and need to prove:

$$\vec{u} \in \underline{A}'.$$

Now our case-hypothesis, $\forall m\ \lambda\vec{u}.e(<m> * \vec{u}) \in \underline{M}$, amounts to:

(b) $\forall m\ \forall\vec{t}\ [\forall\vec{v}\ (e(<m> * \vec{v}) > 0 \to \vec{t} * \vec{v} \in \underline{A}' \to \vec{t} \in \underline{A}']$.

From (a), putting $\vec{w} = <m> * \vec{v}$, we get:

$$\forall m\ \forall\vec{v}\ (e(<m> * \vec{v}) > 0 \to (\vec{u}{}^\frown m) * \vec{v} \in \underline{A}').$$

Hence, putting $\vec{t} = \vec{u}{}^\frown m$ in (b), we have:

$$\forall m\ \vec{u}{}^\frown m \in \underline{A}'.$$

Since by assumption $\underline{R}$ is decidable,

$$\exists\vec{v}(\vec{v} \in \underline{R}\ \&\ \vec{u} \preccurlyeq \vec{v}) \quad \vee \quad \neg\exists\vec{v}(\vec{v} \in \underline{R}\ \&\ \vec{u} \preccurlyeq \vec{v}).$$

If $\exists\vec{v}\ (\vec{v} \in \underline{R}\ \&\ \vec{u} \preccurlyeq \vec{v})$, then, by definition of $\underline{A}'$,

$\vec{u} \in \underline{A}'$, as desired.

If, however, $\neg\exists\vec{v}\ (\vec{v} \in \underline{R}\ \&\ \vec{u} \preccurlyeq \vec{v})$, then, since

$\forall m\ \vec{u}\frown m \in \underline{A}'$, we have, by the definition of $\underline{A}'$:

$$\forall m\ (\vec{u}\frown m \in \underline{A} \vee \vec{u}\frown m \in \underline{R})\ .$$

Since $\underline{R} \subseteq \underline{A}$ by assumption,

$$\forall m\ (\vec{u}\frown m \in \underline{A})\ .$$

But since also $\underline{A}$ is, by assumption, hereditary upwards,

$$\vec{u} \in \underline{A}\ ,$$

whence again, by the definition of $\underline{A}'$,

$$\vec{u} \in \underline{A}'\ .$$

This concludes the proof that BI_D is derivable from KI.

Because K-Induction is equivalent to Bar Induction, one method, followed, e.g., by Troelstra, of formalizing intuitionistic analysis is by assuming it, rather than Bar Induction, axiomatically. The symbol $\underline{K}$ was originally introduced by Kreisel for the species of neighbourhood functions (in Baire space) considered as inductively defined by:

$$(\kappa) \qquad \underline{K}^* \subseteq \underline{K}\ \ \&\ \ \forall\underline{M}\ (\underline{M}^* \subseteq \underline{M} \rightarrow \underline{K} \subseteq \underline{M})\ ,$$

i.e. by the requirement that $\underline{K}$ be the smallest species such that $\underline{K}^* \subseteq \underline{K}$. (The existence of such inductively defined species is not, of course, guaranteed by the predicative comprehension axiom, since the condition for membership of it involves quantification over species of the same type.) Using this notation, $\underline{K} \subseteq \underline{K}_o$ follows at once from the fact that $\underline{K}_o^* \subseteq \underline{K}_o$. Since $\underline{K}^* \subseteq \underline{K}$, K-Induction implies that $\underline{K}_o \subseteq \underline{K}$. KI can thus be expressed in this notation by:

$$\underline{K} = \underline{K}_o\ .$$

Troelstra's actual procedure is to take K(e) as a primitive predicate governed by an axiom schema corresponding to (κ), and then to connect this predicate with the theory of choice sequences by assuming $\forall\alpha\exists n$-continuity in the form:

$$\forall n\ \mathrm{Ext}_{\alpha}\ C(\alpha, n)\ \&\ \forall\alpha\ \exists n\ C(\alpha, n) \rightarrow \exists e_{K(e)}\ \forall\alpha\ C(\alpha, e(\alpha)).$$

From this it is then straightforward to derive BI_M. However, not only is Bar Induction formally independent of the Continuity Principle (since it is classically true while the Continuity Principle is not), but, at least in the weaker form BI_D, it is surely conceptually independent of it also, and it therefore seems advantageous to express K-Induction, as here, in a form from which BI_D may be derived without appeal to the Continuity Principle.

3.6. THE UNIFORM CONTINUITY THEOREM

We now return to the topic of real analysis, bearing in mind the results about choice sequences. However, before we can apply any of these results, we must revise our notion of the reals: instead of starting from constructively given (lawlike) sequences of rationals, we consider choice sequences of rationals, i.e. choice sequences together with an effective correlation law from natural numbers to rationals, and define the reals from these in exactly the same way. We will denote such sequences by 'α', 'β', 'γ', ... and the reals by 'x', 'y', 'z', 'u', 'v', ... as before. None of the proofs of our previous theorems about reals depended on the nature of the sequences from which they were defined, so all the theorems still apply.

Definition. If $\underline{A}$ is a species of reals, then x is a *least upper bound* of $\underline{A}$ iff

$$\forall y_{y \in \underline{A}}\ y \nmid x$$

$$\text{and} \quad \forall k\ \exists y_{y \in \underline{A}}\ y > x - 2^{-k}.$$

The definition of a *greatest lower bound* is analogous.

We use the notation 'f(x)', 'g(x)', ... for functions

of reals, but such a function can only really be considered as a functional from sequences of rationals to sequences of rationals. Clearly, such a functional Φ can represent a real-valued function of reals if and only if, whenever α is a real number generator, so is $\Phi(\alpha)$, and $\sim$ is a congruence with respect to Φ.

Definition. A real-valued function f is *uniformly continuous on an interval* [u,v] iff

$$\forall k\ \exists m\ \forall x_{x \in [u,v]}\ \forall y_{y \in [u,v]}\ (|x - y| < 2^{-m} \rightarrow |f(x) - f(y)| < 2^{-k}) .$$

By the Axiom of Choice, there is a constructive function a which, when applied to k, determines such an m. This function is the *modulus of continuity* of f on [u,v] .

f is *uniformly continuous* iff it is uniformly continuous on every interval.

Definition. x is a *least upper bound for* f *on* [u,v] iff x is a least upper bound of $\{f(y) \mid y \in [u,v]\}$. (Similarly for *greatest lower bound*.)

It is not obvious constructively that every function defined and bounded on [u,v] has a least upper bound (greatest lower bound); however, a sufficient condition is that f be uniformly continuous on the interval. In the proof of this, we make use of the following result.

Lemma. If $x_0 = \max\{x_1,\ldots,x_q\}$ then $\forall k\ \exists i_{0<i\leq q} |x_0 - x_i| < 2^{-k}$.

Proof. Suppose $\alpha_i \in x_i$ for $0 < i \leq q$, and define α_0 by

$$\alpha_0(n) = \max\{\alpha_1(n),\ldots, \alpha_q(n)\} .$$

Then $\alpha_0 \in x_0$.

Since the α_i are real number generators, by the Axiom of Choice there exist functions a_i such that:

$\forall i\ \forall k\ \forall m_{m > a_i(k)}\ |\alpha_i(a_i(k)) - \alpha_i(m)| < 2^{-k-3}$.

Let $a(k) = \max\ \{a_i(k)\ |\ 0 \le i \le q\}$.

Then
$\forall i_{i \le q} \forall k\ \forall m_{m > a(k)}\ |\alpha_i(a(k)) - \alpha_i(m)| < 2^{-k-2}$.

Now, for some $i \neq 0,\ i \le q$,

$$\alpha_o(a(k)) = \alpha_i(a(k)).$$

So, for $m > a(k)$, we have, for this i,

$$|\alpha_o(a(k)) - \alpha_i(m)| < 2^{-k-2}$$

and $|\alpha_o(a(k)) - \alpha_o(m)| < 2^{-k-2}$.

Therefore, $\forall m_{m > a(k)} |\alpha_o(m) - \alpha_i(m)| < 2^{-k-1}$.

Hence, $|x_o - x_i| < 2^{-k}$.

In order to simplify the proof of the theorem, we consider the interval [0,1].

Theorem 1. If f is uniformly continuous on [0,1], then f has a least upper bound and greatest lower bound on [0,1].

Proof. For each k, put $r_{n,k} = n.2^{-k}$ where $0 \le n \le 2^k$, and let

$$\underline{A}_k = \{f(r_{n,k})\ |\ 0 \le n \le 2^k\}.$$

Then $\underline{A}_o \subset \underline{A}_1 \subset \underline{A}_2 \subset \dots$.

Now put $x_k = \max \underline{A}_k$.

Clearly, $m < n \to x_m \nleq x_n$. We claim that $\langle x_k \rangle$ is a Cauchy sequence. Let a be the modulus of continuity of f on [0,1]. Then, given any k and m, by the lemma applied to $\underline{A}_m$, we can find an n such that

$$|f(r_{n,m}) - x_m| < 2^{-k}.$$

Moreover, we can certainly find a j such that

$$|r_{n,m} - r_{j,a(k)}| < 2^{-a(k)} ,$$

whence

$$|f(r_{n,m}) - f(r_{j,a(k)})| < 2^{-k}, \text{ since f is}$$

uniformly continuous on [0,1]. By Theorem 19 (ii) of section 2.3,

$$x_m \nless f(r_{n,m}) \text{ and } x_{a(k)} \nless f(r_{j,a(k)}).$$

Also, if $m > a(k)$, then $x_m \nless x_{a(k)}$. So, for $m > a(k)$, we have

$$|x_m - x_{a(k)}| = x_m - x_{a(k)}$$

$$= (x_m - f(r_{n,m})) + (f(r_{n,m}) -$$

$$f(r_{j,a(k)})) + (f(r_{j,a(k)}) - x_{a(k)}).$$

Therefore,

$$|x_m - x_{a(k)}| < 2^{-k} + 2^{-k} = 2^{-k+1} .$$

Hence,

$$\forall k \ \forall m_{m>a(k+1)} \ |x_m - x_{a(k+1)}| < 2^{-k}.$$

So $\langle x_k \rangle$ is a Cauchy sequence.

By Theorem 24 of section 2.4, $\langle x_k \rangle$ has a limit x, which, we claim, is at least upper bound for f on [0,1].

Suppose that, for some $y \in [0,1]$, $f(y) > x$. Then, for some n,

$$f(y) - x > 2^{-n} .$$

But we can choose k so that

$$|y - r_{k,a(n)}| < 2^{-a(n)} ,$$

whence ,

$$|f(y) - f(r_{k,a(n)})| < 2^{-n} ,$$

and so $f(y) - f(r_{k,a(n)}) < 2^{-n}$.

But $x \nless x_{a(n)} \nless f(r_{k,a(n)})$,

so $f(y) - x < 2^{-n}$, in contradiction to the hypothesis. Therefore,

$$\forall y_{y \in [0,1]}\ f(y) \ngtr x.$$

Finally, for given k,

$$x - x_{a(k+2)} \ngtr 2^{-k-1},$$

and, for some j,

$$x_{a(k+2)} - f(r_{j,a(k+2)}) < 2^{-k-1}.$$

Therefore

$$f(r_{j,a(k+2)}) > x - 2^{-k}.$$

Hence x is a least upper bound for f on [0,1]. Similarly, f has a greatest lower bound on [0,1].

Intuitionistically, we cannot go further and assert that f actually attains its least upper bound, i.e. we cannot necessarily find a $y \in [0,1]$ such that $f(y) = x$. The following weak counter-example illustrates why not.

Let A(n) and B(n) express properties of natural numbers such that A(n) is decidable for each n, but we do not know whether $\exists n\, A(n)$ or not, and, if $\exists n\, A(n)$ we do not know whether B(n) holds of the least such n. Define a real number generator α as follows:

$$\alpha(n) = \begin{cases} 0 & \text{if } \forall m_{m \le n}\ \neg A(m) \\ 2^{-k} & \text{if } k \le n \ \&\ A(k)\ \&\ B(k)\ \& \\ & \qquad \forall m_{m<k}\ \neg A(m) \\ -2^{-k} & \text{if } k \le n \ \&\ A(k)\ \&\ \neg B(k)\ \& \\ & \qquad \forall m_{m<k}\ \neg A(m). \end{cases}$$

Let z_0 be the real number determined by α, and define f by

$$f(y) = y.z_o .$$

Then f is uniformly continuous, and so has a least upper bound x on [0,1]. Suppose f attains x, i.e. for some $y_o \in [0,1]$, $f(y_o) = x$. Now, $0 < y_o \vee y_o < 1$. But if $z_o < 0$, then clearly $x = 0$, and so $y_o = 0$. So $0 < y_o \rightarrow z_o \nless 0$, and likewise $y_o < 1 \rightarrow z_o \ngtr 0$.

So $\quad 0 < y_o \rightarrow \neg \exists k\, [A(k)\ \&\ \neg B(k)\ \&\ \forall m_{m<k}\ \neg A(m)]$

and $\quad y_o < 1 \rightarrow \neg \exists k\, [A(k)\ \&\ B(k)\ \&\ \forall m_{m<k}\ \neg A(m)]$.

But by construction of the predicates A(n) and B(n), neither of these conclusions can be asserted. Hence we cannot assume that f attains its least upper bound.

When dealing with reals, instead of considering any real number generator from which a real may be defined, we can limit our attention to real number generators of a particular kind, known as *canonical real number generators*. These enable us to represent the reals in, say, [0,1] by a dressed fan <s,c>. That is, we can construct a fan <s,c> such that

$$x \in [0,1] \rightarrow \exists \alpha_{\alpha \in \langle s,c\rangle}\ \alpha \in x .$$

The elements of the fan are canonical real number generators. The fan is constructed as follows. Take the naked spread to be the full ternary spread:

$$s(\vec{u}) = 0 \leftrightarrow \forall i_{i \leq \ell h(\vec{u})}\ 0 \leq u_i \leq 2 .$$

If $s(\vec{u}) = 0$ and $\ell h(\vec{u}) = k$, then c is given by

$$c(\vec{u}) = [1 + \sum_{i=0}^{k-1} (u_i . 2^{k-i-1})] . 2^{-k-1} .$$

This describes the fan:

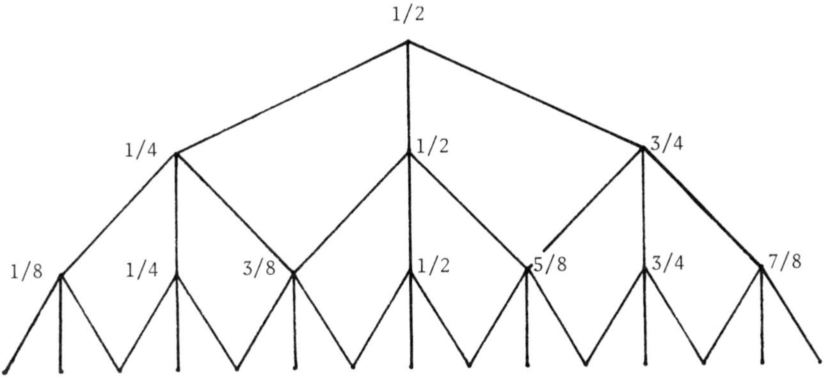

(Notice that c is so defined as to map distinct admissible sequences of s, of the same length, on to the same rational; e.g. $c(<2,0>) = c(<1,2>) = 5/8$.)

We now have to prove that every $x \in [0,1]$ is represented by an element of the fan. We in fact prove something slightly stronger.

Theorem 2. Let $s(\vec{u}) = 0$ and $\ell h(\vec{u}) = \ell$, so that $c(\vec{u}) = p.2^{-\ell-1}$ for some p, $0 < p < 2^{\ell+1}$. Define $<s',c'>$ by:

$$s'(\vec{v}) = 0 \leftrightarrow s(\vec{v}) = 0 \;\&\; (\vec{u} \preccurlyeq \vec{v} \vee \vec{v} \preccurlyeq \vec{u})$$

and c' is the restriction of c to s'.

Now if $x \in [(p-1).2^{-\ell-1},(p+1).2^{-\ell-1}]$, then

$$\exists\alpha_{\alpha \in s'}\ c(\alpha) \in x.$$

Proof. Suppose $\beta \in x$. We want to construct $\alpha \in s'$ such that $c(\alpha) \sim \beta$. Clearly, for $n < \ell$, we can take $\alpha(n) = u_n$. For $n \geq \ell$ we define ℓ inductively: assume $\bar{\alpha}(n)$ has been constructed so that

$$c(\bar{\alpha}(n)) = q.2^{-n-1}$$

and $x \in [c(\bar{\alpha}(n)) - 2^{-n-1}, c(\bar{\alpha}(n)) + 2^{-n-1}]$.

We can find r such that

$$\forall m_{m > r} \; |\beta(r) - \beta(m)| < 2^{-n-4},$$

and so

$$|\beta(r) - x| < 2^{-n-3} \; .$$

Now choose k such that

$$|\beta(r) - k.2^{-n-2}| < 2^{-n-3}.$$

Then

$$|x - k.2^{-n-2}| < 2^{-n-2}.$$

Now we define $\alpha(n)$ by:

$$\alpha(n) = \begin{cases} 0 & \text{if } k \le 2q-1 \\ 1 & \text{if } k = 2q \\ 2 & \text{if } k \ge 2q+1. \end{cases}$$

From the definition of c,

$$|k.2^{-n-2} - c(\bar{\alpha}(n+1))| < 2^{-n-2},$$

and, by hypothesis,

$$(q-1).2^{-n-1} \not> x \not> (q+1).2^{-n-1}.$$

Therefore,

$$|x - c(\bar{\alpha}(n+1))| < 2^{-n-2},$$

i.e.

$$x \in [c(\bar{\alpha}(n+1)) - 2^{-n-2}, c(\bar{\alpha}(n+1)) + 2^{-n-2}] \; .$$

Thus we have successfully defined an element of the fan. Moreover

$$|\beta(r) - c(\bar{\alpha}(n+1))| < 2^{-n-2} + 2^{-n-3},$$

$$\forall m_{m > r} \; |\beta(r) - \beta(m)| < 2^{-n-4}$$

and $\forall m_{m > n} \; |c(\bar{\alpha}(n+1)) - c(\bar{\alpha}(m))| < 2^{-n-2}.$

Therefore,

$$\forall m_{m > \max(n,r)} \ |\beta(m) - c(\bar{\alpha}(m))| < 2^{-n}.$$

Hence

$$c(\alpha) \sim \beta .$$

So far we have not made any actual use of the fact that we are now considering real number generators as choice sequences in a dressed spread rather than as constructive functions from natural numbers to rationals. We now proceed to do so by applying the Extended Fan Theorem to our representation of the interval [0,1] as a fan, thereby obtaining the following strong non-classical result.

<u>Theorem 3</u>. (Uniform Continuity Theorem). If f is everywhere defined on [0,1], then f is uniformly continuous on [0,1].

Proof. Since f is everywhere defined on [0,1], we can consider f as a functional from elements of <s,c> to reals, and write:

$$f(c(\alpha)) = f(x) \text{ whenever } \alpha \in s \text{ and } c(\alpha) \in x.$$

For each n and $x \in [0,1]$, we can approximate f(x) to within 2^{-n-1},

i.e. $\forall\alpha_{\alpha \in s} \ \exists m \ |f(c(\alpha)) - m.2^{-n-1}| < 2^{-n-1}$.

The relation between α and m is plainly extensional, so we can apply the Extended Fan Theorem, for each n, to find an r such that:

$$\forall\alpha_{\alpha\in s} \ \exists m \ \forall\beta_{\beta\in s,\beta\in\bar{\alpha}(r)} \ |f(c(\beta)) - m.2^{-n-1}| < 2^{-n-1}.$$

Suppose x, $y \in [0,1]$ and $|x - y| < 2^{-r-1}$. By Theorem 2, we can assume $c(\alpha_1) \in x$, for some $\alpha_1 \in s$, and we can find $\alpha_2 \in s$ such that $\bar{\alpha}_1(r) = \bar{\alpha}_2(r)$ and $c(\alpha_2) \in y$. For some m we have:

$$\forall\beta_{\beta \in s,\beta \in \bar{\alpha}_1(r)} \ |f(c(\beta)) - m.2^{-n-1}| < 2^{-n-1}.$$

So,

$$|f(x) - f(y)| = |f(c(\alpha_1)) - f(c(\alpha_2))|$$

$$\ngtr |f(c(\alpha_1)) - m.2^{-n-1}| + |f(c(\alpha_2)) - m.2^{-n-1}|$$

$$< 2^{-n} .$$

For each n, we can effectively find such an r, so f is uniformly continuous on [0,1].

It is evident that the representation of the interval [0,1] as a fan provided by Theorem 2 can be adapted to give a similar representation of any interval [u,v], and that Theorem 3 can accordingly be generalized. Theorems 1 and 3 together yield the result that, if f is everywhere defined on [0,1], then f has a least upper bound and greatest lower bound on [0,1], a result which again holds for any closed interval. The Continuity Principle finds many other applications in intuitionistic analysis, usually via the Extended Fan Theorem; sometimes, as above, to prove results which contradict classical analysis, sometimes to prove the analogues of classical theorems which could not otherwise be obtained.

4

THE FORMALIZATION OF INTUITIONISTIC LOGIC

4.1. NATURAL DEDUCTION

A *natural deduction system* is a formalization of logic in which no formulas are axiomatically assumed as valid, but there are only rules of inference. To compensate for the lack of axioms, it is permitted to introduce any formula as a hypothesis at any stage. Most rules of inference will present the conclusion as depending on all hypotheses on which any of the premisses depends; but there will be some rules which discharge hypotheses (in the sense that the conclusion of the rule no longer depends on the hypothesis discharged). Those hypotheses left undischarged at the termination of the derivation serve as the premisses on which the conclusion depends.

It is thus of crucial importance, in a natural deduction system, that we should be able, in the course of a derivation, to keep track of which hypotheses each formula occurring in it depends upon.

One familiar way of doing this is to take a derivation in such a system to be a finite linearly ordered array of formulas, each of which either is introduced as a hypothesis or follows from some preceding formula(s) by one of the rules of the system, and to number the hypotheses and indicate, next to each formula, upon which of these hypotheses it depends.

Alternatively, instead of merely indicating the hypotheses on which a formula rests, we can write out these hypotheses in full each time, and take a derivation to be composed of *sequents* rather than formulas. A sequent $\Gamma : A$ may for present purposes be taken as an ordered pair $\langle \Gamma, A \rangle$, where A is a formula and Γ a finite (possibly empty) set of formulas. We write:

$$\text{'}\Gamma, \Delta : A\text{'} \text{ for } \text{'}\Gamma \cup \Delta : A\text{'},$$

$$\text{'}\Gamma, B : A\text{'} \text{ for } \text{'}\Gamma \cup \{B\} : A\text{'},$$

$$\text{'}B_1, B_2, \ldots, B_n : A\text{'} \text{ for } \text{'}\{B_1, B_2, \ldots, B_n\}: A\text{'},$$

$$\text{and } \text{'} : A\text{'} \text{ for } \text{'}\emptyset : A\text{'} .$$

Γ is called the *antecedent* and A the *succedent* of the sequent Γ : A. A *basic sequent* is a sequent of the form Γ, A : A.

Another variation is obtained by exploiting the two-dimensional character of the page, and presenting the formal derivation, not as a linear array, but as a tree, where each sequent stands immediately below those from which it has been derived by one of the rules of inference. This has the advantage of being more perspicuous, although it will lengthen the derivation whenever any sequent figures as a premiss for more than one rule or application of a rule. We accordingly give the following definitions.

A *proof tree-trunk* is a finite tree with its vertex at the bottom whose nodes are correlated with sequents in such a way that every sequent associated to a node which has nodes above it is a direct consequence by one of the rules of inference of the sequent(s) associated to the node(s) immediately above it. A *proof* or *proof-tree* is a proof tree-trunk in which every topmost node is correlated with a basic sequent. The sequent assigned to the vertex of the tree is the *conclusion* of the proof.

We say that A *is derivable from* Γ *in* N, and write '$\Gamma \vdash_N A$', if there exists a proof in the system N of which Γ_0 : A is the conclusion, for some finite subset Γ_0 of Γ. When $\emptyset \vdash_N A$, we write '$\vdash_N A$', and say that A *is provable in* N.

The rules of inference of the system N of natural deduction for intuitionistic logic which we shall now consider are divided into *structural rules* and *logical rules*. There is only one structural rule, the thinning rule:

$$\frac{\Gamma : B}{\Gamma, A : B} .$$

Since the antecedents of sequents are sets rather than sequences of formulas, no further structural rules are needed to allow for repetition or permutation of formulas within the antecedent of a sequent. By repeated application of the thinning rule, we get the more general version:

$$\frac{\Gamma : B}{\Gamma, \Delta : B} .$$

The logical rules are divided into *introduction rules* and *elimination rules*. We denote an introduction rule for v by 'v+' and an elimination rule by 'v-', and similarly for the other connectives. The rules are as follows:

	+	-
&	$\dfrac{\Gamma : A \qquad \Delta : B}{\Gamma, \Delta : A \,\&\, B}$	$\dfrac{\Gamma : A \,\&\, B}{\Gamma : A} \qquad \dfrac{\Gamma : A \,\&\, B}{\Gamma : B}$
$\vee$	$\dfrac{\Gamma : A}{\Gamma : A \vee B} \qquad \dfrac{\Gamma : B}{\Gamma : A \vee B}$	$\dfrac{\Gamma : A \vee B \quad \Delta, A : C \quad \Theta, B : C}{\Gamma, \Delta, \Theta : C}$
$\rightarrow$	$\dfrac{\Gamma, A : B}{\Gamma : A \rightarrow B}$	$\dfrac{\Gamma : A \qquad \Delta : A \rightarrow B}{\Gamma, \Delta : B}$
$\neg$	$\dfrac{\Gamma, A : B \qquad \Delta, A : \neg B}{\Gamma, \Delta : \neg A}$	$\dfrac{\Gamma : A \qquad \Delta : \neg A}{\Gamma, \Delta : B}$
$\forall$	$\dfrac{\Gamma : A(y)}{\Gamma : \forall x A(x)}$	$\dfrac{\Gamma : \forall x A(x)}{\Gamma : A(t)}$
$\exists$	$\dfrac{\Gamma : A(t)}{\Gamma : \exists x A(x)}$	$\dfrac{\Gamma : \exists x A(x) \qquad \Delta, A(y) : C}{\Gamma, \Delta : C}$

The rules for $\forall$ and $\exists$ hold only under these conditions:

(a) y is a variable and t is a term and both are free for x in A(x);

(b) A(y) and A(t) result from A(x) by replacing every free occurrence of x by y and t respectively;

(c) in $\forall$+, y does not occur free in Γ : $\forall$xA(x), and in $\exists$-, y does not occur free in Γ,Δ : C or in $\exists$xA(x).

y is said to be *free for* x *in* A(x) just in case no free occurrence of x in A(x) stands in the scope of a quantifier binding the variable y. A term t is *free for* x *in* A(x) just in case all the free variables in t are free for x in A(x). Classical logic is obtained from N by replacing the $\neg$ -rule by double negation elimination:

$$\frac{\Gamma : \neg\neg A}{\Gamma : A} \; .$$

There is a certain amount of redundancy in the system as formulated here. The effect of the thinning rule could be obtained by an application of &+ followed by &- :

$$\cfrac{\cfrac{\Gamma : A \qquad B : B}{\Gamma, B : A \,\&\, B}\,\&+}{\Gamma, B : A}\,\&- \; .$$

Conversely, in the presence of the thinning rule, those logical rules with more than one premiss could be weakened by writing 'Γ' in place of 'Δ' and 'Θ', so that, for example, &+ would appear as:

$$\frac{\Gamma : A \qquad \Gamma : B}{\Gamma : A \,\&\, B} \; .$$

Moreover, in the presence of the thinning rule, it is unnecessary to define 'basic sequent' in the general manner given above: basic sequents could be restricted to those of the form A : A. Conversely, with 'basic sequent' defined in the more general way, we could dispense with the thinning rule altogether, and still state the logical rules in the more restricted form given in the last paragraph: any formula which was going to be needed at a later stage could be put into the antecedent of the relevant basic sequent.

A minor awkwardness arises from the formulation of ¬+ : the idea of the rule is plainly that of reductio ad absurdum -- if from A together with other hypotheses Γ we can derive an inconsistent pair of formulas B_1, B_2, then we are entitled to assert ¬A on the basis of Γ. However, the requirement in ¬+ that B_2 be $\neg B_1$ sometimes creates a certain awkwardness in carrying out proofs by reductio, particularly when the application of ¬+ is preceded by a use of the ∨-rule. For example, a proof of the sequent ¬A & ¬B : ¬(A ∨ B) must take the form:

$$\neg+\;\dfrac{\vee-\;\dfrac{A \vee B : A \vee B \qquad \neg-\;\dfrac{\&-\;\dfrac{\neg A \,\&\, \neg B : \neg A \,\&\, \neg B}{\neg A \,\&\, \neg B : \neg A} \qquad A : A}{\neg A \,\&\, \neg B, A : B} \qquad B : B}{\neg A \,\&\, \neg B, A \vee B : B} \qquad \dfrac{\neg A \,\&\, \neg B : \neg A \,\&\, \neg B}{\neg A \,\&\, \neg B : \neg B}\;\&-}{\neg A \,\&\, \neg B : \neg(A \vee B)\ .}$$

This proof does not follow the natural intuitive argument, which is that, since both pairs of assumptions ¬A & ¬B, A and ¬A & ¬B, B lead to a contradiction, so does the pair ¬A & ¬B, A ∨ B. We cannot formulate this idea directly in the system N as it stands, since different contradictions follow from ¬A & ¬B, A and from ¬A & ¬B, B.

On the assumption that the rules of N are sound, all inconsistencies are interderivable; so one way of avoiding awkwardness of this kind would be to add a new primitive '⊥', denoting a constant absurd proposition, and to define '¬A' as 'A → ⊥'. ¬+ now becomes a derived rule; the rule ¬- also becomes derivable if we adopt, as a rule governing ⊥ :

$$\bot- \quad \frac{\Gamma : \bot}{\Gamma : B}$$

Alternatively, we could retain '$\neg$' as primitive and assume the rules:

$$\bot+ \quad \frac{\Gamma : A \qquad \Delta : \neg A}{\Gamma, \Delta : \bot} \qquad\qquad \bot- \quad \frac{\Gamma : \bot}{\Gamma : B}$$

$$\neg+' \quad \frac{\Gamma, A : \bot}{\Gamma : \neg A}$$

$\neg+$ and $\neg-$ are plainly derived rules of this modified system.

An axiomatic formalization of intuitionistic logic, Ax, can be obtained by, in effect, transforming the rules of N (other than $\rightarrow -$, $\forall +$, and $\exists -$) into axiom schemata:

Axioms:
(1) $A \rightarrow (B \rightarrow A)$
(2) $A \rightarrow (B \rightarrow A \,\&\, B)$
(3) $A \,\&\, B \rightarrow A$
(4) $A \,\&\, B \rightarrow B$
(5) $A \rightarrow A \vee B$
(6) $B \rightarrow A \vee B$
(7) $A \vee B \rightarrow ((A \rightarrow C) \rightarrow ((B \rightarrow C) \rightarrow C))$
(8) $(A \rightarrow B) \rightarrow ((A \rightarrow (B \rightarrow C)) \rightarrow (A \rightarrow C))$
(9) $(A \rightarrow B) \rightarrow ((A \rightarrow \neg B) \rightarrow \neg A)$
(10) $A \rightarrow (\neg A \rightarrow B)$
(11) $\forall x A(x) \rightarrow A(t)$
(12) $A(t) \rightarrow \exists x A(x)$

Rules:
(a) $\dfrac{A \qquad A \rightarrow B}{B}$

(b) $\dfrac{C \rightarrow A(y)}{C \rightarrow \forall x A(x)}$

(c) $\dfrac{A(y) \rightarrow C}{\exists x A(x) \rightarrow C}$

In (11) and (12), t is any term that is free for x in A(x), and A(t) is formed by replacing every free occurrence of x in A(x) by t. In (b) and (c), y is free for x in A(x), and does not occur free in C or in A(x); A(y) is formed by replacing every free occurrence of x in A(x) by y.

A *derivation* in Ax of a formula A from a set Γ of closed formulas is a finite linearly ordered sequence of formulas, each of which is either a member of Γ or an instance of one of the axiom schemata (1) - (12) or derived by one of the rules (a) - (c) from some formula or pair of formulas occurring earlier in the sequence, and the last member of which is A. (If we want to allow the more general case when Γ may contain formulas with free variables, we shall need to impose restrictions on the applications of the two quantifier rules (b) and (c).) We write '$\Gamma \vdash_{Ax} A$' to mean that there exists a derivation in Ax of A from Γ, and say that A is *provable* in Ax if $_{Ax}A$ ($\emptyset$ $_{Ax}A$).

<u>Lemma</u> (Deduction Theorem). If $\Gamma, B \vdash_{Ax} A$, then $\Gamma \vdash_{Ax} B \rightarrow A$.

Sketch of proof. The Deduction Theorem in effect states that $\rightarrow +$ is a derived rule in Ax. It should be perfectly familiar from classical logic, and its proof is exactly the same here. If we have a derivation of A from Γ and B, we transform it into a derivation of B → A from Γ as follows. First we replace every line C of the derivation by B → C. We now have to modify those transitions which correspond, in the original derivation, to applications of the rules of inference. An application of rule (a) will now have become:

$$\frac{B \rightarrow C \qquad B \rightarrow (C \rightarrow D)}{B \rightarrow D}$$

and this step can be effected by appeal to axiom schema (8) and two applications of rule (a). An application of rule (b) will now have become:

$$\frac{B \to (C \to D(y))}{B \to (C \to \forall x\, D(x))}\ .$$

This step can be effected by deriving $B \,\&\, C \to D(y)$ from $B \to (C \to D(y))$, applying rule (b), and then deriving $B \to (C \to \forall x\, D(x))$ from $B \,\&\, C \to \forall x\, D(x)$. It is left as an exercise to check that these derivations are possible. Finally an application of rule (c) will have become:

$$\frac{B \to (D(y) \to C)}{B \to (\exists x\, D(x) \to C)}\ .$$

We derive $D(y) \to (B \to C)$ from $B \to (D(y) \to C)$, apply rule (c), and then derive $B \to (\exists x\, D(x) \to C)$ from $\exists x\, D(x) \to (B \to C)$. It is again left as an exercise to check that these derivations are possible. In both the last two cases, a little complication is involved when B is not required to be closed, and in fact contains y free. As a final step, we have to consider those formulas in the original derivation which were not derived by a rule of inference from earlier lines. B itself may have figured as such a formula, in which case it has been transformed into $B \to B$, which is provable in Ax (it is left as an exercise to verify this). Otherwise, such a formula C must be either an instance of an axiom schema or a member of Γ, and in either case can be derived from Γ by means of axiom schema (1) and rule (a).

We can now prove the equivalence of Ax and N.

Theorem. For any set Γ and formula E, $\Gamma \vdash_{Ax} E$ iff $\Gamma \vdash_N E$.

Sketch of proof. (i) If $\Gamma \vdash_{Ax} E$, then $\Gamma \vdash_N E$.

We indicate the inductive argument by which it may be shown that the derivation of E from Γ in Ax can be transformed into a proof-tree of N with conclusion $\Gamma_0 : E$ for some $\Gamma_0 \subseteq \Gamma$. Any line of the derivation consisting of a formula C belonging to Γ is transformed into $C : C$. Any line consisting of an instance of one of the axiom schemas (1)-(12) is replaced by a proof-tree for the sequent having null antecedent and that axiom as succedent. We illustrate with the case of axiom schema (7):

$$
\begin{array}{l}
\vee- \dfrac{A \vee B : A \vee B \qquad \to- \dfrac{A : A \quad A \to C : A \to C}{A \to C,\ A : C} \qquad \to- \dfrac{B : B \quad B \to C : B \to C}{B \to C,\ B : C}}{A \vee B,\ A \to C,\ B \to C : C} \\
\to+ \dfrac{A \vee B,\ A \to C,\ B \to C : C}{A \vee B,\ A \to C : (B \to C) \to C} \\
\to+ \dfrac{A \vee B,\ A \to C : (B \to C) \to C}{A \vee B : (A \to C) \to ((B \to C) \to C)} \\
\to+ \dfrac{A \vee B : (A \to C) \to ((B \to C) \to C)}{A \vee B \to ((A \to C) \to ((B \to C) \to C))} \ .
\end{array}
$$

Now suppose that we can derive in N a pair of sequents $\Gamma' : A$ and $\Gamma'' : A \to B$ whose succedents form the premisses of an application of rule (a) of Ax, where $\Gamma' \subseteq \Gamma$ and $\Gamma'' \subseteq \Gamma$. Then by $\to-$, we can derive in N $\Gamma', \Gamma'' : B$. Next, suppose that we can derive in N the sequent $\Gamma' : C \to A(y)$, where $\Gamma' \subseteq \Gamma$ and the succedent forms the premiss of an application of rule (b). Then we can derive $\Gamma' : C \to \forall x\, A(x)$ as follows:

$$
\begin{array}{l}
\to- \dfrac{C : C \quad \Gamma' : C \to A(y)}{\Gamma', C : A(y)} \\
\forall+ \dfrac{\Gamma', C : A(y)}{\Gamma', C : \forall x\, A(x)} \\
\to+ \dfrac{\Gamma', C : \forall x\, A(x)}{\Gamma' : C \to \forall x\, A(x)} \ .
\end{array}
$$

Finally, suppose that we can derive in N the sequent $\Gamma' : A(y) \to C$, whose succedent forms the premiss of an application of rule (c) and where again $\Gamma' \subseteq \Gamma$. Then we may derive $\Gamma' : \exists x\, A(x) \to C$ as follows:

$$
\begin{array}{l}
\exists- \dfrac{\exists x\, A(x) : \exists x\, A(x) \qquad \to- \dfrac{A(y) : A(y) \quad \Gamma' : A(y) \to C}{\Gamma', A(y) : C}}{\Gamma', \exists x\, A(x) : C} \\
\to+ \dfrac{\Gamma', \exists x\, A(x) : C}{\Gamma' : \exists x\, A(x) \to C} \ .
\end{array}
$$

From this it is clear that the transformation can be carried through.

(ii) If $\Gamma \vdash_N E$, then $\Gamma \vdash_{Ax} E$.

We argue by induction on the maximum length of a path in the proof-tree with conclusion $\Gamma_0 : E(\Gamma_0 \subseteq \Gamma)$.

If the maximum length is 0, $\Gamma_0 : E$ is a basic sequent, and so $E \in \Gamma$, whence $\Gamma \vdash_{Ax} E$ by definition of $\vdash_{Ax}$.

The induction step proceeds by cases, corresponding to the final rule of N applied in the proof-tree. We take one example, that of $\neg$+. Then E is $\neg D$ for some D, and the end of the proof-tree has the form

$$\neg+ \ \frac{\Gamma', D : B \qquad \Gamma'', D : \neg B}{\Gamma', \Gamma'' : \neg D}$$

where $\Gamma' \subseteq \Gamma$ and $\Gamma'' \subseteq \Gamma$. By the induction hypothesis $\Gamma', D \vdash_{Ax} B$ and $\Gamma'', D \vdash_{Ax} \neg B$, whence by the Deduction Theorem for Ax $\Gamma' \vdash_{Ax} D \rightarrow B$ and $\Gamma'' \vdash_{Ax} D \rightarrow \neg B$. From (9), by two applications of (a), we have $D \rightarrow B, D \rightarrow \neg B \vdash_{Ax} \neg D$. Simply by stringing together these derivations, we have at once $\Gamma', \Gamma'' \vdash_{Ax} \neg D$. The other cases are left as an easy exercise.

(Strictly speaking, we were entitled to state the lemma and theorem only for the case that Γ consisted of closed formulas, since we did not give the exact definition of $\vdash_{Ax}$ for when Γ contains open formulas; but it is easy to see how to extend the definition to this case, and there is no difficulty about the theorem in the general case either.)

For those previously quite unfamiliar with intuitionistic logic, practice in trying to see whether a proof of this or that classically valid sequent in N is highly recommended; because it keeps so closely to intuitive reasoning, it gives genuine insight into logical principles.

We may conclude the section by proving the result stated in section 1.3, that if $\neg A$ is provable in classical sentential logic, then it is also provable in intuitionistic logic (remember that this result does not extend to predicate logic).

Just as we may obtain a classical version NK of the system N by replacing ¬- by double negation elimination, so we may obtain an axiomatic formalization of classical logic, AxK, by replacing axiom schema (10) of Ax by:

$$(10^K) \qquad \neg\neg A \rightarrow A.$$

AxK and NK may be shown to be equivalent in just the same way as Ax and N. We now state a lemma.

Lemma. If A and B are any formulas, then

(1) $A \vdash_N \neg\neg A$

(2) $\vdash_N \neg\neg(\neg\neg A \rightarrow A)$

(3) $\neg\neg A, \neg\neg(A \rightarrow B) \vdash_N \neg\neg B$

(4) $\neg\neg\neg A \vdash_N \neg A.$

Proof. We display appropriate proof-trees in N.

For (1):

$$\neg+\ \dfrac{p, \neg p : p \qquad \neg p : \neg P}{p : \neg\neg P\ .}$$

For (2):

$$\neg+\ \dfrac{\neg-\ \dfrac{\neg+\ \dfrac{\rightarrow+\ \dfrac{p, \neg\neg p : p}{p : \neg\neg p \rightarrow p} \qquad p, \neg(\neg\neg p \rightarrow p) : \neg(\neg\neg p \rightarrow p)}{\neg(\neg\neg p \rightarrow p) : \neg p} \qquad \neg\neg p : \neg\neg p}{\rightarrow+\ \dfrac{\neg(\neg\neg p \rightarrow p), \neg\neg p : p}{\neg(\neg\neg p \quad p) : \neg\neg p \rightarrow p}} \qquad \neg(\neg\neg p \rightarrow p) : \neg(\neg\neg p \rightarrow p)}{: \neg\neg(\neg\neg p \rightarrow p)\ .}$$

For (3):

$$\dfrac{\dfrac{\dfrac{\dfrac{p : p \quad p \to q : p \to q}{p \to q,\ p : q}{\scriptstyle \to -} \qquad \neg q,\ p : \neg q}{p \to q,\ \neg q : \neg p}{\scriptstyle \neg +} \qquad \neg\neg p,\ p \to q : \neg\neg p}{\neg\neg p, \neg q : \neg(p \to q)}{\scriptstyle \neg +} \qquad \neg\neg(p \to q), \neg q : \neg\neg(p \to q)}{\neg\neg p, \neg\neg(p \to q) : \neg\neg q}{\scriptstyle \neg +} \ .$$

For (4):

$$\dfrac{\dfrac{p,\ \neg p : p \quad \neg p : \neg p}{p : \neg\neg p}{\scriptstyle \neg +} \qquad \neg\neg\neg p,\ p : \neg\neg\neg p}{\neg\neg\neg p : \neg p}{\scriptstyle \neg +} \ .$$

From this we can derive:

<u>Theorem</u>. If A is a formula of sentential logic, and $\vdash_{NK} A$, then $\vdash_N \neg\neg A$.

Proof. Suppose $\vdash_{NK} A$. Then there is a proof of A in AxK. Each line of the formal proof is either (i) an instance of one of the axiom schemata (1) - (9), or (ii) an instance of (10^K), or (iii) a consequence of two of the earlier lines by rule (a) (modus ponens). We argue, by induction on the position of C in the proof, that, for each line C of the proof, $\vdash_N \neg\neg C$. In case (i) $\vdash_N C$ and hence $\vdash_N \neg\neg C$ by part (1) of the lemma. In case (ii) $\vdash_N \neg\neg C$ by part (2) of the lemma. In case (iii), suppose C is derived from earlier lines B and $B \to C$. By the induction hypothesis $\vdash_N \neg\neg B$ and $\vdash_N \neg\neg(B \to C)$, By the induction hypothesis $\vdash_N \neg\neg B$ and $\vdash_N \neg\neg(B \to C)$, and hence $\vdash_N \neg\neg C$ by part (3) of the lemma.

<u>Corollary</u>. If A is a formula of sentential logic, and $\vdash_{NK} \neg A$, then $\vdash_N \neg A$.

Proof. If $\vdash_{NK} \neg A$, then, by the theorem, $\vdash_{N} \neg\neg\neg A$, whence, by part (4) of the lemma, $\vdash_{N} \neg A$.

4.2. THE SEQUENT CALCULUS

For the sequent calculus L we use a broadened notion of sequent which provides the force of the primitive $\perp$ discussed in the last section. Henceforth, a sequent $\Gamma : C$ is to be an ordered pair $\langle\Gamma, C\rangle$ such that Γ, as before, is a finite set of formulas, and C is either a formula or the empty set. We shall abbreviate '$\Gamma : \emptyset$' as '$\Gamma :$ '. Intuitively, $\Gamma :$ is supposed to be provable just in case Γ is inconsistent.

We preserve the introduction rules of N, but replace the elimination rules by rules of introduction on the left. Also we add a rule for thinning on the right when the succedent is empty. We shall denote, e. g. the rule of $\vee$-introduction on the right by '$:\vee$', and the rule of $\vee$-introduction on the left by '$\vee:$'. The complete set of rules for L is now as follows:

	Right	Left
Thin	$\dfrac{\Gamma :}{\Gamma : A}$	$\dfrac{\Gamma : C}{\Gamma, A : C}$
&	$\dfrac{\Gamma : A \qquad \Delta : B}{\Gamma, \Delta : A \,\&\, B}$	$\dfrac{\Gamma, A, B : C}{\Gamma, A \,\&\, B : C}$
$\vee$	$\dfrac{\Gamma : A}{\Gamma : A \vee B} \qquad \dfrac{\Gamma : B}{\Gamma : A \vee B}$	$\dfrac{\Gamma, A : C \qquad \Delta, B : C}{\Gamma, \Delta, A \vee B : C}$
$\rightarrow$	$\dfrac{\Gamma, A : B}{\Gamma : A \rightarrow B}$	$\dfrac{\Gamma, B : C \qquad \Delta : A}{\Gamma, \Delta, A \rightarrow B : C}$
$\neg$	$\dfrac{\Gamma, A :}{\Gamma : \neg A}$	$\dfrac{\Gamma : A}{\Gamma, \neg A :}$
$\forall$	$\dfrac{\Gamma : A(y)}{\Gamma : \forall x A(x)}$	$\dfrac{\Gamma, A(t) : C}{\Gamma, \forall x A(x) : C}$
$\exists$	$\dfrac{\Gamma : A(t)}{\Gamma : \exists x A(x)}$	$\dfrac{\Gamma, A(y) : C}{\Gamma, \exists x A(x) : C}$

In all cases, C may be either a formula or the empty set. 'Proof tree-trunk' and 'proof-tree' or 'proof' are defined for L just as they were for N. The rules governing the quantifiers are subject to the conditions (a) - (c) of section 4.1.

It might seem more natural to formulate &: as the two rules:

$$\frac{\Gamma, A : C}{\Gamma, A \,\&\, B : C} \qquad \frac{\Gamma, B : C}{\Gamma, A \,\&\, B : C} ,$$

but, using thinning on the left, the two formulations are easily seen to be equivalent. As in N, thinning on the left is derivable when the more general notion of basic sequent is used; thinning on the right, however, is indispensable.

A formalization of classical logic can be obtained by modifying L so as to allow sequents to take the general form $\Gamma : \Delta$, where both Γ and Δ are finite (possibly empty) sets of formulas. A sequent of this general type is intended to be derivable if, under any interpretation under which all the formulas in Γ come out true, at least one formula in Δ comes out true. In this classical system the notion of a basic sequent is also generalized so as to include sequents of the form $\Gamma, A : A, \Delta$. The rules of inference are similarly generalized to allow any number of extraneous formulas in the succedents. Thus the rule of thinning on the right becomes

$$\frac{\Gamma : \Delta}{\Gamma : \Delta, B}$$

and :& takes the form

$$\frac{\Gamma : A, \Delta \qquad \Gamma' : B, \Delta'}{\Gamma, \Gamma' : A \,\&\, B, \Delta, \Delta'} .$$

Moreover, it now becomes possible to formulate :∨ as a dual of :&, namely as:

$$\frac{\Gamma : A,\ B,\ \Delta}{\Gamma : A \vee B,\ \Delta}\ .$$

To verify that this system does indeed yield classical logic, we derive the law of excluded middle in it:

$$\dfrac{\dfrac{A : A}{: A,\ \neg A}\,{:}\neg}{: A \vee \neg A}\,{:}\vee\ .$$

(This derivation uses the rule :∨ in its modified form, but it is evident that two applications of the old :∨ rule would give the same result.)

We can also have an intuitionistic system L' in which the sequents may take this yet more extended form Γ : Δ, that is, where more than one formula can occur in the succedent. In L', however, the rules :→ , :¬ , and :∀ must retain the restricted form which they have in L; i.e. the succedent of the premiss for :→ or for :∀ must consist of a single formula, and that of the premiss for :¬ must be empty. For the classical :¬ rule

$$\frac{\Gamma,\ A : \Delta}{\Gamma : \neg A,\ \Delta}$$

allows the derivation of such intuitionistically invalid sequents as A → B : ¬A ∨ B and : ¬A, A; while the classical :∀ rule

$$\frac{\Gamma : A(y),\ \Delta}{\Gamma : \forall x\ A(x),\ \Delta}$$

allows the derivation of the invalid sequent

$\forall x\,(A(x) \vee B) : \forall x\, A(x) \vee B$. Likewise, the classical $:\rightarrow$ rule

$$\frac{\Gamma, A : B, \Delta}{\Gamma : A \rightarrow B, \Delta}$$

allows the derivation of $: A \rightarrow B, A$. Even the restricted form

$$\frac{\Gamma, A : B_1, \ldots, B_k}{\Gamma : A \rightarrow B_1, \ldots, A \rightarrow B_k}$$

is intuitionistically invalid, since it would allow us to derive the sequent $A \rightarrow (B \vee C) : (A \rightarrow B) \vee (A \rightarrow C)$, which we saw to be invalid in section 1.3. In all the other cases, however, the generalized classical rules for sequents of this extended type are intuitionistically valid. Note that, in the classical system, thinning on the right is derivable, just as thinning on the left is derivable in N and in L: any formula which is introduced into a sequent by thinning on the right could have been put into the succedent of the appropriate basic sequent at the outset. In L', on the other hand, thinning on the right is essential: we cannot introduce everything we need into the basic sequents, for to do so might impede a needed application of one of the restricted rules $:\rightarrow$, $:\neg$, and $:\forall$.

In sum, therefore, the rules of L' are as follows:

	Right	Left
Thin	$\dfrac{\Gamma : \Delta}{\Gamma : A, \Delta}$	$\dfrac{\Gamma : \Delta}{\Gamma, A : \Delta}$
&	$\dfrac{\Gamma : A, \Delta \qquad \Gamma' : B, \Delta'}{\Gamma, \Gamma' : A \,\&\, B, \Delta, \Delta'}$	$\dfrac{\Gamma, A, B : \Delta}{\Gamma, A \,\&\, B : \Delta}$
$\vee$	$\dfrac{\Gamma : A, B, \Delta}{\Gamma : A \vee B, \Delta}$	$\dfrac{\Gamma, A : \Delta \qquad \Gamma', B : \Delta'}{\Gamma, \Gamma', A \vee B : \Delta, \Delta'}$

$\neg$ $$\frac{\Gamma, A :}{\Gamma : \neg A} \qquad\qquad \frac{\Gamma : A, \Delta}{\Gamma, \neg A : \Delta}$$

$\forall$ $$\frac{\Gamma : A(y)}{\Gamma : \forall x\, A(x)} \qquad\qquad \frac{\Gamma , A(t) : \Delta}{\Gamma , \forall x\, A(x) : \Delta}$$

$\exists$ $$\frac{\Gamma : A(t), \Delta}{\Gamma : \exists x\, A(x), \Delta} \qquad\qquad \frac{\Gamma , A(y) : \Delta}{\Gamma, \exists x\, A(x) : \Delta}$$

For present purposes, however, L' may be regarded as a less natural formalization of intuitionistic predicate logic than L, to which we now return.

L does not succeed in mimicking natural reasoning as well as N does; but inspection of the rules shows that the system has the great advantage of having the *subformula property*: in any proof in L of the sequent Γ : A every formula which occurs in any sequent in the proof must be a subformula either of A or of one of the formulas in Γ. ('Subformula' is here to be so understood that every formula is a subformula of itself; if $\neg A$ is a subformula of C, so is A; if A & B, $A \vee B$, or $A \rightarrow B$ is a subformula of C, so are A and B; and if $\forall x\, A(x)$ or $\exists x\, A(x)$ is a subformula of C, so is A(t) for any term t.) Because of the subformula property, we have a much firmer mental grasp on the character of any possible proof of a given sequent in the system L than we have in such a system as N, where, by an application of $\rightarrow$-, a formula may completely disappear from subsequent sequents.

We have now to ask whether L is adequate to capture intuitionistic logic. Since we do not as yet have any semantical notions, the best we can do at this stage is to consider whether N and L are equivalent in the sense that, for all Γ and A, $\Gamma\ {}_{N}A$ iff $\Gamma\ {}_{L}A$.

<u>Theorem</u>. If $\Gamma \vdash_{L} A$, then $\Gamma \vdash_{N} A$.

Proof. It is sufficient to show that all the rules proper to L, namely the left-introduction rules and the right thinning rule, are derived rules of N, where, when the succedent C of a sequent of L is the empty set, we replace it by $\bot$ or by a chosen contradiction $D \,\&\, \neg D$. The case of &: is shown here and the other cases are left as an exercise.

$$\dfrac{\dfrac{\dfrac{\dfrac{\Gamma, A, B : C}{\Gamma, B : A \to C}{\scriptstyle \to+} \qquad \dfrac{A \,\&\, B : A \,\&\, B}{A \,\&\, B : A}{\scriptstyle \&-}}{\Gamma, A \,\&\, B, B : C}{\scriptstyle \to-}}{\Gamma, A \,\&\, B : B \to C}{\scriptstyle \to+} \qquad \dfrac{A \,\&\, B : A \,\&\, B}{A \,\&\, B : B}{\scriptstyle \&-}}{\Gamma, A \,\&\, B : C \,.}{\scriptstyle \to-}$$

We cannot hope to prove the converse of this theorem by the same method, since, in view of the fact that L has the subformula property and N lacks it, the elimination rules of N cannot be derived rules of L in the sense that it is possible to get, by the rules of L, from the premiss or premisses of an elimination rule to its conclusion. An elimination rule can hold in L only in the weaker sense that, if its premiss or premisses can be proved, then so can its conclusions, that is to say, as a derived rule of proof rather than a derived rule of inference. It is this that we aim to establish.

In order to do this, we consider an auxiliary system L^+ which has all the rules of L, and, in addition, the *cut rule*:

$$\text{Cut} \qquad \dfrac{\Gamma : A \qquad \Delta, A : C}{\Gamma, \Delta : C}$$

Clearly, owing to the presence of the cut rule, L^+ does not possess the subformula property. It is now easy to establish the following:

<u>Theorem</u>. If $\Gamma \vdash_N A$, then $\Gamma \vdash_{L^+} A$.

Proof. The rules proper to N, i.e. the elimination rules, are derivable in L^+. We again show this for the case of &, and leave the other cases to the reader.

$$\text{Cut}\ \frac{\Gamma : A \mathbin{\&} B \qquad \dfrac{A, B : A}{A \mathbin{\&} B : A}\ \&:}{\Gamma : A}$$

In fact, N and L^+ are equivalent, since, as we have seen, the rules of L are derived rules of N, and the cut rule is also derivable:

$$\to -\ \frac{\Gamma : A \qquad \dfrac{\Delta, A : B}{\Delta : A \to B}\ \to +}{\Gamma, \Delta : B}$$

It now remains to show that $\Gamma \vdash_{L^+} A$ implies $\Gamma \vdash_L A$, which, in effect, is to show that all applications of the cut rule can be eliminated from proofs in L^+.

4.3. CUT-ELIMINATION

The basic idea of the cut-elimination theorem, which is due to Gentzen, as is the sequent calculus in general, is that a proof which employs the cut rule is not going direct to its objective. The cut formula A has been introduced unnecessarily, only to be removed by the cut: if we attend to the way in which A was introduced in the first place, on the right and on the left, we shall be able to straighten out the loop made by its introduction and eventually to obtain a cut-free proof which goes straight to its conclusion.

Cut-Elimination Theorem. If $\Gamma_o \vdash_{L^+} A_o$, then $\Gamma_o \vdash_L A_o$.

Proof. Given a proof in L^+ of $\Gamma_o : A_o$, it is sufficient to consider any application of the cut rule occurring within it, say

$$\text{Cut}\ \frac{\Gamma : A \qquad \Delta, A : C}{\Gamma, \Delta : C}\ ,$$

which is such that no other application of the cut rule stands above it on any path of the proof-tree, and to show that the sub-proof of Γ, Δ : C can be replaced by a sub-proof in L of the same sequent, i.e. a sub-proof with no cut; iteration of this process of replacement will then lead to a cut-free proof of $\Gamma_0 : A_0$. We say that the *left rank* of the cut is r - 1, where r is the maximum length of a path in the proof-tree proceeding upwards from Γ : A such that the succedent of each sequent associated with a node on that path is A. (Thus if no premiss from which Γ : A was derived has A on the right, the left rank of the cut is 0.) Likewise, the *right rank* of the cut is r - 1, where r is the maximum length of a path in the proof-tree proceeding upwards from Δ, A : C such that the antecedent of each sequent associated with a node on that path contains A. The *rank* of the cut is the sum of its left and right ranks. The *degree* of the cut is the number of logical constants in the cut formula A. (These terms are to be taken as similarly defined for all cuts.)

The method of proof is as follows. If the rank of the original cut is positive, we first replace the given sub-proof of Γ, Δ : C by a sub-proof with the same conclusion which has only applications of the cut rule - perhaps several of them - with the same cut formula A but of lower rank: iteration of this process will lead to a sub-proof containing cuts of rank 0 and of the same degree as the original cut. Next, if the degree of these cuts is positive, each of them is either eliminated or replaced by one or more cuts of lower degree. These may have positive rank, but alternate repetition of these two processes leads to a sub-proof in which every cut is of rank and degree 0; these can be eliminated outright.

Throughout this proof there is a multitude of cases, corresponding to the various rules, to be considered; we shall merely indicate how to deal with them by way of example, rather than go through each one.

We have first to show that, if the rank of the original cut was positive, we can replace it by one or more cuts of lower rank. Suppose, first, that the left rank of the original cut is positive, and that Γ : A was derived from a single sequent $\Gamma' : A$, thus:

$$\text{Cut}\ \dfrac{\dfrac{\overset{\vdots}{\Gamma' : A}}{\Gamma : A} \qquad \overset{\vdots}{\Delta, A : C}}{\Gamma, \Delta : C} \ .$$

We rearrange the sub-proof so that $\Gamma' : A$ now becomes the left-hand premiss of a new cut, the right-hand premiss again being $\Delta, A : C$, and apply to the conclusion of this cut the rule by which $\Gamma : A$ was obtained from $\Gamma' : A$, as follows:

$$\dfrac{\text{Cut}\ \dfrac{\overset{\vdots}{\Gamma' : A} \qquad \overset{\vdots}{\Delta, A : C}}{\Gamma', \Delta : C}}{\Gamma, \Delta : C} \ .$$

If the left rank of the original cut was positive, but $\Gamma : A$ was obtained by a rule with two premisses, then this rule must be either $\rightarrow:$ or $\vee:$. If it was $\rightarrow:$, then only one of its two premisses had A as antecedent, and we can again replace the given cut by just one of lower rank. If, however, the rule was $\vee:$, then A was the antecedent of both premisses, and in this case each of these two sequents must serve as the left-hand premiss of a new cut, the original cut thus being replaced by two new cuts of lower rank.

If the left rank of the original cut was 0, but the right rank was positive, we have to make a transformation to reduce the right rank. In most cases, this can be done in an exactly analogous way: where $\Delta, A : C$ was obtained from a single sequent, thus:

$$\text{Cut}\ \dfrac{\overset{\vdots}{\Gamma : A} \qquad \dfrac{\overset{\vdots}{\Delta', A : C'}}{\Delta, A : C}}{\Gamma, \Delta : C}$$

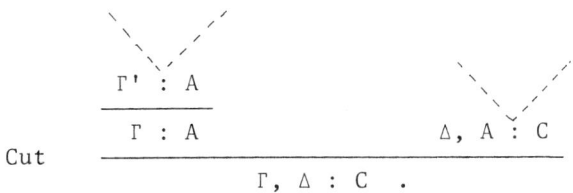

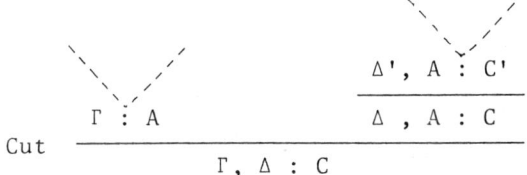

we can transform this into:

$$\text{Cut}\ \dfrac{\begin{matrix}\vdots\\ \Gamma : A\end{matrix} \qquad\qquad \begin{matrix}\vdots\\ \Delta', A : C'\end{matrix}}{\dfrac{\Gamma, \Delta' : C'}{\Gamma, \Delta : C}}$$

and similarly for the cases when Δ, A : C was obtained from two sequents. In one type of case, however, this transformation will not produce the desired result: namely, whenever Δ, A : C was obtained by means of a left-introduction rule in which A was the principal formula. For instance, A might be D & E, the original fragment of the proof taking this form:

$$\text{Cut}\ \dfrac{\begin{matrix}\vdots\\ \Gamma : D \,\&\, E\end{matrix} \qquad\qquad \&{:}\ \dfrac{\begin{matrix}\vdots\\ \Delta, D, E, D \,\&\, E : C\end{matrix}}{\Delta, D \,\&\, E : C}}{\Gamma, \Delta : C}$$

The above transformation would then lead to:

$$\text{Cut}\ \dfrac{\begin{matrix}\vdots\\ \Gamma : D \,\&\, E\end{matrix} \qquad\qquad \begin{matrix}\vdots\\ \Delta, D, E, D \,\&\, E : C\end{matrix}}{\&{:}\ \dfrac{\Gamma, \Delta, D, E : C}{\Gamma, \Delta, D \,\&\, E : C}}$$

In the same way, in each case of this kind, we shall by this means obtain a sub-proof, not of Γ, Δ : C, but of Γ, Δ, A : C. We therefore have, in addition, to insert a further cut, namely:

$$\text{Cut}\ \dfrac{\begin{matrix}\vdots\\ \Gamma : A\end{matrix} \qquad\qquad \begin{matrix}\vdots\\ \Gamma, \Delta, A : C\end{matrix}}{\Gamma, \Delta : C}$$

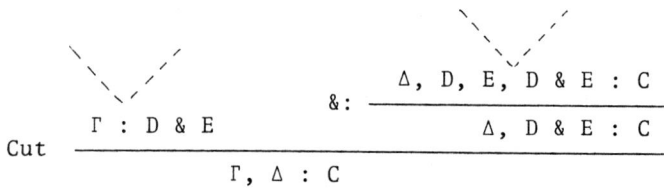

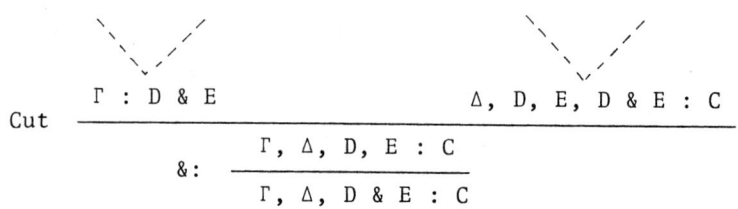

We can assume without loss of generality that A does not belong to Γ or to Δ(if it does, then $\Gamma, \Delta, A : C$ is identical with $\Gamma, \Delta : C$, and the additional cut is unnecessary). Hence the right rank of the additional cut may be taken to be 0, while the right rank of the new cut which stands above it will be lower by 1 than that of the original cut.

So far the cut formula in all the cuts of the resultant sub-proof is A. We now aim to reduce the degree of the cut formula. Without loss of generality we can assume that the original cut was of zero rank, and we claim that the sub-proof can be replaced by one in which every cut is of lower degree than the given one.

If $\Gamma : A$ is a basic sequent, then $\Gamma = \Gamma' \cup \{A\}$ for some Γ', and the cut is of the form:

$$\text{Cut}\ \frac{\Gamma', A : A \qquad\qquad \Delta, A : C}{\Gamma', A, \Delta : C}$$

In this case, the conclusion $\Gamma', \Delta, A : C$ can be derived from the right-hand premiss $\Delta, A : C$ alone by repeated applications of the thinning rule, the cut being eliminated altogether.

If $\Delta, A : C$ is a basic sequent, it is either of the form $\Delta, A : A$ or of the form $\Delta', C, A : C$. In the former case, the cut rule can be eliminated, and the conclusion $\Gamma, \Delta : A$ derived by thinning from the left-hand premiss $\Gamma : A$ alone; in the latter case, the conclusion $\Gamma, \Delta', C : C$ is itself a basic sequent.

If the left-hand premiss $\Gamma : A$ was obtained by thinning on the right, the sequent standing above it is $\Gamma :$, from which the conclusion $\Gamma, \Delta : C$ of the cut could have been obtained directly by thinning, without use of the cut rule. Likewise, if A was introduced into $\Delta, A : C$ by thinning on the left, the cut can be replaced by repeated use of thinning.

It remains to show that, if A was introduced into both premisses of the cut by means of a logical rule; we can arrive at the same conclusion by means of one or more cuts of lower degree. The transformations are shown here for $\to$ and $\forall$. The remaining cases are left as an exercise.

(i) Suppose that A is $B \to D$. Then the proof must have the form:

$$\text{Cut}\ \dfrac{{:}\!\to\ \dfrac{\Gamma, B : D}{\Gamma : B \to D} \qquad \dfrac{\Delta', D : C \quad \Delta'' : B}{\Delta, B \to D : C}\ \to\!{:}}{\Gamma, \Delta : C}$$

We transform this into:

$$\text{Cut}\ \dfrac{\text{Cut}\ \dfrac{\Delta'' : B \qquad \Gamma, B : D}{\Gamma, \Delta'' : D} \qquad \Delta', D : C}{\Gamma, \Delta : C}\ .$$

(ii) Suppose that A is $\forall x A(x)$. In this case, the proof must be as follows:

$$\text{Cut}\ \dfrac{{:}\forall\ \dfrac{\Gamma : A(y)}{\Gamma : \forall x A(x)} \qquad \dfrac{\Delta, A(t) : C}{\Delta, \forall x A(x) : C}\ \forall{:}}{\Gamma, \Delta : C}$$

To reduce the degree here, we first transform the proof of $\Gamma : A(y)$ into a proof of $\Gamma : A(t)$, and then replace the above by:

$$\text{Cut}\ \dfrac{\Gamma : A(t) \qquad \Delta, A(t) : C}{\Gamma, \Delta : C}.$$

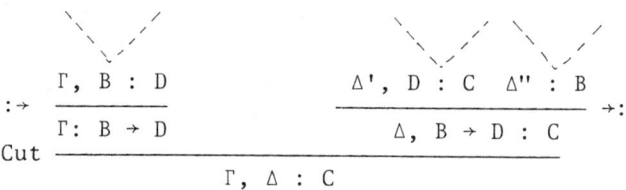

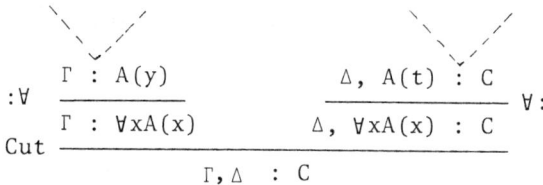

To complete the elimination of the cut altogether we continue to reduce the ranks and degrees of the cuts in the proofs until we arrive at a proof of the original conclusion in which all the cuts are of rank and degree zero. But in any cut of degree zero, the cut formula must be atomic, and so must have been introduced either by thinning or in a basic sequent, and, as we have seen, in any such case no application of the cut rule is needed.

We can now conclude from this theorem and the two theorems of section 4.2 that $\Gamma \vdash_N A$ iff $\Gamma \vdash_L A$.

Cut-elimination is directly connected with establishing consistency, and was so intended by Gentzen. Given the equivalence of N and L, acceptance of ex falso quodlibet in the form of $\neg$- in N or thinning on the right in L makes the consistency of N equivalent to the non-derivability of the empty sequent '$\emptyset$: $\emptyset$' in L. But simply by examination of the rules of L, it is clear that there is no rule by which this sequent could possibly be derived. Recognition of $\neg$- as a correct principle thus amounts to recognizing the consistency of N. Likewise, if a cut-elimination theorem can be proved for some formalized theory of arithmetic, it will follow that only numerical equations can occur in a proof of a numerical equation, and hence that there can be no proof of $0 = 1$; and from this the consistency of the theory follows. This was Gentzen's original strategy for a consistency proof for arithmetic.

However, it should also be noted that, just because the cut-elimination theorem does not establish derived rules in the strict sense, it emphatically cannot be presumed to carry through for extensions of a given system for which it has been proved. The operation of cut-elimination consists in taking each cut which has no cut above it, and successively pushing it upwards until it disappears at the top. We have to show that this pushing upwards can be done through any rule or basic sequent that can occur in the proof; so if we add new kinds of basic sequents or new rules of inference, as we add axioms and induction in the case of arithmetic, the possibility of cut-elimination has to be verified afresh for each of these. For arithmetic as ordinarily formalized, the cut-elimination theorem does not go through, as Gentzen found, and he had accordingly to modify his original strategy.

4.4. DECIDABILITY OF INTUITIONISTIC SENTENTIAL LOGIC

Apart from its relevance to questions of consistency, the cut-free system has great power as an instrument of proof theory, just because it yields so much sharper a conception of what a proof of a given sequent can be like. This is illustrated by the following theorems due to Gentzen.

Now that we have established the equivalence of N and L, we shall use '$\vdash$' with no subscript when we wish to say that something is intuitionistically derivable without reference to a specific formalization. When we are also interested in classical results, we shall use '$\vdash_{IC}$' for intuitionistic logic and '$\vdash_{PC}$' for classical logic.

Theorem. $\vdash A \vee B$ iff $\vdash A$ or $\vdash B$.

Proof. Let $: A \vee B$ be the conclusion of a formal proof in L. Then it can only have been derived by $:\vee$, either from $: A$ or from $: B$. The converse is obvious.

A similar argument is used to prove:

Theorem. $\vdash \exists x A(x)$ iff $\vdash A(t)$ for some term t.

This method of looking upwards to see from where a given sequent could possibly have been derived in one step provides a decision procedure for sentential logic.

Theorem. Intuitionistic sentential logic is decidable.

Proof. We want to discover whether $\Gamma \vdash A$, for given A and finite Γ, i.e. whether there is a proof of $\Gamma : A$ in L. Because of the subformula property, since there are only finitely many subformulas of formulas in $\Gamma \cup \{A\}$, there are only finitely many sequents which could occur in a proof of $\Gamma : A$. Let us call a proof *irredundant* if no sequent appears more than once on any path of it. Then, evidently, if there is a proof of $\Gamma : A$, there is an irredundant proof of it. A finite number of trials will suffice to establish whether or not there exists an irredundant proof of $\Gamma : A$.

In order to exhibit this procedure perspicuously, it is preferable to reformulate some of the rules of L slightly. We remarked for N that, taking basic sequents to be of the form '$\Gamma, A : A$', we could, without weakening the system, restrict those rules which have more than one premiss (:&, v:, and →:) by requiring the Δ occurring in the statement of the rules to be identical with the Γ. This also applies to L, and we shall impose this restriction. If $\Gamma, A \to B : C$ is derived by →:, we cannot exclude the possibility that $A \to B$ appeared in the antecedent of the right-hand premiss; but we shall not weaken the system if we require that it did so appear. Thus we assume →: in the form:

$$\frac{\Gamma, B : C \qquad \Gamma, A \to B : A}{\Gamma, A \to B : C} .$$

Plainly, we do not weaken the system by assuming that $A \to B$ does not occur in Γ (i.e. in the antecedent of the left-hand premiss), and we accordingly make that assumption. :& and v: take the form:

$$\frac{\Gamma : A \qquad \Gamma : B}{\Gamma : A \,\&\, B} \quad \text{and} \quad \frac{\Gamma, A : C \qquad \Gamma, B : C}{\Gamma, A \vee B : C}$$

and in the latter we assume that $A \vee B$ does not occur in Γ. In ¬:, however, we again cannot rule out the possibility that $\neg A$ occurs in the antecedent of the premiss, but lose no generality by assuming that it does, and we accordingly adopt the rule in the form:

$$\frac{\Gamma, \neg A : A}{\Gamma, \neg A :} .$$

These simplifications are made possible by our maintaining the most general notion of a basic sequent, and this makes it possible also to drop the rule of thinning on the left.

Definition. A *proof tree-trunk of level n* is a proof tree-trunk none of whose paths is of length greater than n, and such that the sequent associated with the topmost node of any path of length less than n is a basic sequent. A proof tree-trunk *for* Γ : A has 'Γ : A' associated with the vertex.

The procedure now consists in constructing all possible irredundant proof tree-trunks for Γ : A. That is to say, at stage n we have constructed all possible proof tree-trunks for Γ : A of level n, and at stage n+1 we extend these to proof tree-trunks of level n+1. A given proof tree-trunk of level n may have more than one extension, but always only finitely many, since each sequent can be inferred, by the rules of L as revised, from only finitely many sequents or pairs of sequents, according to the various introduction rules for the principal operators of the constituent formulas. However, some proof tree-trunks will have no extensions: these are the ones containing a sequent which is not a basic sequent but all of whose constituent formulas are sentence letters. Others will have no irredundant extension. The procedure can be shortened if we do not bother to extend proof tree-trunks which contain classically invalid sequents.

It is evident that, if an irredundant proof exists, it will constitute an irredundant proof tree-trunk. It is also clear that, if no proof exists, the procedure will terminate at some stage when no proof tree-trunk can be extended without redundancy. This is because there are only finitely many sequents that can occur anywhere within a proof tree-trunk for Γ : A.

Example.

Consider the sequent

$$(\neg\neg p \rightarrow p) \rightarrow p \vee \neg p : \neg\neg p \rightarrow p.$$

This sequent is invalid, so the procedure will terminate without our finding a proof. There are three proof tree-trunks of level 1:

$$:\text{Thin}\quad \frac{(\neg\neg p \to p) \to p \vee \neg p :}{(\neg\neg p \to p) \to p \vee \neg p : \neg\neg p \to p}$$

$$:\to\quad \frac{(\neg\neg p \to p) \to p \vee \neg p, \neg\neg p : p}{(\neg\neg p \to p) \to p \vee \neg p : \neg\neg p \to p}$$

$$\to:\quad \frac{p \vee \neg p : \neg\neg p \to p \qquad (\neg\neg p \to p) \to p \vee \neg p : \neg\neg p \to p}{(\neg\neg p \to p) \to p \vee \neg p : \neg\neg p \to p}\ .$$

The first of these contains a classically invalid sequent, and the third is redundant, the same sequent occurring both as a premiss and as conclusion. The second has two possible extensions of level 2. The sequent

$$(\neg\neg p \to p) \to p \vee \neg p, \neg\neg p : p$$

could either have been derived by thinning on the right or by $\to:$. However, the extension got by considering thinning yields the classically invalid sequent

$$(\neg\neg p \to p) \to p \vee \neg p, \neg\neg p : \ ,$$

so we need only consider the other possibility, namely the following proof tree-trunk:

$$:\to\quad \frac{\to:\quad \dfrac{\neg\neg p, p \vee \neg p : p \qquad \neg\neg p, (\neg\neg p \to p) \to p \vee \neg p : \neg\neg p \to p}{(\neg\neg p \to p) \to p \vee \neg p, \neg\neg p : p}}{(\neg\neg p \to p) \to p \vee \neg p : \neg\neg p \to p}\ .$$

The left premiss of this application of $\to:$ is intuitionistically valid, so, extending, we should get a proof of it:

$$
\dfrac{\neg\neg p,\ p : p \qquad \dfrac{\dfrac{\dfrac{\neg p : \neg p}{\neg\neg p, \neg p :}\ \neg :}{\neg\neg p, \neg p : p}\ : \text{Thin}}{\neg\neg p,\ p \vee \neg p : p}\ \vee : \quad .
$$

However, the right-hand premiss leaves open only one irredundant extension, and that yields a classically invalid sequent. For if

$$\neg\neg p,\ (\neg\neg p \rightarrow p) \rightarrow p \vee \neg p : \neg\neg p \rightarrow p$$

was derived by $: \rightarrow$, it must have come from a sequent identical with the one at level 1. If obtained by $\rightarrow :$, it would itself appear as one of the premisses. The only possibility left is that it was obtained by thinning on the right, but then it must have come from

$$(\neg\neg p \rightarrow p) \rightarrow p \vee \neg p, \neg\neg p :$$

which is not classically valid. The proof tree-trunk therefore has no irredundant extensions and has not yielded a proof of the given sequent, so the sequent is not provable.

So far, we have dealt only with sentential logic. The cut-free system also provides a decision procedure for prenex formulas of predicate logic. As was remarked in section 1.3, not every formula can be put into prenex form, owing to the failure intuitionistically of the converses of the following laws:

$$
\begin{aligned}
\exists x \neg A(x) &\vdash \neg \forall x A(x)\\
\forall x A(x) \vee B &\vdash \forall x (A(x) \vee B)\\
\exists x (B \rightarrow A(x)) &\vdash B \rightarrow \exists x A(x)\\
\exists x (A(x) \rightarrow B) &\vdash \forall x A(x) \rightarrow B \quad .
\end{aligned}
$$

The procedure is as before, using the rules for the quantifiers. Once we reach a stage at which all formulas are quantifier-free, we can apply the sentential decision procedure.

Example. Consider a sequent of the form

$$: \forall u\ \exists v\ \forall x\ \exists y\ A(x,y,u,v)\ ,$$

where A is quantifier-free and has no free variables other than x, y, u, and v. For convenience we shall use the letters a, b, c for free variables, reserving x, y, u, v for bound variables, and assume that there are no function symbols or individual constants in the language.

Apart from inessential variants with a different free variable, the only first-level proof tree-trunk is:

$$:\forall\ \frac{:\ \exists v\ \forall x\ \exists y\ A(x,y,a,v)}{:\ \forall u\ \exists v\ \forall x\ \exists y\ A(x,y,u,v)}\ .$$

The sequent at level 1 must have been derived by :∃ either from :∀x ∃y A(x,y,a,a) or from :∀x ∃y A(x,y,a,b). However, we can disregard the latter possibility since, plainly, if there is a proof of :∀x ∃y A(x,y,a,b) there is also a proof of :∀x ∃y A(x,y,a,a). At stage 4 the proof tree-trunk is:

$$:\forall\ \frac{:\exists\ \dfrac{:\forall\ \dfrac{:\ \exists y\ A(c,y,a,a)}{:\ \forall x\ \exists y\ A(x,y,a,a)}}{:\ \exists v\ \forall x\ \exists y\ A(x,y,a,v)}}{:\ \forall u\ \exists v\ \forall x\ \exists y\ A(x,y,u,v)}\ .$$

Again, we can ignore the possibility that the topmost sequent was derived from : A(c,b,a,a), leaving two alternatives: either it was obtained from : A(c,c,a,a) or from : A(c,a,a,a). These can now be treated as sentential expressions.

4.5. NORMALIZATION

The sequent calculus was derived by Gentzen as being, in view of its cut-free character, a more powerful proof-theoretic tool than the natural deduction system. However, in more recent times the work of Prawitz and others has established that similar results can be obtained by reducing proofs within a natural deduction system to a normal form. As with cut-elimination, the fundamental idea is the avoidance of unnecessary detours within the proof. Such a detour takes place whenever a sequent occurs as the conclusion of an introduction rule and, simultaneously, as the major premiss of an elimination rule (the major premiss being that which, in the schematic representation of the rule, contains the logical constant to be eliminated). An occurrence of a sequent of this kind is called *maximal*, since the logical complexity of the formula serving as succedent attains a local maximum. It is easy to see that the introduction of a maximal sequent was unnecessary. This enables us to specify certain reduction steps which eliminate such maximal sequents. In doing so, it is convenient to take the system as lacking a thinning rule, and as having basic sequents only of the form $A : A$; we must then construe the rules which discharge hypotheses ($\vee$-, $\exists$-, $\rightarrow$+, and $\neg$+) as not requiring the presence of the hypothesis to be discharged in the antecedent of the relevant premiss (e.g. $\Gamma : A \rightarrow B$ may be validly inferred by $\rightarrow$+ from $\Gamma : B$). The result of the reduction will be a (sub-)proof the conclusion of which is a sequent whose succedent coincides with that of the original, and whose antecedent is contained in that of the original. The reduction step is distinguished by cases according to the logical constant involved.

(&)

$$\dfrac{\dfrac{\Gamma : A \qquad \Delta : B}{\Gamma, \Delta : A \,\&\, B}\,\&+}{\Gamma, \Delta : A}\,\&- \qquad \text{reduces to} \qquad \Gamma : A \quad .$$

The subcase in which the other &- rule is used, to yield the conclusion $\Gamma, \Delta : B$, is dealt with similarly.

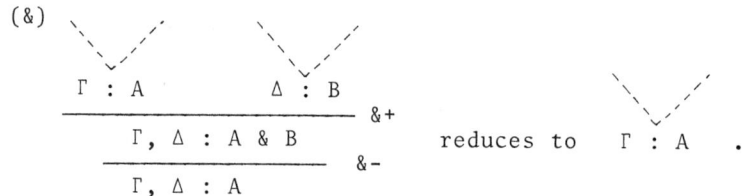

(∨)

$$\dfrac{\dfrac{\begin{array}{c}\vdots\\ \Gamma : A\end{array}}{\Gamma : A \vee B}\,\vee+ \qquad \begin{array}{c}A : A\\ \vdots\\ A, \Delta : C\end{array} \qquad \begin{array}{c}\vdots\\ B, \Theta : C\end{array}}{\Gamma, \Delta, \Theta : C}\,-$$

reduces to

$$\begin{array}{c}\vdots\\ \Gamma : A\\ \vdots\\ \Gamma, \Delta : C .\end{array}$$

This reduction relates to the subcase in which the formula A actually occurs in the antecedent of the first minor premiss of the ∨- inference, in which case the basic sequent A : A must have occurred (in one or more places) above it; the idea is to obtain the effect of a cut by replacing that basic sequent by Γ : A and its proof, to obtain the proof of a sequent in whose antecedent Γ replaces A. This idea will be used repeatedly in the other cases. In the second subcase, in which the first minor premiss takes the form Δ : C, the reduction is even simpler; we simply omit everything but the proof of this minor premiss. Similar subcases may occur in the other cases to be treated below, but will not be explicitly mentioned. When the major premiss Γ : A ∨ B of the ∨- inference was obtained by the other ∨+ rule from Γ : B, the reduction is carried out similarly.

(→)

$$\dfrac{\begin{array}{c}\vdots\\ \Gamma : A\end{array} \qquad \dfrac{\begin{array}{c}A : A\\ \vdots\\ A, \Delta : B\end{array}}{\Delta : A \to B}\,\to+}{\Gamma, \Delta : B}\,\to-$$

reduces to

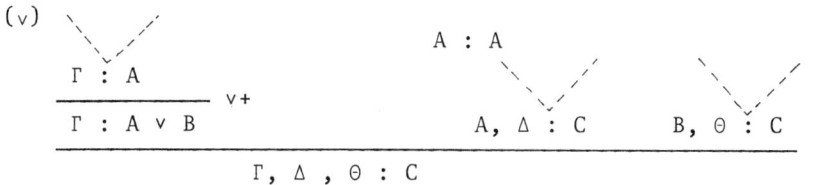

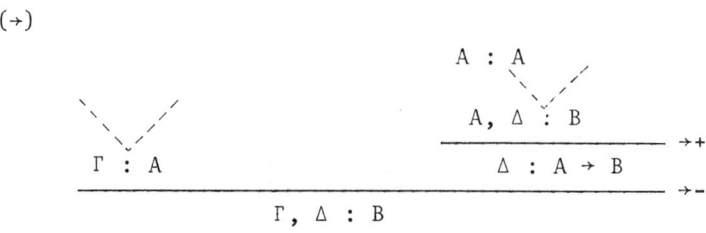

$$\begin{array}{c} \Gamma : A \\ \Gamma\, ,\, \Delta : B\ . \end{array}$$

$(\neg)$

$$\dfrac{\Gamma : A \qquad \dfrac{\begin{array}{c}A : A\\ A,\ \Delta : B\end{array} \qquad \begin{array}{c}A : A\\ A,\ \Theta : \neg B\end{array}}{\Delta,\ \Theta : \neg A}\ \neg +}{\Gamma,\ \Delta,\ \Theta : C}\ \neg -$$

reduces to

$$\dfrac{\begin{array}{c}\Gamma : A\\ \Gamma,\ \Delta : B\end{array} \qquad \begin{array}{c}\Gamma : A\\ \Gamma,\ \Theta : \neg B\end{array}}{\Gamma,\ \Delta,\ \Theta : C\ .}\ \neg -$$

$(\forall)$

$$\dfrac{\dfrac{\Gamma : A(y)}{\Gamma : \forall x\ A(x)}\ \forall +}{\Gamma : A(t)}\ \forall - \qquad \text{reduces to} \qquad \Gamma : A(t)\quad .$$

Here the free variable y, which by assumption does not occur free in Γ, is replaced throughout the proof of $\Gamma : A(y)$ by the term t, which by assumption is free for x in A(x), to obtain a proof of $\Gamma : A(t)$. In the process, it may be necessary to change some of the bound variables occurring earlier in the proof, but it is easy to satisfy oneself that this can be done.

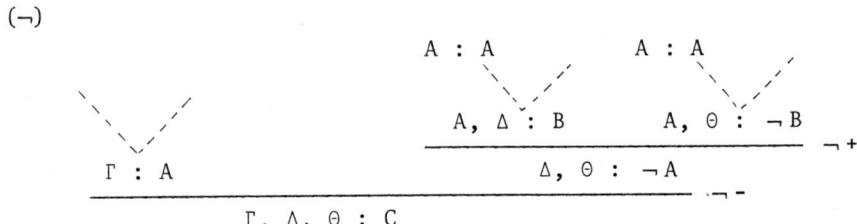

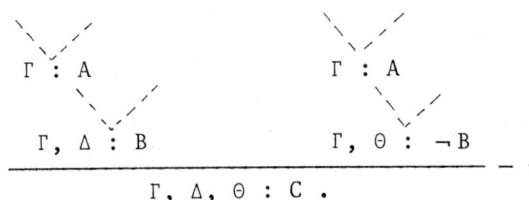

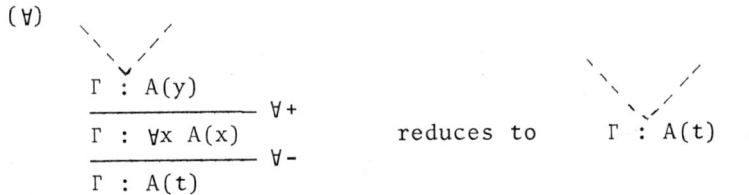

(∃)

$$\dfrac{\dfrac{\begin{array}{c}\vdots\\ \Gamma : A(t)\end{array}}{\Gamma : \exists x\, A(x)}\ \exists+ \qquad \begin{array}{c}A(y) : A(y)\\ \vdots\\ A(y),\ \Delta : C\end{array}}{\Gamma,\ \Delta : C}\ \exists-$$

reduces to

$$\begin{array}{c}\vdots\\ \Gamma : A(t)\\ \vdots\\ \Gamma,\ \Delta : C\ .\end{array}$$

Here every free occurrence of y in the proof of A(y), Δ : C is replaced by the term t, if necessary after some changes of bound variables, while the basic sequent A(y) : A(y) is replaced by Γ : A(t) and its proof; this will not affect Δ or C, which by assumption do not contain y free.

The above reduction steps, called *proper reductions*, embody the basic idea of normalization; but further reductions may also be carried out, as follows. The most important are *permutative reductions*. To see the point of them, consider the following fragment of a proof:

$$\dfrac{\dfrac{\begin{array}{c}\vdots\\ \Gamma : A \vee B\end{array} \qquad \dfrac{\begin{array}{c}\vdots\\ A,\ \Delta : C\end{array} \quad \begin{array}{c}\vdots\\ \Theta : D\end{array}}{A,\ \Delta,\ \Theta : C \,\&\, D}\ \&+ \qquad \begin{array}{c}\vdots\\ B,\ \Lambda : C \,\&\, D\end{array}}{\Gamma,\ \Delta,\ \Theta,\ \Lambda : C \,\&\, D}\ \vee-}{\Gamma,\ \Delta,\ \Theta,\ \Lambda : C\ .}\ \&-$$

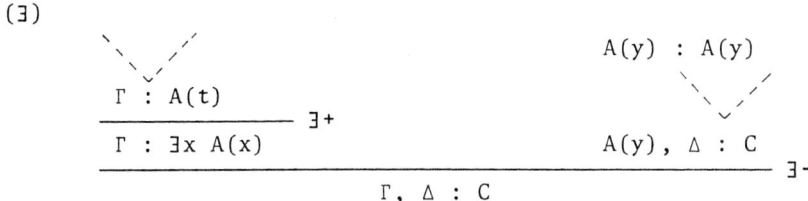

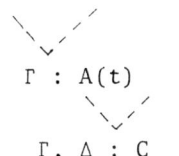

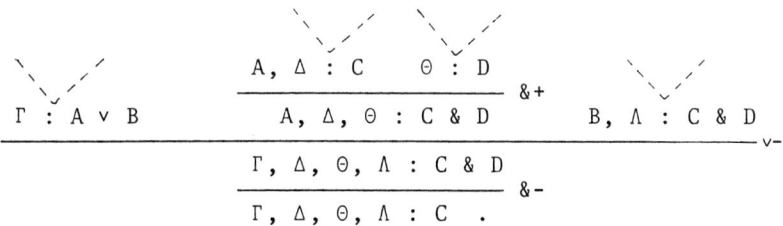

Evidently, the use of the &+ rule to obtain C & D as the succedent of the first minor premiss of the ∨- rule was unnecessary; but it cannot be eliminated immediately by a proper reduction, since the ∨-rule intervenes between the &+ rule and the &- rule. In order to be able to carry out a proper reduction, we therefore first transform the proof into the following:

$$\dfrac{\Gamma : A \vee B \qquad \dfrac{\dfrac{A, \Delta : C \qquad \Theta : D}{A, \Delta, \Theta : C \,\&\, D}\,\&+}{A, \Delta, \Theta : C}\,\&- \qquad \dfrac{B, \Lambda : C \,\&\, D}{B, \Lambda : C}\,\&-}{\Gamma, \Delta, \Theta, \Lambda : C}\,\vee- \quad .$$

We can now eliminate the succession of the &+ and &- rules in the proof of the first minor premiss; we made this possible by permuting the order of the application of the ∨- rule and the &- rule. A permutative reduction is possible whenever the conclusion of an application of the ∨- or ∃- rule serves as the major premiss of any elimination rule. The general form of the reduction in the case of the ∨- rule is as follows.

$$\dfrac{\dfrac{\Gamma : A \vee B \qquad A, \Delta : C \qquad B, \Theta : C}{\Gamma, \Delta, \Theta : C}\,\vee- \qquad \Lambda : D}{\Gamma, \Delta, \Theta, \Lambda : E}\,\underline{R}$$

reduces to

$$\dfrac{\Gamma : A \vee B \qquad \dfrac{A, \Delta : C \qquad \Lambda : D}{A, \Delta, \Lambda : E}\,\underline{R} \qquad \dfrac{B, \Theta : C \qquad \Lambda : D}{B, \Theta, \Lambda : E}\,\underline{R}}{\Gamma, \Delta, \Theta, \Lambda : E}\,\vee- \quad .$$

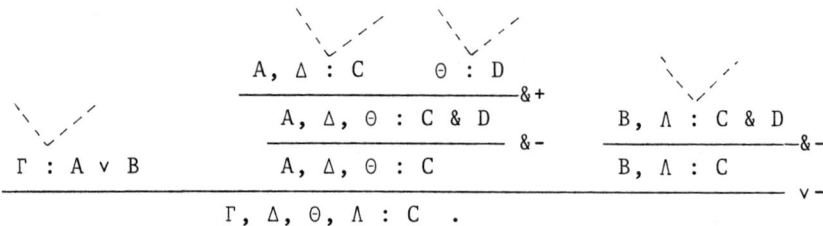

Here R is some elimination rule of which, in the original application, Λ : D (which may be missing) is the minor premiss. Similarly, the general form of a permutative reduction in the case of the ∃- rule is as below.

$$\cfrac{\cfrac{\begin{matrix}\vdots\\ \Gamma : \exists x\, A(x)\end{matrix} \qquad \begin{matrix}\vdots\\ A(y),\, \Delta : C\end{matrix}}{\Gamma,\, \Delta : C}\,\exists\text{-} \qquad \begin{matrix}\vdots\\ \Theta : D\end{matrix}}{\Gamma,\, \Delta,\, \Theta : E}\,\underline{R}$$

reduces to

$$\cfrac{\begin{matrix}\vdots\\ \Gamma : \exists x\, A(x)\end{matrix} \qquad \cfrac{\begin{matrix}\vdots\\ A(y),\, \Delta : C\end{matrix} \qquad \begin{matrix}\vdots\\ \Theta : D\end{matrix}}{A(y),\, \Delta,\, \Theta : E}\,\underline{R}}{\Gamma,\, \Delta,\, \Theta : E}\,\exists\text{-} \quad .$$

The conclusion of an application of the ∨- or ∃- rule has the same succedent as the minor premiss or premisses. Consider, in any proof, a sequent which is not itself the conclusion of a ∨- or ∃- rule, but forms the minor premiss of an application of one of those rules. The conclusion of that rule might again be a minor premiss of one or other of the two rules. If we proceed down the path of the proof-tree leading to the conclusion of the proof, and stop as soon as we come to a sequent which is not the minor premiss of an application of the ∨- or ∃- rule, we shall have picked out a section of that path on which all the sequents have the same succedent. A sequence $\Gamma_1 : A, \ldots, \Gamma_n : A$ of consecutive sequents so obtained is called a *segment*: if $\Gamma_1 : A$ was the conclusion of an introduction rule, and $\Gamma_n : A$ is the major premiss of an elimination rule, then the segment is called *maximal*, by analogy with a maximal sequent. Repeated applications of the permutative reduction steps, followed by a proper reduction, eliminate a maximal segment from a proof.

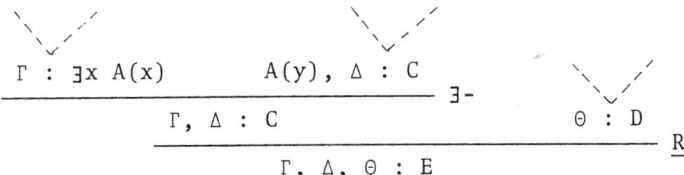

Another type of reduction, called an *immediate simplification*, eliminates redundant applications of the ∨- or ∃- rule. An application of one of those rules is redundant when the antecedent of the conclusion wholly contains the antecedent of one of the minor premisses (intuitively, when none of the hypotheses of that premiss is discharged); for instance, if the major premiss of a $\vee$- is $\Gamma : A \vee B$, and one of the minor premisses is $\Delta : C$. It is obvious that the application of $\vee$- was superfluous, and can be eliminated.

Finally, a $\neg$-reduction lowers the degree of the succedent of the conclusion of an application of $\neg$-. For instance, we may replace

$$\frac{\Gamma : A \qquad \Delta : \neg A}{\Gamma, \Delta : B \;\&\; C}\;\neg-$$

by

$$\frac{\dfrac{\Gamma : A \qquad \Delta : \neg A}{\Gamma, \Delta : B}\;\neg- \qquad \dfrac{\Gamma : A \qquad \Delta : \neg A}{\Gamma, \Delta : C}\;\neg-}{\Gamma, \Delta : B \;\&\; C}\;\&+$$

and similarly for other logical constants. Repeated application of reductions of this type has the effect of making the succedent of the conclusion of every instance of the $\neg$-rule atomic.

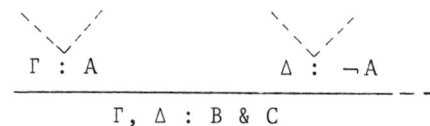

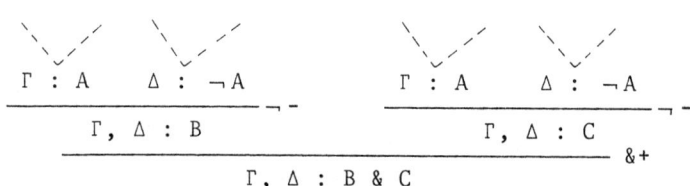

A proof is said to be in *normal form* (with respect to a given class of reductions) if no reduction (of that class) can be applied to it. A *normal form theorem* states that every provable sequent can be derived by means of a proof in normal form. A *normalization theorem* states that every proof can be brought to normal form by an appropriate sequence of reduction steps. A *strong normalization theorem* states that every sequence of reduction steps, starting with a given proof, terminates in a proof in normal form, and that the normal form corresponding to each proof is unique. The normalization theorem for our system of natural deduction with respect to proper and permutative reductions can be established by induction on the pair $\langle d, \ell \rangle$, where d is the highest degree of a maximal segment occurring in the given proof, and ℓ is the sum of the lengths of maximal segments of degree d in the proof; a maximal formula constitutes a maximal segment of length 1, and the degree of a segment is the number of logical constants occurring in the formula which serves as the common succedent of the sequents composing it. We take $\langle d, \ell \rangle$ to precede $\langle d', \ell' \rangle$ if either $d < d'$ or $d = d'$ and $\ell < \ell'$. We now choose, for an application of a reduction step, some maximal segment of degree d such that no maximal segment of degree d occurs above it in the proof-tree, and which is also such that the elimination rule of which the last sequent in the segment is the major premiss does not have as a minor premiss a sequent which either belongs to or stands below another maximal segment of degree d. If the segment we have chosen is composed of a single formula, we apply a proper reduction, which either reduces d or reduces ℓ by 1. If the length of the segment is greater than 1, we apply a permutative reduction, which, in view of the condition on the selection of the segment, must reduce ℓ. The process of reduction therefore terminates. It is now easy to see that the process of applying $\neg$-reductions also terminates, and trivial that that of applying immediate simplifications does so. A more complicated induction will yield the strong normalization theorem.

A normal proof cannot be quite so simply characterized as a cut-free proof in the sequent calculus, but we can describe its structure with the aid of the concept of a *track* in the proof-tree. (Prawitz uses the word 'path', but, since what is intended is not in general a path in the ordinary sense, an unusual term serves to avoid confusion.) A track is a sequence of sequents which begins with a basic sequent the antecedent of which is not subsequently discharged by an application of $\vee$- or $\exists$-, and proceeds down a path of the proof-tree, passing from each sequent to the one which stands immediately below, until it reaches (a) the major premiss of an $\vee$- or $\exists$- inference, or (b) the minor premiss of a $\rightarrow$- or $\neg$- inference, or (c) the conclusion of the proof. In case (b) and, of course, case (c) it terminates; but in case (a) it continues with a basic sequent whose antecedent is eventually discharged by that ∨- or ∃-inference whose major premiss it had reached, and then proceeds downwards as before (passing through the ∨- or ∃-inference to its conclusion if and when it reaches the minor premiss). (Strictly speaking, we need to provide in our definition of a track for the occurrence of redundant ∨- or ∃- inferences, but we can ignore this since we are concerned only with tracks in a normal proof.) Any such track splits up into segments; and, in a normal proof, there will be a minimal segment, dividing the rest of the track into an E-part above the minimal segment and an I-part below it; either of these may be empty. The minimal segment is characterized by the fact that (i) the last sequent in each segment in the E-part is the major premiss of an elimination rule, while (ii) the first sequent in each segment in the I-part is the conclusion of an introduction rule; furthermore only the sequent immediately above the minimal segment can be the major premiss of an application of the ¬-rule. That the track can be so divided may be seen as follows. By the definition of 'segment', the last sequent of a segment cannot be the minor premiss of an ∨- or ∃- inference, nor can its first sequent be the conclusion of such an inference; the track terminates at minor premisses of →- and ¬-inferences, while &- and ∀- inferences do not have minor premisses. It follows that a track which cannot be divided into an E-part and an I-part must contain a segment whose first sequent is the conclusion of an introduction rule and whose last sequent is the major premiss of an elimination rule. But such a segment is maximal, and cannot occur in a normal proof. The fact that there can be at most one application, on the track, of the ¬- rule, and that at the end of the E-part, follows from the fact that, in a normal proof, the succedent of the conclusion of such an inference must be atomic; it can be seen that the violation of this condition would again involve the existence of a maximal segment.

Let us call the succedent common to all sequents in a

segment the 'segment formula'. It is easily seen that if B is the formula of a segment in the I-part of a track in a normal proof, and A the formula of the immediately preceding segment, A is a subformula of B; and, further, that if B is the formula of a segment in the E-part, and A the formula of the immediately succeeding segment, then again A is a subformula of B, save in the case when the later segment is the minimal segment and the last sequent in the earlier one was the major premiss of a $\neg$- inference. From this we can derive the result that normal proofs possess the subformula property, just as cut-free proofs do: every formula occurring in a normal proof of Γ : A is a subformula of A or of some formula in Γ .

The normalization theorem, and its corollaries as just stated, can be used to derive the same results as the cut-elimination theorem, for instance the disjunction property for first-order logic, i.e. the fact that $\vdash A \vee B$ if and only if $\vdash A$ or $\vdash B$. (This theorem can be strengthened by appeal to the relation of being a *strictly positive part* (s.p.p.) of a formula, namely the smallest transitive and reflexive relation such that A and B are s.p.p.s of $A \vee B$ and of A & B, A(t) is a s.p.p. of $\forall x\, A(x)$ and of $\exists x\, A(x)$, and B is a s.p.p. of $A \rightarrow B$. If, now, $\Gamma \vdash A \vee B$, and no formula in Γ has a s.p.p. of the form $C \vee D$, then either $\Gamma \vdash A$ or $\Gamma \vdash B$.) Instead of going over old ground, we may conclude this section with a brief sketch of another result derivable from the normalization theorem, the interpolation theorem for intuitionistic first-order logic. In order to state this theorem, we need a formulation of first-order logic with a constant sentence, so that we can construct formulas containing no schematic letters (sentence-letters, predicate-letters, individual constants, or function symbols). It is most convenient for this purpose to use the symbol $\bot$ for the constant false sentence, mentioned earlier; we treat $\neg A$ as an abbreviation for $A \rightarrow \bot$; we replace the $\neg$+ and $\neg$- rules by the single rule $\bot$- :

$$\frac{\Gamma : \bot}{\Gamma : B}\ \bot\text{-}$$

The normalization theorem still holds good, with $\bot$- playing the role previously played by $\neg$-.

Interpolation Theorem. If $\Gamma \vdash A$, then there exists a formula F, called the *interpolated formula* for the pair $\langle \Gamma, A \rangle$, such that $\Gamma \vdash F$ and $F \vdash A$, and such that every schematic letter occurring in F occurs both in A and in a formula of Γ.

Sketch of proof. We actually prove a slightly more general result, namely that if $\Gamma \cup \Delta \vdash A$, where Γ and Δ are disjoint, then there exists an interpolated formula for the triple $\langle \Gamma, \Delta, A \rangle$, viz. a formula F such that $\Gamma \vdash F$ and $\Delta \cup \{F\} \vdash A$, and every schematic letter in F occurs both in some formula of Γ and in some formula of $\Delta \cup \{A\}$. This then reduces to the theorem when Δ is empty. The proof proceeds by induction on the length of a normal proof. The induction basis relates to a one-line proof, consisting of the basic sequent A : A. A is an interpolated formula for $\langle \{A\}, \emptyset, A \rangle$, and $\bot \rightarrow \bot$ is an interpolated formula for $\langle \emptyset, \{A\}, A \rangle$, For the induction step, there are two cases, according as, in a normal proof, the conclusion Γ, Δ : A is inferred by an introduction rule or by an elimination rule. If by an introduction rule, we apply the induction hypothesis to the premisses of that final inference: by considering each rule separately, we see that it is possible to construct an interpolated formula for $\langle \Gamma, \Delta, A \rangle$ from the interpolated formula(s) associated with the premiss(es). (By the interpolated formula associated with a sequent Θ : B is meant the interpolated formula for the triple $\langle \Gamma \cap \Theta, \Delta \cap \Theta, B \rangle$.) For example, suppose that A is of the form B & C, so that the final inference of the proof has the form:

$$\frac{\Theta : B \qquad \Lambda : C}{\Gamma, \Delta : B \,\&\, C}\,\&+ \quad .$$

By the induction hypothesis, there exists an interpolated formula G for $\langle \Gamma \cap \Theta, \Delta \cap \Theta, B \rangle$ and an interpolated formula H for $\langle \Gamma \cap \Lambda, \Delta \cap \Lambda, C \rangle$. It is then easily seen that G & H is an interpolated formula for $\langle \Gamma, \Delta, B \,\&\, C \rangle$, as desired.

The other case is that in which the conclusion $\Gamma, \Delta : A$ of the proof was obtained by means of an elimination rule. In this case, we consider a path of the proof-tree leading upwards from the conclusion and always going through the major premiss of each elimination rule encountered. It is easily shown that, in a normal proof, only elimination rules can occur on such a path (which is therefore unique). We now consider that inference which is applied to the basic sequent which stands at the head of this path, which will, of course, be an elimination rule. We modify the proof by removing everything standing above the conclusion of this elimination rule, replacing that conclusion by the basic sequent with the same succedent, and making all the consequent alterations to the antecedents of sequents occurring below it on the path. We now have a shorter normal proof, and can apply the induction hypothesis to the conclusion of this modified proof; from the interpolated formula so yielded, together with the interpolated formula associated with the minor premiss, if any, of the elimination rule of which the basic sequent originally at the head of the path was the major premiss, we can construct an interpolated formula for Γ, Δ, A . For example, suppose that the basic sequent at the head of the path is $B \& C : B \& C$, and that the conclusion of the &- rule applied to it is $B \& C : B$. We suppress the original basic sequent, and replace $B \& C : B$ by $B : B$, at the same time replacing, in the antecedents of sequents below it on the path, any occurrences of $B \& C$ that were inherited from the original basic sequent, by B. The result is a normal proof of $\Gamma', \Delta', B : A$, where Γ' and Δ' are like Γ and Δ save for possibly not containing $B \& C$. If $B \& C$ belonged to Γ, then $\Delta' = \Delta$, and we put $\Gamma^* = \Gamma' \cup \{B\}$, and take G as the interpolated formula for $\langle\Gamma^*, \Delta, A\rangle$. Since $\Gamma^* \vdash G$, also $\Gamma \vdash G$, and G is an interpolated formula for $\langle\Gamma, \Delta, A\rangle$ as well. If, on the other hand, $B \& C$ belonged to Δ, then, similarly, $\Gamma' = \Gamma$, we put $\Delta^* = \Delta' \cup \{B\}$ and take G as the interpolated formula for $\langle\Gamma, \Delta^*, A\rangle$; it is again evident that G will serve as an interpolated formula for $\langle\Gamma, \Delta, A\rangle$. The other subcases, some of which are considerably harder, are left as an exercise.

5
THE SEMANTICS OF INTUITIONISTIC LOGIC

5.1. VALUATION SYSTEMS

For the next four sections we consider only sentential logic. By a *sentential language* $\mathbb{L}$ we mean a finite sequence of sentential operators, $\langle O_1,\ldots,O_n\rangle$, where, for each i, O_i is of fixed finite degree d_i. We assume that there is a stock of denumerably many *sentence-letters*, p_0, p_1, $p_2,\ldots$, used in every sentential language. A *formula* of $\mathbb{L}$ is then defined inductively as follows:

(i) Each sentence-letter is a formula;

(ii) If O_i is a sentential operator of $\mathbb{L}$ and $A_1,\ldots,A_{d_i}$ are formulas of $\mathbb{L}$ then $O_iA_1\ldots A_{d_i}$ is a formula.

A *sentential logic* $\mathcal{L}$ is an ordered pair $\langle\mathbb{L},\vdash_{\mathcal{L}}\rangle$, where $\vdash_{\mathcal{L}}$ is a relation (which we call the *derivability relation* for $\mathcal{L}$) defined between sets of formulas of $\mathbb{L}$ and formulas of $\mathbb{L}$, such that:

(i) If $A \in \Gamma$, then $\Gamma \vdash_{\mathcal{L}} A$;

(ii) if $\Gamma \subseteq \Delta$ and $\Gamma \vdash_{\mathcal{L}} A$, then $\Delta \vdash_{\mathcal{L}} A$;

(iii) if $\Gamma \vdash_{\mathcal{L}} A$ and $\Delta \cup \{A\} \vdash_{\mathcal{L}} B$, then $\Gamma \cup \Delta \vdash_{\mathcal{L}} B$;

(iv) if $\Gamma \vdash_{\mathcal{L}} A$ and * is any substitution, then $\Gamma^* \vdash_{\mathcal{L}} A^*$;

(v) If $\Gamma \vdash_{\mathcal{L}} A$, then $\Gamma_0 \vdash_{\mathcal{L}} A$ for some finite subset Γ_0 of Γ.

In (iv) a *substitution* is a mapping from formulas to formulas such that $[O_iA_1\ldots A_{d_i}]^* = O_iA_1^*\ldots A_{d_i}^*$; $\Gamma^* = \{B^* \mid B \in \Gamma\}$. When $\emptyset \vdash A$ we write '$\vdash_{\mathcal{L}} A$', and say that A is *provable* in $\mathcal{L}$. We put $V_{\mathcal{L}} = \{A \mid \vdash_{\mathcal{L}} A\}$.

If $\mathbb{L}$ is $\langle O_1,\ldots,O_n\rangle$, a *valuation system* $\mathfrak{M}$ for $\mathbb{L}$ is a structure $\langle \underline{M}, \underline{D}, f_1,\ldots,f_n\rangle$, where:

(i) $\underline{M}$ is a set with at least two elements;

(ii) $\underline{D}$ is a proper subset of $\underline{M}$ with at least one element;

(iii) for each i, $1 \leq i \leq n$, f_i is a function mapping $\underline{M}^{d_i}$ into $\underline{M}$.

For example, the truth-tables for the classical connectives form a valuation system with $\underline{M} = \{\top, \bot\}$, $\underline{D} = \{\top\}$, and f_1, f_2, f_3, f_4 as the functions defined by the truth-tables for &, ∨, →, ¬.

An *assignment* relative to a valuation system $\mathfrak{m}$ for a language $\mathbb{L}$ is a function ϕ from sentence-letters to $\underline{M}$. ϕ induces a *valuation* ν_ϕ which maps all formulas of $\mathbb{L}$ into $\underline{M}$:

$$\nu_\phi(p_i) = \phi(p_i) \qquad \nu_\phi(O_i A_1 \ldots A_{d_i}) = f_i(\nu_\phi(A_1), \ldots \nu_\phi(A_{d_i})).$$

If Γ is a set of formulas and A is a formula of $\mathbb{L}$, and $\mathfrak{m}$ is a valuation system for $\mathbb{L}$, we define:

$\Gamma \vDash_{\mathfrak{m}} A$ (A *is entailed by* Γ *in* $\mathfrak{m}$) iff, for every assignment ϕ, whenever $\nu_\phi(B) \in \underline{D}$ for all $B \in \Gamma$, then $\nu_\phi(A) \in \underline{D}$. When $\emptyset \vDash_{\mathfrak{m}} A$, we write '$\vDash_{\mathfrak{m}} A$', and say that A is *valid in* $\mathfrak{m}$. We put $V_{\mathfrak{m}} = \{A \mid \vDash_{\mathfrak{m}} A\}$.

Note that the properties (i)-(iv), but not the property (v), required for a derivability-relation $\vdash_{\mathcal{L}}$ automatically hold for an entailment-relation $\vDash_{\mathfrak{m}}$.

<u>Definition</u>. $\mathfrak{m}$ is *faithful* to $\mathcal{L} = \langle \mathbb{L}, \vdash_{\mathcal{L}} \rangle$ iff, for every Γ and A, if $\Gamma \vdash_{\mathcal{L}} A$, then $\Gamma \vDash_{\mathfrak{m}} A$.

<u>Definition</u>. $\mathfrak{m}$ is (*finitely*) *strictly characteristic* for $\mathcal{L}$ iff, for every (finite) Γ and A,

$$\Gamma \vdash_{\mathcal{L}} A \text{ iff } \Gamma \vDash_{\mathfrak{m}} A.$$

Valuation systems have been widely used to provide semantics for non-classical languages, and semantic definitions of logical consequence for non-classical logics. The most direct use is to consider the elements of $\underline{M}$ as truth-values, each sentence being assumed to take exactly one of these truth-values, independently of our recognition of its truth-value. Then the assertion of a sentence amounts to a claim that it has a designated value, viz. a value belonging to $\underline{D}$.

On one intuitive interpretation of 'true', 'is true' can then be taken to mean 'has a designated value' and 'is false' to mean 'has an undesignated value'. The different individual designated values are then to be taken not as degree of truth, but, rather, as corresponding to different ways in which a sentence might be true. We cannot determine the truth or falsity of a complex sentence just from the truth or falsity of its constituents; to do this we must know the particular ways in which they are true or false.

Since intuitionistic logic is founded on a rejection of the whole notion of objectively determined truth-values independent of our capacity for recognizing them, it does not appear as if valuation systems are going to be of any help towards formulating notions of completeness for intuitionistic logic. However, there are other intuitive interpretations of valuation systems more fruitful for our purposes.

One large and extremely important class of such intuitive interpretations of a valuation system proceeds by taking $\underline{M}$ as some set of subsets of a space $\underline{S}$, so that to atomic formulas are assigned subsets of $\underline{S}$ and, by extending to valuations, all formulas are mapped on to subsets. We can then say:

A is *true at* a point x in $\underline{S}$ just in case $x \in \nu_\phi(A)$.

A valuation system for the language $\langle \&, \vee, \rightarrow, \neg \rangle$ will be of the form $\langle \underline{M}, \underline{D}, \cap, \cup, \Rightarrow, - \rangle$. So, for example, if $\cap$ is the usual operation of set intersection, we shall have:

A & B is true at x iff A is true at x and B is true at x, corresponding to

$$\nu_\phi(A \& B) = \nu_\phi(A) \cap \nu_\phi(B).$$

If all the operators are taken as the usual Boolean set operations, and each formula can be shown (or is assumed) either to be or not to be true at each point, the logic of the resultant valuation system will be classical. However, under an intuitive interpretation of this class, a semantics can be given for a (sentential) language in such a way that the truth or falsity of a complex sentence at a point x does not depend only on the truth-values of its constituents at x. For example, the necessity operator for the modal logic

S5 is given by:

$\Box A$ is true at x iff $\forall y_{y \in \underline{S}}$ A is true at y;

and for S4 by:

$\Box A$ is true at x iff $\forall y_{y \in \underline{S}, y \leq x}$ A is true at y,

where $\leq$ is a quasi-ordering on $\underline{S}$. Thus, under an intuitive interpretation belonging to this class, the fundamental notion becomes that of a relativized truth-value, that of a formula's being or not being true at a point of the space $\underline{S}$, rather than that of its having one of many absolute truth-values. To give clear intuitive substance to the interpretation, the space $\underline{S}$ must be, not just any mathematically suitable structure, but a set of elements to which it makes intuitive sense to relativize the notion of truth: in the case of a tense logic, $\underline{S}$ will be a set of times; in that of a modal logic, a set of possible worlds. In sections 5.3 and 5.4, we shall consider intuitive interpretations of valuation systems of this kind for intuitionistic logic. For the time being, however, we consider only the mathematics of the question rather than its intuitive significance.

The study of valuation systems is one of the oldest parts of modern mathematical logic, having been pursued in the early years by the Polish school, including Tarski and Łukasiewicz. This was due not only to their applications to semantics, but also to their use as a technical instrument for obtaining results about logical systems. If, for example, someone accepted classical logic, but rejected the two-valued semantics on which it is ordinarily based, the two-element valuation system would nevertheless remain for him a powerful tool in proving facts about classical logic. In the same way, valuation systems for non-classical logics may be used merely to give algebraic characterizations of them, and hence without any thought of a connection with the intended meanings of the logical constants. In fact, the original investigations of valuation systems for intuitionistic logic, by Jaśkowski, Tarski, McKinsey, Rasiowa, and Sikorski, had exactly this motivation.

Such algebraic techniques were introduced originally to overcome the clumsiness, for establishing proof-theoretic results, of the axiomatic formalizations of logic given by Frege, Russell, Hilbert, and others. In this respect, valuation systems have, to a large extent, been superseded by the stronger proof-theoretic techniques based on applications of Gentzen's sequent calculus and cut-free systems.

The use, before Gentzen, of axiomatic formalizations of logic had other, unfortunate, effects. It was, in the first place, conceptually misleading. It diverted attention from the obvious fact that, whereas in a *theory* our concern is to establish true statements, and the derivation of statements from others is only a means to this end, in *logic* the process of derivation is itself the object of study. Axiomatic formalizations, unlike natural deduction, tempt us to regard logic as a search for logical truths.

One effect of this was that a logic $\mathcal{L}$ tended to be thought of as characterized not, as above, by a derivability relation $\vdash_{\mathcal{L}}$, but merely by the set of provable formulas. In the same way, in place of the notions of a valuation system being faithful to or strictly characteristic for a logic, the relations considered tended to be those of the inclusion of the set of provable formulas in the set of valid formulas, and their coincidence. We now introduce some further relations between valuation systems and logics, with a view to establishing results about strictly characteristic valuation systems.

Definition. $\mathcal{M}$ is *weakly characteristic* for $\mathcal{L}$ iff $\mathcal{M}$ is faithful to $\mathcal{L}$ and $V_{\mathcal{M}} = V_{\mathcal{L}}$.

We have defined the notion of a logic $\mathcal{L}$ by reference to the structural properties of its derivability relation $\vdash_{\mathcal{L}}$, without regard to the manner in which it may be formalized. Since by condition (iv) on $\vdash_{\mathcal{L}}$, $\vdash_{\mathcal{L}}$ is closed under substitution, we may regard a (primitive or derived) rule of inference of a logic $\mathcal{L}$ as holding just in case the corresponding derivability relation obtains: e.g. the adjunction rule (&+)

$$\frac{A \qquad B}{A \,\&\, B}$$

holds in $\mathcal{L}$ just in case $\{p,q\} \vdash_{\mathcal{L}} p \,\&\, q$. In some formalizations, however, rules are used which are intended only for obtaining provable formulas from provable formulas; e.g. the modal rule of necessitation

$$\frac{A}{\Box A}$$

is intended to allow us to conclude, for a modal logic $\mathcal{L}$, that, if $\vdash_{\mathcal{L}} A$, then $\vdash_{\mathcal{L}} \Box A$, but not to make $\Box p$ derivable from p. Such a rule is called a *rule of proof*. It is possible, in terms only of the relation $\vdash_{\mathcal{L}}$ and the operation of substitution, to characterize the fact that a (primitive or derived) rule of proof holds for a logic $\mathcal{L}$ as the existence of a relation $\Vdash_{\mathcal{L}}$ between a set Γ of formulas and a formula A, thus:

Definition. $\Gamma \Vdash_{\mathcal{L}} A$ iff, for any substitution * such that $\Gamma^* \subseteq V_{\mathcal{L}}$, $\vdash_{\mathcal{L}} A^*$.

Thus, in a modal logic $\mathcal{L}$ for which the above rule of necessitation holds, $\{p\} \Vdash_{\mathcal{L}} \Box p$. We can then state the

Theorem. If $\mathfrak{M}$ is weakly characteristic for $\mathcal{L}$ and $\Gamma \models_{\mathfrak{M}} A$, then $\Gamma \Vdash_{\mathcal{L}} A$.

Proof. Suppose $\Gamma^* \subseteq V_{\mathcal{L}}$, for some substitution *. Since $\mathfrak{M}$ is weakly characteristic for $\mathcal{L}$, $V_{\mathfrak{M}} = V_{\mathcal{L}}$, so $\Gamma^* \subseteq V_{\mathfrak{M}}$. But $\Gamma \models_{\mathfrak{M}} A$, and so, clearly, $\Gamma^* \models_{\mathfrak{M}} A^*$. Therefore $\models_{\mathfrak{M}} A^*$, and, since $V_{\mathfrak{M}} = V_{\mathcal{L}}$, $\vdash_{\mathcal{L}} A^*$. Hence $\Gamma \Vdash_{\mathcal{L}} A$.

Definition. A logic $\mathcal{L}$ is *smooth* iff, whenever $\Gamma \Vdash_{\mathcal{L}} A$, $\Gamma \vdash_{\mathcal{L}} A$. If, for some Γ and A, $\Gamma \Vdash_{\mathcal{L}} A$ but $\Gamma \nvdash_{\mathcal{L}} A$, $\mathcal{L}$ is said to be *rough*. ('$\Gamma \nvdash_{\mathcal{L}} A$' is used to abbreviate 'not $\Gamma \vdash_{\mathcal{L}} A$', and similarly with the symbols for the other relations.)

It is an obvious corollary of the above theorem that if $\mathfrak{M}$ is weakly characteristic for $\mathcal{L}$ and $\mathcal{L}$ is smooth, then $\mathfrak{M}$ is strictly characteristic for $\mathcal{L}$.

Classical logic is smooth, but intuitionistic logic is rough, for we have, for example,

$$\{\neg p \to q \vee r\} \Vdash_{IC} (\neg p \to q) \vee (\neg p \to r)$$

but not

$$\{\neg p \to q \vee r\} \vdash_{IC} (\neg p \to q) \vee (\neg p \to r).$$

So there exist weakly characteristic valuation systems for IC which are not strictly characteristic.

A special case of a weakly characteristic valuation system is singled out by the following definition:

Definition. $\mathcal{M}$ is (*finitely*) *strongly characteristic* for $\mathcal{L}$ iff for every (finite) Γ and A,

$$\Gamma \models_{\mathcal{M}} A \quad \text{iff } \Gamma \Vdash_{\mathcal{L}} A.$$

Clearly, if $\mathcal{L}$ is rough, no strongly characteristic valuation system can be strictly characteristic, whereas if $\mathcal{L}$ is smooth there is no distinction between weakly, strongly and strictly characteristic valuation systems.

Theorem (Lindenbaum). For every logic $\mathcal{L}$ there exists a valuation system $\mathcal{M}$ which is strongly characteristic for $\mathcal{L}$, such that $\underline{M}$ is denumerable.

Proof. Take $\underline{M}$ as the set of formulas of $\mathbb{L}$ and put $\underline{D} = V_{\mathcal{L}}$. $f_i(A_1,\ldots,A_{d_i})$ is to be the formula '$O_iA_1\ldots A_{d_i}$'. $\underline{M}$ is denumerable since there are countably many formulas of $\mathbb{L}$. An assignment ϕ maps each p_i on to some formula A_i, so the induced valuation ν_ϕ is in fact a substitution. Conversely, every substitution is a valuation. So, trivially,

$$\Gamma \Vdash_{\mathcal{L}} A \text{ iff } \Gamma \models_{\mathcal{M}} A.$$

Since $\Gamma \vdash_{\mathcal{L}} A$ implies $\Gamma \Vdash_{\mathcal{L}} A$, it follows from this that $\mathcal{M}$ is faithful to $\mathcal{L}$. Finally, if $A \in V_{\mathcal{M}}$, then $A^* \in V_{\mathcal{L}}$ for every substitution *; so, in particular $A \in V_{\mathcal{L}}$. Hence $V_{\mathcal{M}} = V_{\mathcal{L}}$.

The valuation system defined in this theorem cannot be used to extract any information about $\mathcal{L}$, since we already have to know what is provable in $\mathcal{L}$ in order to know what are the elements of $\underline{D}$. The effective content of the theorem is that a formula cannot be provable in $\mathcal{L}$ and yet invalid in every valuation system faithful to $\mathcal{L}$. The result can be strengthened by taking equivalence classes under interderivability.

Definition. $A \dashv\vdash_{\mathcal{L}} B$ iff $\{A\} \vdash_{\mathcal{L}} B$ and $\{B\} \vdash_{\mathcal{L}} A$.

From the definition of a logic, it follows immediately that $\dashv\vdash_{\mathcal{L}}$ is an equivalence relation.

Corollary. If $\dashv\vdash_{\mathcal{L}}$ is a congruence relation with respect to the sentential operators of $\mathbb{L}$, then there exists a valuation system with just one designated element which is strongly characteristic for $\mathcal{L}$, and in which $\underline{M}$ is denumerable.

Proof. Define $\mathfrak{M}_{\mathcal{L}}$ by taking the domain, $\underline{M}$, as the set of equivalence classes of formulas under $\dashv\vdash_{\mathcal{L}}$. $\underline{D}$ is the set of equivalence classes of valid formulas. But there is obviously only one of these, namely $V_{\mathcal{L}}$ itself, since $\vdash_{\mathcal{L}} A$ and $\vdash_{\mathcal{L}} B$ implies $A \dashv\vdash_{\mathcal{L}} B$. The functions f_i are well-defined by

$$f_i(|A_1|,\ldots,|A_{d_i}|) = |O_i A_1,\ldots,A_{d_i}|,$$

where $|A|$ is the equivalence class to which A belongs, since $\dashv\vdash$ is a congruence relation with respect to the O_i.

$\mathfrak{M}_{\mathcal{L}}$ is called the *Lindenbaum algebra* for $\mathcal{L}$.

We can now generalize our notions of characteristic systems to families of valuation systems. If $\mathbb{H}$ is a family of valuation systems for $\mathbb{L}$, we define: $\Gamma \models_{\mathbb{H}} A$ iff $\Gamma \models_{\mathfrak{M}} A$ for every $\mathfrak{M} \in \mathbb{H}$, and $V_{\mathbb{H}} = \{A | \models_{\mathbb{H}} A\}$.

Definition. $\mathbb{H}$ is *(finitely) strictly characteristic* for $\mathcal{L}$ iff, for every (finite) Γ and A,

$$\Gamma \vdash_{\mathcal{L}} A \quad \text{iff } \Gamma \models_{\mathbb{H}} A.$$

The weaker notions are extended similarly.

From the point of view of characterizing a logic $\mathcal{L}$ by means of valuation systems, the best thing that can happen is that there exists a finite valuation system strictly characteristic for $\mathcal{L}$. The next best thing is that there should be a family $\mathbb{H}$ of finite valuation systems, such that $\mathbb{H}$ is strictly characteristic for $\mathcal{L}$. If there is a family $\mathbb{F}$

of finite valuation systems which is at least finitely strictly characteristic for $\mathcal{L}$, we shall say that $\mathcal{L}$ has the *finite model property*. We conclude this section by showing that there is no finite valuation system even weakly characteristic for IC. In the next two sections we shall show that IC has the finite model property.

Theorem (Gödel). There is no finite valuation system weakly characteristic for IC.

Proof. Suppose $\mathcal{M}$ is such a valuation system in which $\underline{M}$ has n elements. Consider the formula A, consisting of the disjunction,

$$\bigvee_{0 \leq i < j \leq n} p_i \leftrightarrow p_j \quad .$$

Let ϕ be any assignment relative to $\mathcal{M}$. Then, since A contains n + 1 sentence-letters, and $\underline{M}$ has only n elements, for some i,j, $0 \leq i < j \leq n$, $\phi(p_i) = \phi(p_j)$. Since $\mathcal{M}$ is assumed faithful to IC and $\vdash_{IC} p \leftrightarrow p$, we have $\models_{\mathcal{M}} p \leftrightarrow p$, whence $v_\phi(p_i \leftrightarrow p_j) \in \underline{D}$. But also $p_i \leftrightarrow p_j \vdash_{IC} A$, and so $p_i \leftrightarrow p_j \models_{\mathcal{M}} A$, whence $v_\phi(A) \in \underline{D}$. But ϕ was arbitrary, and so $v_\phi(A) \in \underline{D}$ for every ϕ, i.e. $\models_{\mathcal{M}} A$. Now if $\mathcal{M}$ is weakly characteristic, $\vdash_{IC} A$. But, by the decision procedure, not $\vdash_{IC} A$. Hence $\mathcal{M}$ is not weakly characteristic for IC.

5.2. LATTICES AND THE FINITE MODEL PROPERTY

By a lattice is meant a structure $\langle \underline{M}, \cap, \cup \rangle$, where $\cap$ and $\cup$ are binary functions called 'meet' and 'join' satisfying, for all a,b,c $\in \underline{M}$:

$$a \cap a = a \cup a = a$$

$$a \cap b = b \cap a, \ a \cup b = b \cup a$$

$$a \cap (b \cap c) = (a \cap b) \cap c, \ a \cup (b \cup c) = (a \cup b) \cup c$$

$$a \cap (a \cup b) = a \cup (a \cap b) = a \ .$$

Theorem. If $\langle \underline{M}, \cap, \cup \rangle$ is a lattice, and we define '$\leq$' by: $a \leq b$ iff $a \cup b = b$ (equivalently, $a \cap b = a$), then $\leq$ is a partial ordering on $\underline{M}$, $a \cap b$ is the greatest lower bound of $\{a,b\}$, and $a \cup b$ is the least upper bound of $\{a,b\}$.

The proof is left as an exercise. Given a lattice, $\leq$ will always be taken as defined here.

Theorem. If $\leq$ is a partial ordering on a set $\underline{M}$, such that, for any a, b $\in \underline{M}$, the greatest lower bound and least upper bound of $\{a,b\}$ exist, and are denoted by '$a \cap b$' and '$a \cup b$' respectively, then $\langle \underline{M}, \cap, \cup \rangle$ is a lattice.

The verification that $\cap$ and $\cup$ satisfy the conditions for a lattice is left to the reader.

Definition. A lattice $\langle \underline{M}, \cap, \cup \rangle$ is *distributive* iff, for every a, b, c $\in \underline{M}$:

$$a \cap (b \cup c) = (a \cap b) \cup (a \cap c)$$

and $$a \cup (b \cap c) = (a \cup b) \cap (a \cup c) .$$

Either of these laws is derivable from the other.

If a lattice has a least element, it is called the *zero* of the lattice, and denoted by '0'; we have $a \cup 0 = a$ and $a \cap 0 = 0$. If it has a greatest element, this is called the *unit* of the lattice, and denoted by '1'; we have $a \cup 1 = 1$ and $a \cap 1 = a$. A finite lattice always has a zero and unit.

Definition. A lattice $\langle \underline{M}, \cap, \cup \rangle$ is called a *Heyting lattice* if it has a zero and there exists a binary operation '$\Rightarrow$' such that, for all a, b, c $\in \underline{M}$:

$$c \leq a \Rightarrow b \text{ iff } a \cap c \leq b.$$

Evidently, if such an operation exists, it is unique.

Theorem. Any Heyting lattice is distributive.

Proof. In any lattice, $b \leq b \cup c$ and $c \leq b \cup c$,

so $a \cap b \leq a \cap (b \cup c)$

and $a \cap c \leq a \cap (b \cup c)$.

Hence $(a \cap b) \cup (a \cap c) \leq a \cap (b \cup c)$.

Further, $a \cap b \leq (a \cap b) \cup (a \cap c)$; so, in a Heyting lattice, $b \leq a \Rightarrow [(a \cap b) \cup (a \cap c)]$, and similarly

$$c \leq a \Rightarrow [(a \cap b) \cup (a \cap c)].$$

Hence $b \cup c \le a \Rightarrow [(a \cap b) \cup (a \cap c)]$.
Therefore,

$$a \cap (b \cup c) \le (a \cap b) \cup (a \cap c).$$

Thus $a \cap (b \cup c) = (a \cap b) \cup (a \cap c)$.

Theorem. Any finite distributive lattice is a Heyting lattice.

Proof. Since the lattice is finite, it has a zero. Given a, b, let $\underline{C} = \{c \mid a \cap c \le b\} = \{c_1, \dots, c_k\}$. Let c_o be the least upper bound of $\underline{C}$, so $c_o = c_1 \cup \dots \cup c_k$.
Obviously, if $a \cap c \le b$, then $c \le c_o$.
Conversely, if $c \le c_o$,

$$a \cap c \le a \cap c_o = a \cap (c_1 \cup \dots \cup c_k)$$
$$= (a \cap c_1) \cup \dots \cup (a \cap c_k),$$

since the lattice is distributive.
But $a \cap c_i \le b$ for each i, $1 \le i \le k$. Therefore $a \cap c_o \le b$. Hence $a \cap c \le b$. Thus c_o satisfies the conditions on $a \Rightarrow b$. So the lattice is a Heyting lattice.

If $\langle \underline{M}, \cap, \cup \rangle$ is a Heyting lattice, $\mathfrak{M} = \langle \underline{M}, \{1\}, \cap, \cup, \Rightarrow, - \rangle$ is the *associated valuation system* in which, for each $a \in \underline{M}$, $-a = a \Rightarrow 0$, and $1 = -0$, being the unit of the lattice.

Theorem. $\mathfrak{M} = \langle \underline{M}, \{1\}, \cap, \cup, \Rightarrow, - \rangle$ is faithful to IC iff $\mathfrak{M}$ is the associated valuation system of a Heyting lattice $\langle \underline{M}, \cap, \cup \rangle$.

Proof. It is straightforward to show that if $\mathfrak{M}$ is faithful to IC, then the underlying lattice $\langle \underline{M}, \cap, \cup \rangle$ is a Heyting lattice of which $\mathfrak{M}$ is the associated valuation system. For, the converse, we wish to show, by induction on the length of proof in L, that if a sequent $A_1, \dots, A_n : B$ is derivable, then $\{A_1, \dots, A_n\} \models_{\mathfrak{M}} B$. In order to do this, we establish the stronger result that, if the sequent is derivable, then, for any assignment ϕ with respect to $\mathfrak{M}$,

$$\nu_\phi(A_1) \cap \dots \cap \nu_\phi(A_n) \le \nu_\phi(B).$$

The details are left as an exercise.

With the next two theorems we make an approach to showing that IC has the finite model property by establishing that the family of valuation systems assocaited with finite Heyting lattices is weakly characteristic for IC.

Theorem. If $\mathfrak{M} = \langle \underline{M}, \{1\}, \cap, \cup, \Rightarrow, - \rangle$ is faithful to IC, and $\not\models_{\mathfrak{M}} A$ for some formula A, then there exists a finite subset $\underline{M}_o$ of $\underline{M}$ such that for some $\Rightarrow_o$ and $-_o$, $\mathfrak{M}_o = \langle \underline{M}_o, \{1\}, \cap, \cup, \Rightarrow_o, -_o \rangle$, is faithful to IC and $\not\models_{\mathfrak{M}_o} A$.

Proof. By the preceding theorem we can assume $\mathfrak{M}$ is the associated valuation system of a Heyting lattice. Since $\not\models_{\mathfrak{M}} A$, there is some assignment ϕ with respect to $\mathfrak{M}$ such that $v_\phi(A) \neq 1$.

Let $\underline{N}_o = \{v_\phi(B) \mid B \text{ is a subformula of } A\} \cup \{0, 1\}$, and let $\underline{M}_o$ be the closure of $\underline{N}_o$ under $\cap$ and $\cup$ in $\langle \underline{M}, \cap, \cup \rangle$. $\underline{N}_o$ is finite. As is quite easily shown, in a distributive lattice, a sublattice generated by a finite set of elements is itself finite. So $\underline{M}_o$ is finite.

Since $\langle \underline{M}, \cap, \cup \rangle$ is a Heyting lattice, it is distributive; so $\langle \underline{M}_o, \cap, \cup \rangle$ is a finite distributive lattice, and therefore a Heyting lattice. Hence, $\mathfrak{M}_o$, the associated valuation system, is faithful to IC.

If ϕ_o is the restriction of ϕ to those sentence-letters occurring in A, then ϕ_o is an assignment with respect to $\mathfrak{M}_o$, and v_{ϕ_o} is the corresponding valuation. We claim that $v_\phi(B) = v_{\phi_o}(B)$ for every subformula B of A, and so, in particular, that $v_{\phi_o}(A) = v_\phi(A) \neq 1$. By definition, $v_{\phi_o}(p) = v_\phi(p)$, for sentence-letters p occurring in A. Further, since $\underline{M}_o$ is closed under the operations $\cap$ and $\cup$ of $\langle \underline{M}, \cap, \cup \rangle$, we certainly have $v_{\phi_o}(B) = v_\phi(B)$, whenever B is a subformula whose only connectives are & and v. However, $\underline{M}_o$ may not be closed under $\Rightarrow$; if we had chosen it to be so, we could not have ensured that it was finite. So, in general, $\Rightarrow_o$ is not the same as $\Rightarrow$. However, since we are only concerned with subformulas of A, it suffices to show:

if a, b, $a \Rightarrow b \in \underline{M}_0$, then $a \Rightarrow_0 b = a \Rightarrow b$.

For, if $C \rightarrow D$ is a subformula, then $\nu_\phi(C)$, $\nu_\phi(D)$, and $\nu_\phi(C \rightarrow D)$, i.e. $\nu_\phi(C) \Rightarrow \nu_\phi(D)$, are in $\underline{M}_0$. To establish this, note that:

(1) for any $c \in \underline{M}_0$, $c \leq a \Rightarrow_0 b$ iff $a \cap c \leq b$

(2) for any $c \in \underline{M}$, $c \leq a \Rightarrow b$ iff $a \cap c \leq b$.

By (2), $a \cap (a \Rightarrow b) \leq b$, whence by (1) $a \Rightarrow b \leq a \Rightarrow_0 b$. By (1), $a \cap (a \Rightarrow_0 b) \leq b$, whence by (2) $a \Rightarrow_0 b \leq a \Rightarrow b$. Hence $a \Rightarrow b = a \Rightarrow_0 b$.

Thus, for any subformula B of A, $\nu_{\phi_0}(B) = \nu_\phi(B)$.

So $\nu_{\phi_0}(A) \neq 1$, i.e. $\not\models_{\mathfrak{M}_0} A$.

$\dashv\vdash$ is easily shown to be a congruence relation over IC, so, by Lindenbaum's Theorem, the Lindenbaum algebra $\mathfrak{M}_{IC}$ exists and is a valuation system weakly characteristic for IC. So we have the following result:

Theorem. For any formula A such that $\not\vdash_{IC} A$, we can find a finite valuation system $\mathfrak{M}_0$ faithful to IC such that $\not\models_{\mathfrak{M}_0} A$.

Example Consider the formula

$$(p \rightarrow q) \vee (q \rightarrow p).$$

The valuation with respect to the Lindenbaum algebra for IC which assigns to each formula A its equivalence class, $|A|$, under $\dashv\vdash_{IC}$, gives us:

$\underline{N}_0 = \{|p|, |q|, |p \rightarrow q|, |q \rightarrow p|, |(p \rightarrow q) \vee (q \rightarrow p)|, 0, 1\}$,

where 0 is the equivalence class of inconsistent formulas and 1 is V_{IC}.

By Gentzen's decision procedure, $(p \rightarrow q) \vee (q \rightarrow p)$ is obviously not provable in L; so $|(p \rightarrow q) \vee (q \rightarrow p)| \neq V_{IC}$. Taking the closure of $\underline{N}_0$ with respect to $\cap$ and $\cup$, we get the finite sublattice $\langle \underline{M}_0, \cap, \cup \rangle$:

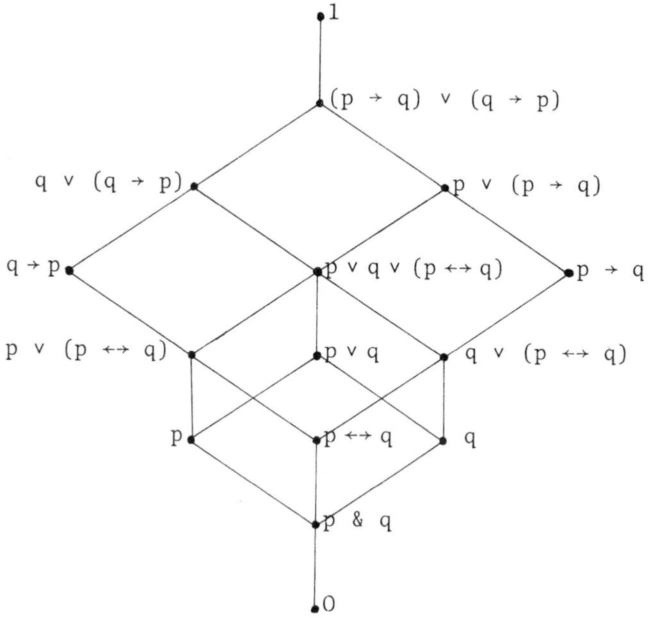

5.3. TOPOLOGICAL SPACES: PO-SPACES

As remarked in section 5.1, it is fruitful to take the domain of a valuation system to be some family of subsets of a space $\underline{S}$. We now consider this possibility when $\underline{S}$ is a topological space.

Definition. A *topological space* $\mathcal{T}$ is a structure $\langle \underline{S}, \mathfrak{I} \rangle$, where is a set and $\mathfrak{I}$ an operation carrying subsets of $\underline{S}$ to subsets of $\underline{S}$ satisfying, for all $\underline{A}$, $\underline{B} \subseteq \underline{S}$:

$$\mathfrak{I}\underline{A} \subseteq \underline{A}$$

$$\mathfrak{I}\mathfrak{I}\underline{A} = \mathfrak{I}\underline{A}$$

$$\mathfrak{I}\underline{S} = \underline{S}$$

$$\mathfrak{I}(\underline{A} \cap \underline{B}) = \mathfrak{I}\underline{A} \cap \mathfrak{I}\underline{B}.$$

A subset $\underline{A}$ of $\underline{S}$ is *open* just in case $\mathfrak{I}\underline{A} = \underline{A}$. Classically, we also define $\mathfrak{C}\underline{A} = -\mathfrak{I}-\underline{A}$, where $-\underline{A}$ denotes the complement of $\underline{A}$ in $\underline{S}$, and say that $\underline{A}$ is *closed* iff $\mathfrak{C}\underline{A} = \underline{A}$. $\mathfrak{I}\underline{A}$ is called the *interior* of $\underline{A}$, and $\mathfrak{C}\underline{A}$ is its *closure*. Evidently $\mathfrak{I}\emptyset = \emptyset$, so $\emptyset$ and $\underline{S}$ are always both open and closed.

Theorem. Let $\langle \underline{S}, \mathcal{I} \rangle$ be a topological space, and let $\underline{\mathcal{O}}$ be the family of open subsets of $\underline{S}$. Then $\langle \underline{\mathcal{O}}, \cap, \cup \rangle$ is a Heyting lattice, where $\cap, \cup$ are Boolean intersection and union, $\emptyset$ is its zero, and the operation $\Rightarrow$ is given by

$$\underline{A} \Rightarrow \underline{B} = \mathcal{I}\{x \mid \text{if } x \in \underline{A}, \text{ then } x \in \underline{B}\}.$$

Proof. For any open sets $\underline{A}$ and $\underline{B}$, $\underline{A} \cap \underline{B}$ and $\underline{A} \cup \underline{B}$ are open. So $\langle \underline{\mathcal{O}}, \cap, \cup \rangle$ is clearly a distributive lattice with a zero. The ordering relation on this lattice is set-inclusion. But, for any sets $\underline{A}$, $\underline{B}$, $\underline{C}$, we have

$$\underline{C} \subseteq \{x \mid x \in \underline{A} \rightarrow x \in \underline{B}\} \text{ iff } \underline{A} \cap \underline{C} \subseteq \underline{B}.$$

For suppose, first, that $\underline{C} \subseteq \{x \mid x \in \underline{A} \rightarrow x \in \underline{B}\}$, and assume that $y \in \underline{A} \cap \underline{C}$. Then $y \in \underline{A}$ and, if $y \in \underline{A}$, then $y \in \underline{B}$. Hence $y \in \underline{B}$. This shows that $\underline{A} \cap \underline{C} \subseteq \underline{B}$. Now suppose that $\underline{A} \cap \underline{C} \subseteq \underline{B}$, and assume that $y \in \underline{C}$. If $y \in \underline{A}$, then $y \in \underline{A} \cap \underline{C}$, and so $y \in \underline{B}$. Thus $y \in \{x \mid x \in \underline{A} \rightarrow x \in \underline{B}\}$. This shows that $\underline{C} \subseteq \{x \mid x \in \underline{A} \rightarrow x \in \underline{B}\}$.
When $\underline{C}$ is open, however, for any $\underline{D}$, $\underline{C} \subseteq \underline{D}$ iff $\underline{C} \subseteq \mathcal{I}\underline{D}$. Hence, taking $\underline{D} = \{x \mid x \in \underline{A} \rightarrow x \in \underline{B}\}$, we have that, for any open sets $\underline{A}$, $\underline{B}$, $\underline{C}$

$$\underline{C} \subseteq \underline{A} \Rightarrow \underline{B} \text{ iff } \underline{A} \cap \underline{C} \subseteq \underline{B}.$$

(Regarded classically, $\underline{A} \Rightarrow \underline{B}$ is $\mathcal{I}(-\underline{A} \cup \underline{B})$.) Thus $\langle \underline{\mathcal{O}}, \cap, \cup \rangle$ is a Heyting lattice.

Such a lattice is called a *topological Heyting lattice*.

There is a representation theorem, due to Tarski and McKinsey, to the effect that every Heyting lattice can be embedded in a topological Heyting lattice by a map which is a morphism with respect to zero and $\Rightarrow$. However, we need to establish this only for the finite case: every finite Heyting lattice is isomorphic to a finite topological Heyting lattice.

A relation is a *quasi-ordering* iff it is reflexive and transitive. Given a quasi-ordered set $\langle \underline{S}, \leq \rangle$ define, for $\underline{A} \subseteq \underline{S}$,

$$\mathcal{I}\underline{A} = \{a \mid b \in \underline{A} \text{ for all } b \leq a\} .$$

$\langle \underline{S}, \mathcal{I} \rangle$ is called a *QO-space*. It is left as an exercise to show that it is a topological space. Note that a subset $\underline{A}$ is open in $\langle \underline{S}, \mathcal{I} \rangle$ iff, whenever $a \in \underline{A}$ and $b \leq a$, then $b \in \underline{A}$.

If, in addition, $\leq$ is antisymmetric (i.e. is a partial ordering), then $\langle \underline{S}, \mathcal{I} \rangle$ is called a *PO-space*. We can restrict our attention to PO-spaces; for, if $\leq$ is a quasi-ordering and $a \leq b$ and $b \leq a$, then, for any open set $\underline{A}$ of the corresponding QO-space, $a \in \underline{A}$ iff $b \in \underline{A}$.

We remark that given any finite topological space $\langle \underline{S}, \mathcal{I} \rangle$ we can define a quasi-ordering on $\underline{S}$ classically by

$$a \leq b \text{ iff } b \in \mathcal{C}\{a\} ,$$

so that $\langle \underline{S}, \mathcal{I} \rangle$ is isomorphic to the corresponding QO-space.

In order to prove the representation theorem, we show that for any finite Heyting lattice $\langle \underline{M}, \cap, \cup \rangle$ there is some partially ordered set $\langle \underline{S}, \leq \rangle$ such that the Heyting lattice of open sets of the corresponding PO-space is isomorphic to the original lattice. We cannot simply take $\underline{S}$ to be $\underline{M}$ and $\leq$ to be the partial ordering of the lattice, since this PO-space will in general have more open sets than there are elements of $\underline{M}$.

Definition. An element a of a lattice $\langle \underline{M}, \cap, \cup \rangle$ is *join-irreducible* iff $a \neq 0$ and, for all $b, c \in \underline{M}$, if $a = b \cup c$ then $a = b$ or $a = c$.

Lemma. If a is a join-irreducible element of a distributive lattice $\langle \underline{M}, \cap, \cup \rangle$, then whenever $a \leq b \cup c$, $a \leq b$ or $a \leq c$.

Proof. Suppose $a \leq b \cup c$, then

$$\begin{aligned} a &= (b \cup c) \cap a \\ &= (b \cap a) \cup (c \cap a) , \text{ by distributivity.} \end{aligned}$$

Therefore, $a = b \cap a$ or $a = c \cap a$, since a is join-irreducible. Hence $a \leq b$ or $a \leq c$.

<u>Theorem</u>. If $\langle \underline{M}, \cap, \cup \rangle$ is a finite Heyting lattice, then there is an isomorphism ϑ from $\langle \underline{M}, \cap, \cup \rangle$ to the lattice $\langle \underline{\mathcal{O}}, \cap, \cup \rangle$ of open sets of the PO-space on the set $\underline{J}$ of join-irreducible elements of $\langle \underline{M}\ \cap, \cup \rangle$ under the restriction of the lattice ordering $\leq$ to $\underline{J}$. Moreover, for all $a, b \in \underline{M}$, $\vartheta(a \Rightarrow b) = \vartheta(a) \Rightarrow \vartheta(b)$.

Proof. Define ϑ on $\underline{M}$ by:

$$\vartheta(a) = \{d \in \underline{J} \mid d \leq a\} .$$

ϑ is clearly well-defined, and $\vartheta(a) \in \underline{\mathcal{O}}$ since, if $c \in \vartheta(a)$, $d \in \underline{J}$ and $d \leq c$, then $d \leq a$, so $d \in \vartheta(a)$. It is obvious that $\vartheta(a \cap b) = \vartheta(a) \cap \vartheta(b)$ and $\vartheta(a \cup b) \supseteq \vartheta(a) \cup \vartheta(b)$. So, to show ϑ is a morphism with respect to $\cap$ and $\cup$, it only remains to show $\vartheta(a \cup b) \subseteq \vartheta(a) \cup \vartheta(b)$. Suppose $d \in \vartheta(a \cup b)$. Then $d \leq a \cup b$ and d is join-irreducible, so, by the lemma, $d \leq a$ or $d \leq b$. Therefore $d \in \vartheta(a) \cup \vartheta(b)$. It is left as an exercise to show that $\vartheta(a \Rightarrow b) = \vartheta(a) \Rightarrow \vartheta(b)$.

To show ϑ maps $\underline{M}$ on to $\underline{\mathcal{O}}$, consider any $\underline{D} \in \underline{\mathcal{O}}$. Since $\underline{J}$ is finite, we can let $\underline{D} = \{d_1, \ldots, d_n\}$. Let $d = d_1 \cup \ldots \cup d_n$; then, for $1 \leq i \leq n$, $d_i \leq d$ and d_i is join-irreducible. So $d_i \in \vartheta(d)$, whence $\underline{D} \subseteq \vartheta(d)$. Conversely, if $c \in \vartheta(d)$, then c is join-irreducible and $c \leq d_1 \cup \ldots \cup d_n$. Therefore, by the lemma, $c \leq d_i$ for some i, $1 \leq i \leq n$. But $d_i \in \underline{D}$ and $\underline{D}$ is open; so $c \in \underline{D}$, whence $\vartheta(d) \subseteq \underline{D}$. Thus, for each open set $\underline{D}$, there exists $d \in \underline{M}$ such that $\vartheta(d) = \underline{D}$.

To prove ϑ is one-one, we show that, for every $a \in \underline{M}$, $a = \bigcup\vartheta(a)$ where $\bigcup\vartheta(a)$ is the join, i.e. the least upper bound, of all the elements of $\vartheta(a)$, and $\bigcup\emptyset = 0$. Define the *level* of a in $\langle \underline{M}, \cap, \cup \rangle$, $\ell(a)$, as the least number of elements on any path leading from a to 0. We proceed by induction. If $\ell(a) = 0$ then $a = 0$ and $\vartheta(a) = \emptyset$; so $a = \bigcup\vartheta(a)$. Suppose

$\ell(a) \geq 1$, and assume, as induction hypothesis, that for all b with $\ell(b) < \ell(a)$, $b = U\vartheta(b)$. If $a \in \underline{J}$, then $a \in \vartheta(a)$, so $a = U\vartheta(a)$. If $a \notin \underline{J}$, then $a = b \cup c$ for some b, c such that $b < a$ and $c < a$. Then $\ell(b) < \ell(a)$ and $\ell(c) < \ell(a)$, so, by the induction hypothesis, $b = U\vartheta(b)$ and $c = U\vartheta(c)$. Therefore,

$$\begin{aligned} U\vartheta(a) = U\vartheta(b \cup c) &= U(\vartheta(b) \cup \vartheta(c)) \\ &= U\vartheta(b) \cup U\vartheta(c) \\ &= b \cup c = a. \end{aligned}$$

Hence, for all a, $a = U\vartheta(a)$. It follows immediately that ϑ is one-one, since $\vartheta(a) = \vartheta(b)$ implies $U\vartheta(a) = U\vartheta(b)$.

This concludes the proof that ϑ is a lattice isomorphism.

Example. Consider the finite Heyting lattice $\langle \underline{M}, \cap, \cup \rangle$:

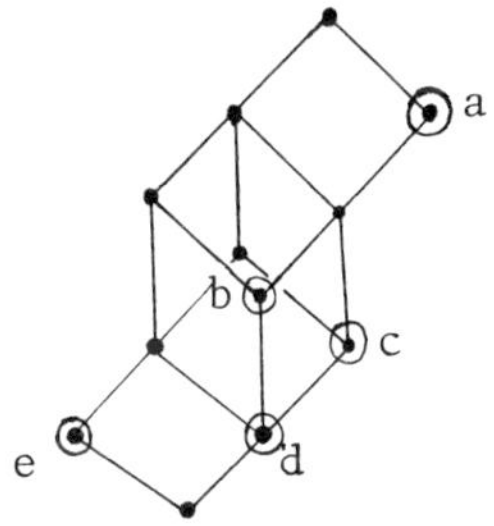

The join-irreducible elements of $\underline{M}$ are circled and form the partially ordered set $\langle \underline{J}, \leq \rangle$:

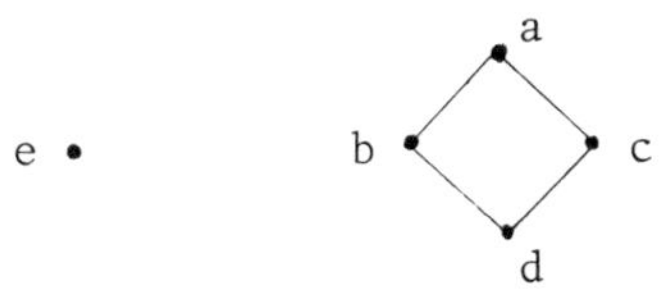

The open subsets of J are thus:

∅, {e}, {d}, {b,d}, {c,d}, {b,c,d}, {a,b,c,d}, {d,e}, {b,d,e}, {c,d,e}, {b,c,d,e}, and J.

The theorem now states that the lattice generated by these open sets is isomorphic to ⟨M, ∩, ∪⟩.

Given a PO-space ⟨S, 𝔍⟩ on a finite partially ordered set ⟨S, ≤⟩ and an assignment ϕ relative to the valuation system associated with the lattice of open sets, we can regard the points of S as states of information and say, as in section 5.1, that for each a ∈ S and sentence-letter p,

p is *true at* a iff a ∈ ϕ(p).

Given any set of formulas, the sentence-letters occurring in them represent unanalysed constituent statements: we are considering states of information only in so far as they bear on the verification of these constituent statements. A state of information consists in a knowledge of two things: which of the constituent statements have been verified; and what future states of information are possible. That the constituent statement represented by a sentence-letter p has been verified in the state of information represented by a point a is itself represented by the fact that a ∈ ϕ(p). That the state of information represented by a may subsequently be improved upon by achieving the state represented by a point b is represented by the fact that b ≤ a. Note that there is no assumption that, at any point, we shall actually every acquire more information: if our present information is represented by a, there is no guarantee that we shall ever advance to a state represented by a point b < a.

The requirement that ϕ(p) be an open set is the requirement that, for all a and b,

if b ≤ a and a ∈ ϕ(p), then b ∈ ϕ(p).

This requirement corresponds intuitively to the assumption that, once a constituent statement has been verified, it remains verified; i.e. that we do not forget what we have verified.

The extension of the assignment ϕ to a valuation ν_ϕ may now be interpreted as supplying an inductively defined sense for saying that a complex statement, represented by a formula A, has been verified in a state of information represented by a point a. This is represented by the fact that $a \in \nu_\phi(A)$, which we also express by saying that A is *true at* a under ϕ. The recursive conditions under which this holds are determined by the definition of ν_ϕ as follows:

A & B is true at a under ϕ iff $a \in \nu_\phi(A \mathbin{\&} B)$
iff $a \in \nu_\phi(A) \cap \nu_\phi(B)$
iff $a \in \nu_\phi(A)$ and $a \in \nu_\phi(B)$
iff A is true at a under ϕ and B is true at a under ϕ.

$A \vee B$ is true at a under ϕ iff $a \in \nu_\phi(A \vee B)$
iff $a \in \nu_\phi(A) \cup \nu_\phi(B)$
iff $a \in \nu_\phi(A)$ or $a \in \nu_\phi(B)$
iff A is true at a under ϕ or B is true at a under ϕ.

$A \rightarrow B$ is true at a under ϕ iff $a \in \nu_\phi(A \rightarrow B)$
iff $a \in \mathcal{I}\{c \mid \text{if } c \in \nu_\phi(A), \text{ then } c \in \nu_\phi(B)\}$
iff for all $b \leq a$, if $b \in \nu_\phi(A)$, then $b \in \nu_\phi(B)$
iff for all $b \leq a$, if A is true at b under ϕ, then B is true at b under ϕ.

$\neg A$ is true at a under ϕ iff $a \in \nu_\phi(\neg A)$
iff $a \in \mathcal{I}(-\nu_\phi(A))$
iff for all $b \leq a$, $b \notin \nu_\phi(A)$
iff for all $b \leq a$, A is not true under ϕ.

Note that, provided that the assignment ϕ is taken to be effective, i.e. each statement of the form $a \in \phi(p)$ (p is true at a under ϕ) is taken to be decidable, then every statement of the form $a \in \nu_\phi(A)$ (A is true at a under ϕ) is also decidable, since $\langle \underline{S}, \leq \rangle$ is finite.

We now complete the proof that there is a family of finite valuation systems which is finitely strictly characteristic for IC. From the representation theorem together with the fact that the family of finite Heyting lattices is weakly characteristic for IC, we can immediately derive:

__Theorem.__ The family $\mathbb{F}$ of valuation systems associated with the topological Heyting lattices generated by finite PO-spaces is weakly characteristic for IC.

It remains only to show that, for any formula A and finite set of formulas Γ, if $\Gamma \models_{\mathbb{F}} A$, then $\Gamma \vdash_{IC} A$. Since it is decidable whether or not $\Gamma \vdash_{IC} A$, we can argue by assuming $\Gamma \models_{\mathbb{F}} A$ and $\Gamma \nvdash_{IC} A$. Suppose Γ is the set $\{C_1,\dots,C_k\}$ and let C be the conjunction $C_1 \,\&\, \dots \,\&\, C_k$.

Then

$$\nvdash_{IC} C \to A .$$

Since $\mathbb{F}$ is weakly characteristic, this means we can find $\mathfrak{M} \in \mathbb{F}$ such that

$$\not\models_{\mathfrak{M}} C \to A .$$

So there is some assignment ϕ relative to $\mathfrak{M}$ such that, for some point a of the PO-space $\langle \underline{J}, \mathcal{J} \rangle$ underlying $\mathfrak{M}$,

$$C \to A \text{ is not true at a under } \phi.$$

But $C \to A$ is true at a iff for every $b \leq a$, if C is true at b then A is true at b. So, since ϕ is effective and $\underline{J}$ is finite, we can find a point $b \leq a$ at which C is true but A is not. Let $\underline{J}_o$ be $\{c \in \underline{J} \mid c \leq b\}$ and $\leq_o$ be the restriction of $\leq$ to $\underline{J}_o$. Let $\mathfrak{M}_o$ be the valuation system associated with the lattice of open sets of the PO-space on $\langle \underline{J}_o, \leq_o \rangle$, and define an assignment ϕ_o relative to $\mathfrak{M}_o$ by

$$\phi_o(p) = \phi(p) \cap \underline{J}_o .$$

This ensures that ϕ_o does map all sentence-letters on to open sets of the PO-space and, as is easily seen, that, for any formula B,

$$b \in \nu_{\phi_o}(B) \text{ iff } b \in \nu_{\phi}(B).$$

Hence, $b \in \nu_{\phi_o}(C)$ and $b \notin \nu_{\phi_o}(A)$; and so $\nu_{\phi_o}(C_i) = \underline{J}_o$ for $1 \leq i \leq k$, while $\nu_{\phi_o}(A) \neq \underline{J}_o$. Therefore $\Gamma \not\models_{\mathcal{M}_o} A$, and $\Gamma \not\models_{\mathbb{F}} A$.

We can thus state:

Theorem. The family $\mathbb{F}$ of the preceding theorem is finitely strictly characteristic for IC.

Notice that, by the above argument, if $\not\vdash_{IC} B \to A$, we can find $\mathcal{M} \in \mathbb{F}$ and an assignment ϕ relative to $\mathcal{M}$ such that $\nu_{\phi}(B)$ is designated and $\nu_{\phi}(A)$ is not. This will always be so when, as in our case, $\mathbb{F}$ is a decidable family of finite valuation systems which is finitely strictly characteristic for a logic $\mathcal{L}$, provided that $\mathcal{L}$ satisfies the Deduction Theorem, i.e. that, if $\Gamma \cup \{B\} \vdash_{\mathcal{L}} A$, then $\Gamma \vdash_{\mathcal{L}} B \to A$; for then if $\not\vdash_{\mathcal{L}} B \to A$, also $\{B\} \not\vdash_{\mathcal{L}} A$.

Kripke Trees. Instead of considering the valuation systems obtained from any finite PO-spaces we can, in fact, restrict ourselves to the family $\mathbb{K}$ consisting of those valuation systems associated with the topological Heyting lattices generated by PO-spaces on finite trees. The valuation system associated with the topological Heyting lattice generated by a PO-space $\langle \underline{S}, \mathcal{J} \rangle$ such that $\langle \underline{S}, \leq \rangle$ is a tree is called a *Kripke tree*: $\mathbb{K}$ is thus the family of all finite Kripke trees. (A *tree* is a partially ordered set $\langle \underline{T}, \leq \rangle$ with a maximal element $a \in \underline{T}$ such that, for any $b \in \underline{T}$, there is a unique finite chain $c_o, \ldots, c_{k-1}$ such that $c_o << c_1 << \ldots << c_{k-1}$, $b = c_o$ and $c_{k-1} = a$, where $b << d$ iff $b < d \;\&\; \neg\exists c \;\; b < c < d$.) Clearly, since $\mathbb{K}$ is a subfamily of $\mathbb{F}$, $\Gamma \models_{\mathbb{F}} A$ implies $\Gamma \models_{\mathbb{K}} A$. To prove the converse implication, we must show that every situation which can be represented on a finite PO-space under the intuitive interpretation can be equally well represented on a

PO-space obtained from a finite tree.

If $\langle \underline{J}, \leq \rangle$ is any partially ordered set, define $\langle \underline{J}^+, \leq^+ \rangle$ by:

$$\underline{J}^+ = \{\langle a_0, \ldots, a_k \rangle \mid k \geq 0 \text{ and, for } 0 \leq i \leq k,\ a_i \in \underline{J} \text{ and } a_i \gg a_{i+1}, \text{ and } a_0 \text{ is maximal}\},$$

where a >> b iff b << a; and

$$\langle a_0, \ldots, a_n \rangle \leq^+ \langle b_0, \ldots, b_m \rangle \text{ iff } m \leq n \text{ and } a_i = b_i \text{ for } 0 \leq i \leq m.$$

Thus $\underline{J}^+$ is just the set of initial segments of paths from maximal elements of $\underline{J}$. For each maximal element, the set so defined obviously forms a tree; so $\langle \underline{J}^+, \leq^+ \rangle$ is either a tree or a finite disjoint union of trees.

Example. Consider the partially ordered set $\langle \underline{J}, \leq \rangle$:

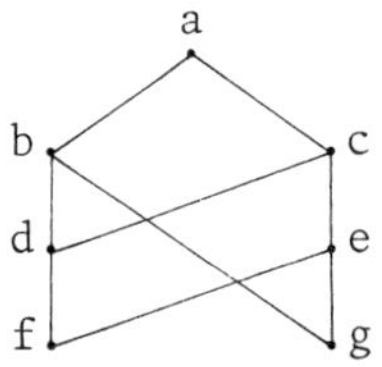

Since $\langle \underline{J}, \leq \rangle$ has only one maximal element, $\langle \underline{J}^+, \leq^+ \rangle$ is a tree:

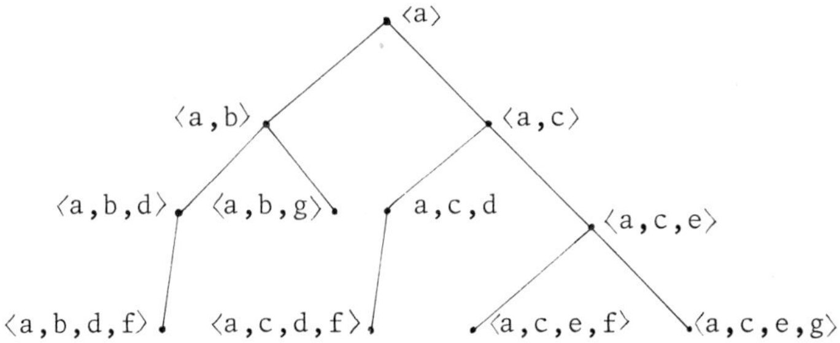

Suppose given a valuation system $\mathcal{M}$ with underlying PO-space $\langle \underline{J}, \mathcal{J} \rangle$ on $\langle \underline{J}, \leq \rangle$. Let A be a formula and Γ a set of formulas. Then given any assignment ϕ relative to $\mathcal{M}$, there is a natural corresponding assignment ϕ^+ relative to $\mathcal{M}^+$, the valuation system obtained from $\langle \underline{J}^+, \leq^+ \rangle$, such that A is true at every point of $\underline{J}$ under ϕ iff A is true at every point of $\underline{J}^+$ under ϕ^+. This is expressed by the following:

Theorem. Let $\langle \underline{\mathcal{O}}, \cap, \cup \rangle, \langle \underline{\mathcal{O}}^+, \cap, \cup \rangle$ be the topological Heyting lattices on $\langle \underline{J}, \leq \rangle$ and $\langle \underline{J}^+, \leq^+ \rangle$ respectively. Then $\langle \underline{\mathcal{O}}, \cap, \cup \rangle$ is isomorphic to a sublattice of $\langle \underline{\mathcal{O}}^+, \cap, \cup \rangle$.

Proof For $\underline{D} \in \underline{\mathcal{O}}$, define ψ by

$$\psi(\underline{D}) = \{\langle a_o, \ldots, a_k \rangle \in \underline{\mathcal{O}}^+ \mid a_k \in \underline{D}\} .$$

ψ is easily seen to be an isomorphism. The assignment ϕ^+ corresponding to ϕ is just the composite map $\psi\phi$. It follows that if $\Gamma \nvdash_{IC} A$, i.e. $\Gamma \nvDash_{\mathbb{IF}} A$, then for some $\underline{J}$ and some assignment ϕ, every formula of Γ is true at every point of $\underline{J}^+$ under ϕ^+, but A is not. If $\underline{J}^+$ is a tree, then we have shown $\Gamma \nvDash_{\mathbb{IK}} A$. However, as remarked, $\underline{J}^+$ will in general be a disjoint union of trees. Suppose $\underline{J}^+ = \underline{J}_1 \cup \ldots \cup \underline{J}_k$, where $\langle \underline{J}_i, \leq_i \rangle$ is a tree, $\leq_i$ being the restriction of $\leq^+$ to $\underline{J}_i$, for $1 \leq i \leq k$. Since ϕ^+ is effective and A is not true at every point of $\underline{J}^+$ under ϕ^+, we can find i such that A is not true at every point of $\underline{J}_i$ under ϕ^+, although all of Γ is. But then ϕ', defined by $\phi'(p) = \phi^+(p) \cap \underline{J}_i$, is an assignment, relative to the valuation system associated with the topological Heyting lattice on $\langle \underline{J}_i, \leq_i \rangle$, such that, for each $B \in \Gamma$, $\nu_{\phi'}(B) = \underline{J}_i$, but $\nu_{\phi'}(A) \neq \underline{J}_i$. Hence $\Gamma \nvDash_{\mathbb{IK}} A$. We have now shown that $\Gamma \vDash_{\mathbb{IK}} A$ iff $\Gamma \vDash_{\mathbb{IF}} A$, from which it follows that $\mathbb{IK}$ is finitely strictly characteristic for IC.

We have thus attained a semantic completeness proof for intuitionistic sentential logic with respect to Kripke trees. It is to be noted that this completeness proof is itself intuitionistically valid. This results from the fact that we established decidability of the property of provability and the relation of derivability for IC, and, further, that we have been able to confine our attention to finite structures. The extent to which the intuitive interpretation of Kripke trees approximates the intended meanings of the logical constants will be discussed later.

The use of finite distributive lattices to yield

valuation systems faithful to IC was initiated by Jaśkowski, who proved, for a particular infinite sequence of such lattices, that the family of valuation systems associated with them was weakly characteristic for IC. The sequence of lattices is that generated by the following sequence of trees, considered as PO-spaces:

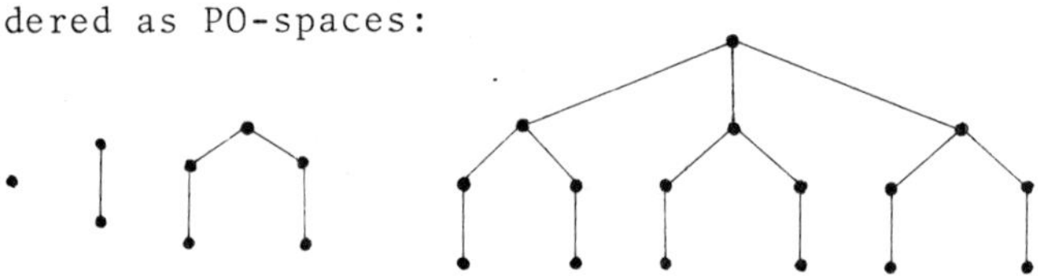
......

The corresponding lattices are:

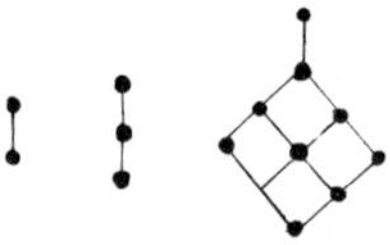
......

Jaśkowski did not characterize his sequence of lattices in terms of the trees, nor did he consider the topological representation of Heyting lattices: his proof involves quite different ideas from those we have been considering. However, to show that the family is weakly characteristic for IC, it is sufficient, in the light of what we have shown, to prove that every finite tree can be embedded in a tree in the Jaśkowski sequence.

A modified version of this family of valuation systems is the family of valuation systems associated with the lattices generated by the sequence of trees:

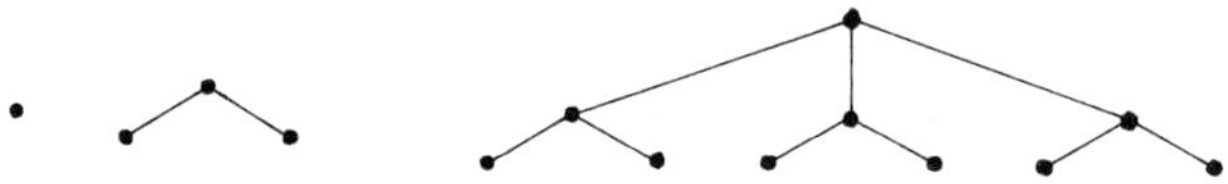
......

The sequences of lattices so generated begins:

......

Finally, we exhibit a single PO-space which generates a topological Heyting lattice, the valuation system associated with which is strictly characteristic for IC. By Godel's Theorem in section 5.1, this lattice is necessarily infinite. This PO-space is not properly speaking a tree, since it has no vertex: however, given any finite tree, it is easy to show that the topological Heyting lattice which it generates is a sublattice of the lattice on this PO-space. The details are left as an exercise.

We take the points of the space to be all ordered pairs of natural numbers, and put:

$\langle n,m\rangle \leq \langle i,j\rangle$ iff $n \leq i$ and $(i+1)!j \leq (n+1)!m < (i+1)!(j+1)$.

A fragment of the partially ordered set so obtained looks like this:

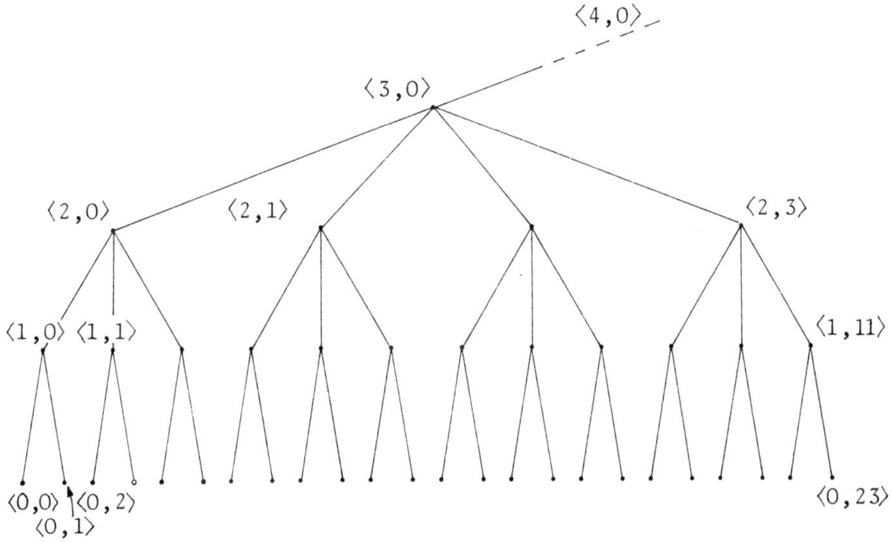

Note that, for this PO-space also, if Γ is finite and $\Gamma \nvdash_{IC} A$, we can effectively find an assignment ϕ such that, for every $B \in \Gamma$, $\nu_\phi(B)$ is the whole space, while $\nu_\phi(A)$ is not.

5.4. BETH TREES

The first person to make a connection between the topological interpretation of intuitionistic logic and the intended meanings of the logical constants was Beth. We have again to consider interpretations of formulas relative to trees, called Beth trees, which can, like the Kripke trees, be viewed as generating topological Heyting lattices, although we cannot, in this case, restrict consideration to finite ones. We shall, however, approach the subject in the reverse direction, describing first the intuitive interpretation, and then the topological representation of it.

A *Beth tree* $\langle \underline{T}, \leq \rangle$ will not, in general, be finite; although for most purposes we can confine our attention to finitary trees (those with only finitely many nodes immediately below any given node), we do not impose this restriction. In order to interpret a formula on a given Beth tree $\langle \underline{T}, \leq \rangle$, we have again to specify a relation between sentence-letters and nodes of $\underline{T}$, which will be extended by definition to a relation between formulas and nodes. We shall not, as before, speak of an 'assignment' to the sentence-letters and a 'valuation' of the formulas, since, in this case, to do so would cause confusion when we come to view the Beth trees as topological spaces. Rather, we shall speak of an *interpretation* of the sentence-letters on the tree, and of a formula's being true at a node under such an interpretation.

There are two possible approaches. On the first approach, we take an interpretation χ on a tree $\langle \underline{T}, \leq \rangle$ as consisting of an association of each sentence-letter p with a set $\chi(p)$ of nodes of $\underline{T}$; we write

p is *true* (*in* $\langle \underline{T}, \leq \rangle$) *at* a (*under* χ) iff $a \in \chi(p)$

(the phrases in brackets will be omitted when there is no ambiguity), and require χ to satisfy

(i_a) if $b \leq a$ and p is true at a, then p is true at b;

(ii_a) if $\underline{S}$ is a set of nodes which bars a, and, for every $b \in \underline{S}$, p is true at b, then p is true at a.

We may or may not also require that $\chi(p)$ be decidable, i.e.

that

(iii_a) p is true at a or p is not true at a.

Intuitively, we take the nodes, as before, as representing states of information. However, in the case of the Beth trees, we must make the assumption that there is necessarily an advance from each non-terminal node to one immediately below it. We suppose time to be divided into successive intervals which we call 'days': if a given non-terminal node a represents our state of information on a certain day, the set $\{b \mid b \ll a\}$ represents all the various possible states to which we can advance by the next day, and to one of which we shall advance. A terminal node represents a state of information which cannot be improved.

On this first approach, the truth of a sentence-letter at a node represents our having, in the corresponding state of information, verified the corresponding atomic statement (which may be atomic only relatively to the statements being considered). Condition (i_a) thus embodies the requirement that we do not forget which of these atomic statements we have verified. Condition (ii_a) embodies the principle that if we are in a position to recognize that a given atomic statement will be verified within a finite time, then we may regard it as already verified. Condition (iii_a) says, of course, that we know whether we have verified an atomic statement or not.

We can now extend the interpretation to all formulas by an inductive stipulation which tallies, save for the clause relating to $\vee$, with the way we extended an assignment to a valuation in the case of a PO-space (or Kripke tree), as follows:

A & B is true at a iff A is true at a and B is true at a

A $\vee$ B is true at a iff for some $\underline{S} \subseteq \underline{T}$, $\underline{S}$ bars a and, for every $b \in \underline{S}$, A is true at b or B is true at b

A $\rightarrow$ B is true at a iff for every $b \leq a$, if A is true at b, then B is true at b

$\neg$A is true at a iff for every b $\leq$ a, A is not true at b.

(Where it is necessary to avoid ambiguity, we must qualify 'A is true at a' by 'in $\langle \underline{T}, \leq \rangle$' or 'under X'.) It is easily established that we have:

(i) if b $\leq$ a and A is true at a, then A is true at b;

(ii) if $\underline{S}$ bars a, and, for every b $\in$ $\underline{S}$, A is true at b, then A is true at a.

The clause for $\vee$ shows it to be useless to consider Beth trees all paths of which are finite, since A $\vee$ $\neg$A is automatically true at the vertex of any such tree under any interpretation. Hence, even if (iii_a) was satisfied, the relation expressed by 'A is true at a' is not in general decidable, since a Beth tree will usually be infinite.

On the second approach, we read 'a $\in$ X(p)', not as 'p is true at a', but as 'p is *verified at* a', and require:

(i_b) if b $\leq$ a and p is verified at a, then p is verified at b;

(iii_b) p is verified at a or p is not verified at a,

without any requirement corresponding to (ii_a). We now give an inductive definition, for all formulas, of 'A is true at a', by taking as the basis clause:

p is *true at* a iff for some $\underline{S} \subseteq \underline{T}$, $\underline{S}$ bars a and, for every b $\in$ $\underline{S}$, p is verified at b,

and using the same inductive clauses as on the first approach. On this definition, we can again establish (i) and (ii); but truth at a node will not in general be decidable even for sentence-letters, i.e. we cannot establish (iii_a).

On this approach, we are distinguishing between the *verification* of an atomic statement in a given state of information, and its being *assertible*; the latter notion is represented by truth at a node, and is defined, for all statements, in terms of the verification of atomic statements. The knowledge that a given atomic statement will be verified within a finite time does not itself constitute a verification of it, but is sufficient ground to entitle us to assert it.

There is an effective means of constructing a Beth tree $\langle \underline{T}, \leq_{\underline{T}} \rangle$ from a given Kripke tree $\langle \underline{J}, \leq_{\underline{J}} \rangle$ so that any formula A is true at the vertex of $\underline{J}$ under some assignment iff A is true at the vertex of $\underline{T}$ under a corresponding interpretation. Take as the elements of $\underline{T}$ (its nodes) all non-empty finite sequences $\langle a_o, \ldots, a_k \rangle$ of elements of $\underline{J}$ such that a_o is the vertex of $\langle \underline{J}, \leq_{\underline{J}} \rangle$ and, for $0 \leq i < k$, a_i is non-terminal and either $a_i = a_{i+1}$ or $a_i \gg_{\underline{J}} a_{i+1}$. We define $\leq_{\underline{T}}$ to be the relation $\preccurlyeq$ on finite sequences (the relation of being an extension). An assignment ϕ relative to the valuation system on the Kripke tree is converted into an interpretation χ on the Beth tree by setting:

p is true (verified) at $\langle a_o, \ldots, a_k \rangle$ under χ iff p is true at a_k under ϕ.

If we construe χ in accordance with the first approach, it is straightfoward to show that (i_a) and (ii_a) are satisfied, and that χ is therefore genuinely an interpretation; provided that the relation expressed by 'p is true at a under ϕ' is decidable, (iii_a) is satisfied also, and hence we could just as well construe χ in accordance with the second approach. We can easily see also that, if $\langle \underline{J}, \leq_{\underline{J}} \rangle$ is finitary, so is $\langle \underline{T}, \leq_{\underline{T}} \rangle$. Moreover, it is apparent that the interpretation does not fully exploit the clause for $\vee$ in the definition of truth at a node: $A \vee B$ is true at a node of $\underline{T}$ under just in case either A or B is true at that node. It follows that, for any A and any $\langle a_o, \ldots, a_k \rangle \in \underline{T}$, A is true at $\langle a_o, \ldots, a_k \rangle$ under χ iff A is true at a_k under ϕ; and, in particular, A is true at the vertex a_o of $\underline{T}$ under χ iff A is true at the vertex a_o of $\underline{J}$ under ϕ.

Examples.

Kripke tree | Beth tree

(1)

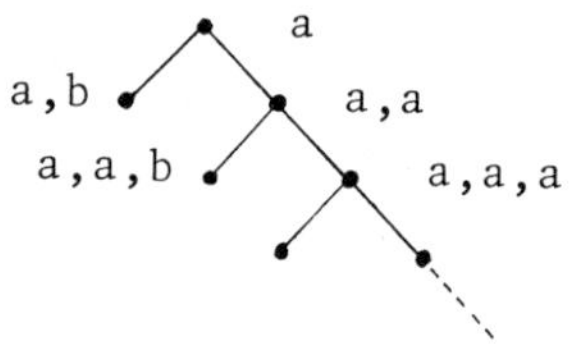

(2)

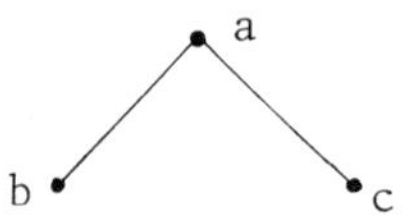

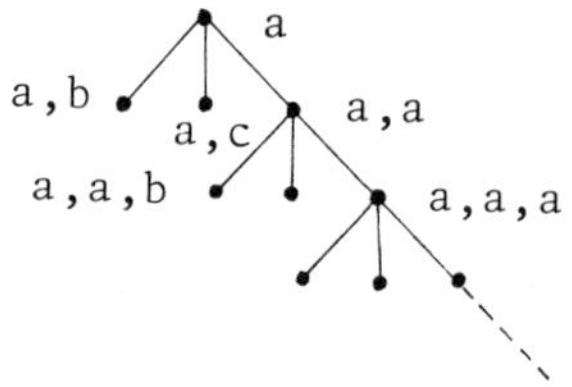

This construction enables us to assert that, for any Γ and A, if there is a Kripke tree in which, under some assignment, all of Γ is true at the vertex, but A is not, then there is a Beth tree of which the same holds. Let $\mathbb{B}$ be the family of Beth trees. Then, taking '$\Gamma \models_{\mathbb{B}} A$' to mean that, for every Beth tree, whenever all of Γ is true at the vertex, A is also true there, we can assert:

> if $\Gamma \models_{\mathbb{B}} A$, then $\Gamma \models_{\mathbb{IF}} A$; and so, provided that Γ is finite, $\Gamma \vdash_{IC} A$.

We have not yet shown that $\mathbb{B}$ is faithful to IC. This is done by representing the Beth trees as topological spaces and relying on our theorem in section 5.2 that every valuation system associated with a Heyting lattice is faithful to IC.

Let $\langle \underline{T}, \leq \rangle$ be a Beth tree. In carrying out this representation, it is plain that we cannot take the points of the space to be the nodes of the tree, at least if assignments are extended to valuations in the standard way for topological Heyting lattices: for then we should always have, as in the Kripke trees,

> $A \vee B$ is true at a under χ iff A is true at a under χ or b is true at a under χ,

which does not hold for Beth trees. Instead, we take the points of the space $\underline{S}$ to be the paths in the tree. The open subsets of $\underline{S}$ are to be all sets of the form

$$\{\alpha \mid \exists a_{a \in \underline{U}}\ \alpha \in a\},$$

for any subset $\underline{U}$ of $\underline{T}$. We will denote the open set associated with $\underline{U}$ by $[\underline{U}]$. In terms of the interior operator, if $\underline{A}$ is a subset of $\underline{S}$, then the topology is determined by:

$$\mathfrak{I}\underline{A} = [\{a \mid \forall\alpha_{\alpha \in a}\ \alpha \in \underline{A}\}] = \{\alpha \mid \exists a_{\alpha \in a}\ \forall\beta_{\beta \in a}\ \beta \in \underline{A}\} .$$

(Here, of course, '$\alpha \in a$' means that the path α goes through the node a, or, synonymously, that a is on α.)

Now suppose given an interpretation χ with respect to the Beth tree $\langle \underline{T}, \leq \rangle$, and let ν_χ map each formula A on to the set $\{a \mid A \text{ is true at } a \text{ under } \chi\}$. (For simplicity, we construe χ in accordance with the first approach.) Let $\mathfrak{M}$ be the valuation system associated with the topological Heyting lattice generated by the topology defined above on $\underline{S}$; then we define an assignment ϕ relative to $\mathfrak{M}$ by:

$$\phi(p) = [\chi(p)] .$$

We then claim that for every formula A:

$$\nu_\phi(A) = [\nu_\chi(A)],$$

where ν_ϕ is the standard extension of ϕ. The proof is by induction on the number of connectives occurring in A; the basis for the induction is given by the definition of ϕ. Assuming the claim for all formulas shorter than A, we argue the cases as follows:

Case 1. A is B & C. Then $\nu_\phi(A) = \nu_\phi(B) \cap \nu_\phi(C)$,

and $\nu_\chi(A) = \nu_\chi(B) \cap \nu_\chi(C)$.

Clearly $[\nu_\chi(A)] \subseteq [\nu_\chi(B)] \cap [\nu_\chi(C)]$.

Also, using condition (i), it is easy to see that

$$[\nu_\chi(A)] \supseteq [\nu_\chi(B)] \cap [\nu_\chi(C)] .$$

Hence $[\nu_\chi(A)] = [\nu_\chi(B)] \cap [\nu_\chi(C)]$
$= \nu_\phi(A)$, by induction hypothesis.

Case 2. A is $B \vee C$. Then $\nu_\phi(A) = \nu_\phi(B) \cup \nu_\phi(C)$,

and $\nu_\chi(A) = \{a \mid \nu_\chi(B) \cup \nu_\chi(C) \text{ bars } a\}$.

$$\text{So } [\nu_\chi(A)] = \{\alpha \mid \exists a_{\alpha \in a}\ \nu_\chi(B) \cup \nu_\chi(C) \text{ bars } a\}$$
$$= \{\alpha \mid \exists b_{\alpha \in b}\ b \in \nu_\chi(B) \cup \nu_\chi(C)\}$$
$$= [\nu_\chi(B)] \cup [\nu_\chi(C)]$$
$$= \nu_\phi(A), \text{ by induction hypothesis.}$$

Case 3. A is $B \to C$.

$$\text{Then } \nu_\phi(A) = \mathfrak{I}\{\beta \mid \beta \in \nu_\phi(B) \to \beta \in \nu_\phi(C)\}$$
$$= \{\alpha \mid \exists a_{\alpha \in a} \forall \beta_{\beta \in a}\ (\beta \in \nu_\phi(B) \to \beta \in \nu_\phi(C))\}$$

and $\nu_\chi(A) = \{a \mid \forall b_{b \leq a}(b \in \nu_\chi(B) \to b \in \nu_\chi(C))\}$.

Hence $[\nu_\chi(A)] = \{\alpha \mid \exists a_{\alpha \in a}\ \forall b_{b \leq a}(b \in \nu_\chi(B) \to b \in \nu_\chi(C))\}$.

In order to show that $\nu_\phi(A) = [\nu_\chi(A)]$, therefore, it is enough to show the equivalence of the conditions:

For every β through a, if $\beta \in \nu_\phi(B)$, then $\beta \in \nu_\phi(C)$

and

For every $b \leq a$, if $b \in \nu_\chi(B)$, then $b \in \nu_\chi(C)$.

By the induction hypothesis, the former condition is equivalent to:

For every β through a, if $\beta \in [\nu_\chi(B)]$, then $\beta \in [\nu_\chi(C)]$,

i.e. to:

For every β through a, if, for some b on β, $b \in \nu_\chi(B)$, then, for some c on β, $c \in \nu_\chi(C)$.

But this is equivalent to:

For every $b \leq a$, if $b \in \nu_\chi(B)$, then $\nu_\chi(C)$ bars b.

By condition (ii), however, $\nu_\chi(C)$ bars b iff $b \in \nu_\chi(C)$, whence the condition is equivalent to:

For every $b \leq a$, if $b \in \nu_\chi(B)$, then $b \in \nu_\chi(C)$,

as claimed.

Case 4. A is $\neg B$. This case is similar to case 3, and is left as an exercise.

It follows that $\nu_{\phi}(A) = \underline{S}$ iff $\nu_{\chi}(A) = \underline{T}$, and so the Beth trees are faithful to IC, since any valuation system associated with a Heyting lattice is faithful to IC.

5.5. THE SEMANTICS OF INTUITIONISTIC PREDICATE LOGIC

We have established the completeness (in relation to finite sets of formulas) of intuitionistic sentential logic with respect to Kripke trees and therefore to Beth trees. It should be noted that our argument was entirely constructive, i.e. intuitionistically valid, because of the decidability of intuitionistic sentential logic. Given an intuitionistically invalid formula B, we can effectively construct the finite sublattice of the Lindenbaum algebra whose elements consist of the 0 and 1 of the algebra together with the equivalence classes of the subformulas of B; this will then constitute a finite Heyting lattice on which B assumes a value < 1 when each sentence-letter p of B is given the value $|p|$ (the equivalence class of the formula p). The procedures for finding from this a finite PO-space, a finite Kripke tree, and a Beth tree, on each of which B is invalid, are then completely effective. In relation to predicate logic, however, the matter is not so straightforward. In the discussion of sentential logic, it is hoped that, not only the completeness proof itself, but all the definitions and arguments have been intuitionistically acceptable (witness the form adopted for the definition of $\Rightarrow$ in the topological interpretation, i.e. for topological Heyting lattices). Terminologically, however, we retained the classical term 'set' in place of the intuitionistic 'species', because, while we were aiming to avoid non-constructive arguments, we were not attempting to present an intuitionistic theory of valuation systems or of lattices. In what follows, on the other hand, we shall need to take greater care to avoid non-intuitionistic reasoning, and hence we shall use the intuitionistic notion of a species rather than the classical notion of a set, save that we shall continue to speak of sets of formulas, and of subsets of a topological space.

In order to inquire into the completeness of intuitionistic predicate logic, we have first to extend our conception of a valuation system to that of a Q-valuation system, relative to which we can define a valuation of a formula of predicate logic. (For simplicity, we shall ignore function-symbols.) A Q-valuation system requires not only a species $\underline{A}$ of values, and a subspecies $\underline{D}$ of $\underline{A}$ containing the designated values, but also a domain $\underline{R}$ for the individual variables. Since we now wish to use the word 'assignment' for a mapping of the individual variables into the domain, $\underline{R}$, we shall use the word 'interpretation' for what corresponds to what we formerly called an 'assignment'. Any particular interpretation ϕ will associate (i) with each individual constant a of the language an element $\phi(a) \in \underline{R}$, (ii) with each sentence-letter p a value $\phi(p) \in \underline{A}$, and (iii) with each n-place predicate-letter F a function $\phi(F)$ which maps n-tuples of elements of $\underline{R}$ into $\underline{A}$. An assignment θ, on the other hand, is now to be taken to be a mapping of all individual variables into $\underline{R}$. The *denotation* of a term t, relative to an interpretation ϕ and an assignment θ, is $\phi(t)$ if t is an individual constant and $\theta(t)$ if t is an individual variable, and can be written $d_{\phi,\theta}(t)$. (If we were admitting function-symbols, this notion of denotation could be extended to complex terms in an obvious way.) Then if B is an atomic formula $Ft_1 \ldots t_n$, the valuation $\nu_{\phi,\theta}(B)$ of B relative to an interpretation ϕ and assignment θ is, plainly, $\phi(F)(d_{\phi,\theta}(t_1), \ldots, d_{\phi,\theta}(t_n))$, which is, of course, an element of $\underline{A}$.

A Q-valuation system for a language with the same logical constants as those of intuitionistic logic is a system

$$\mathfrak{M} = \langle \underline{A}, \underline{D}, \underline{R}, \cap, \cup, \Rightarrow, -, \bigwedge, \bigvee \rangle,$$

where $\underline{A}$ is a species with at least two elements, $\underline{D}$ an inhabited proper subspecies of $\underline{A}$, $\underline{R}$ an inhabited species, $\cap$, $\cup$, and $\Rightarrow$ binary functions from $\underline{A}$ into $\underline{A}$, $-$ a unary function from $\underline{A}$ into $\underline{A}$, and $\bigwedge$ and $\bigvee$ infinitary functions from $\underline{A}$ into $\underline{A}$. Specifically, $\bigwedge$ and $\bigvee$ each map any subspecies of $\underline{A}$ of cardinality less than or equal to that of $\underline{R}$, on to an element of $\underline{A}$. Just as $\cap$, $\cup$, $\Rightarrow$, and $-$ serve to interpret the logical constants &, $\vee$, $\rightarrow$, and $\neg$, so $\bigwedge$ and $\bigvee$ serve to interpret the quantifiers $\forall$ and $\exists$.

Now suppose that we have succeeded, for given formulas B and C, in defining the valuations $\nu_{\phi,\theta}(B)$ and $\nu_{\phi,\theta}(C)$ of those formulas relative to any interpretation ϕ and assignment θ. Then, just as before, we set

$$\nu_{\phi,\theta}(B \,\&\, C) = \nu_{\phi,\theta}(B) \cap \nu_{\phi,\theta}(C)$$

$$\nu_{\phi,\theta}(B \vee C) = \nu_{\phi,\theta}(B) \cup \nu_{\phi,\theta}(C)$$

$$\nu_{\phi,\theta}(B \rightarrow C) = \nu_{\phi,\theta}(B) \Rightarrow \nu_{\phi,\theta}(C)$$

$$\nu_{\phi,\theta}(\neg B) = -\nu_{\phi,\theta}(B).$$

Now let x be any individual variable. Then we may consider the species $\underline{B} \subseteq \underline{A}$ given by

$$\underline{B} = \{\nu_{\phi,\theta'}(B) \mid \theta' \text{ agrees with } \theta \text{ except possibly on } x\} \quad .$$

We then set

$$\nu_{\phi,\theta}(\forall x\, B) = \bigwedge \underline{B}$$

$$\nu_{\phi,\theta}(\exists x\, B) = \bigvee \underline{B} \quad .$$

We have now successfully specified $\nu_{\phi,\theta}(B)$ for any formula B, interpretation ϕ and assignment θ, relative to a Q-valuation system $\mathfrak{M}$; if B is a closed formula, then evidently

$$\nu_{\phi,\theta}(B) = \nu_{\phi,\theta'}(B)$$

for any assignments θ and θ', and we may write simply $\nu_{\phi}(B)$. The notions of entailment with respect to a valuation system, and of a valuation system's being faithful to or strictly, weakly, or strongly characteristic for a logic, may now be extended to Q-valuation systems in the obvious way. (For simplicity, it is easiest to regard $\Gamma \models_{\mathfrak{M}} A$ as defined only when the formulas in $\Gamma \cup \{A\}$ are all closed.)

If $\mathfrak{M} = \langle \underline{A}, \{1\}, \underline{R}, \cap, \cup, \Rightarrow, -, \Lambda, V \rangle$ is a Q-valuation system faithful to IC, then, by our previous results, $\langle \underline{A}, \cap, \cup \rangle$ will be a Heyting lattice of which 1 is the unit and $\Rightarrow$ the arrow operation. Λ and V will then be the operations of infinite meet and join, for species of cardinality $\leq$ the cardinal of $\underline{R}$, which operations must, therefore, exist for this lattice. For example, $\underline{R}$ may be taken as the species $\underline{N}$ of natural numbers: in this case, the lattice must be closed under denumerable meet and join. In particular, suppose that $\langle \underline{A}, \cap, \cup \rangle$ is a topological Heyting lattice, so that $\underline{A}$ is the family of all open subsets of some space $\underline{S}$, and $\underline{D} = \{\underline{S}\}$. In this case, where $\underline{B}$ is a subfamily of $\underline{A}$, $V\ \underline{B}$ will simply be the set-theoretic union of $\underline{B}$ (i.e. the union of all the open sets belonging to $\underline{B}$), that is

$$V\ \underline{B} = \bigcup_{\underline{X} \in \underline{B}} \underline{X},$$

where the $\bigcup$ on the right-hand side is interpreted set-theoretically. An infinite intersection of open sets is not, of course, in general open, so that $\Lambda\ \underline{B}$ must be taken as the interior of the set-theoretic intersection of the members of $\underline{B}$, i.e.

$$\Lambda\ \underline{B} = \bigcap_{\underline{X} \in \underline{B}} \underline{X}.$$

Using classical arguments, we can prove the completeness of intuitionistic predicate logic with respect to Q-valuation systems generated by Heyting lattices, and, in particular, by topological Heyting lattices, where $\underline{N}$ is the domain of the variables; that is, we can show that the family of all Q-valuation systems

$$\langle \underline{A}, \{1\}, \underline{N}, \cap, \cup, \Rightarrow, -, \Lambda, V \rangle,$$

where $\langle \underline{A}, \cap, \cup \rangle$ is a topological Heyting lattice, 1 is its unit, $\Rightarrow$ its arrow operation, $-a = a \Rightarrow 0$, and $\bigwedge$ and $\bigvee$ are $\mathfrak{I}\bigcap$ and $\bigcup$, is strictly characteristic for IC. Indeed, we can find a single strictly characteristic Q-valuation system. Such results are, however, interesting primarily for those concerned to find a non-intuitionistic interpretation of intuitionistic logic. From an intuitionistic point of view, a completeness proof which uses intuitionistically invalid methods of reasoning is, at best, a curiosity; we shall therefore not pursue this topic further here.

What *cannot* be assumed, even classically, is that, if a given family of valuation systems is strictly characteristic for intuitionistic sentential logic, the corresponding family of Q-valuation systems is strictly characteristic for intuitionistic predicate logic. A counter-example is provided by the Kripke trees themselves. We take the species $\underline{N}$ of natural numbers as the domain for our variables. To simplify exposition, we add numerals as individual constants to the language, and consider only interpretations which map each numeral $\bar{n}$ on to the corresponding number n. Then, in order to specify a particular interpretation ϕ relative to a given Kripke tree, we need only say at which nodes of the tree each closed atomic formula is true, where, as before, we require that, if B is an atomic formula true at a under ϕ, and $b \leq a$, then B is true at b under ϕ. If we consider the Kripke tree as a PO-space, then the general requirement on the operation $\bigvee$, as applied within a topological space, yields the result that

$\exists x\, B(x)$ is true at a under ϕ iff, for some n, $B(\bar{n})$ is true at a under ϕ.

In a PO-space, even if infinite, the intersection of open sets is always open, so that in this case $\bigwedge$ actually corresponds to set-theoretic intersection, and we therefore likewise have

$\forall x\, B(x)$ is true at a under ϕ iff, for every n, $B(\bar{n})$ is true at a under ϕ.

Now consider the invalid formula B:

$$\forall x\ (p \vee Fx) \rightarrow (p \vee \forall x\ Fx)\ .$$

Suppose that the antecedent $\forall x\ (p \vee Fx)$ is true at some node a of a Kripke tree. Then, for each n, $p \vee F\bar{n}$ is true at a. Assuming it to be decidable whether a given sentence-letter is true at a given node, there are two cases, according as p is or is not true at a. If p is true at a, then $p \vee \forall x\ Fx$ is also true at a. If, on the other hand, p is not true at a, then since, for each n, $p \vee F\bar{n}$ is true at a, $F\bar{n}$ must be true at a for every n; it follows that $\forall x\ Fx$ is true at a, and hence, again, that $p \vee \forall x\ Fx$ is. We conclude that the invalid formula B is valid on every Kripke tree, and hence that the family of Kripke trees, considered as Q-valuation systems with the fixed domain $\underline{N}$, is not strictly characteristic for intuitionistic predicate logic. Note that this argument does *not* depend upon our restricting ourselves to finite, or even to finitary, Kripke trees, nor to the particular selection of $\underline{N}$ as the fixed domain.

This can be remedied for the Kripke trees, at the cost of changing them into something other than Q-valuation systems in the sense explained above. The remedy is not to have a fixed domain for the variables, but to associate a different inhabited domain $\underline{R}_a \subseteq \underline{N}$ with each node a, subject to the requirement that, if $a \leq b$, then $\underline{R}_b \subseteq \underline{R}_a$. The intuitive explanation of this makes perfectly good sense intuitionistically: we are taking the variables as ranging over an undecidable domain, of which it is required only that we know of at least one element belonging to it. The state of knowledge represented by a node includes knowledge about the elements of the domain: the species $\underline{R}_a$ consists of those objects of which we know, when we are at the stage represented by the node a, that they belong to the domain. For instance, consider a Kripke tree with three nodes, a, b, and c, with $b < a$ and $c < a$ and b and c incomparable, $\underline{R}_a = \underline{R}_c = \{0\}$ and $\underline{R}_b = \{0,1\}$. Then this represents the situation in which we are taking our variables as ranging over such a species as

$$\underline{R} = \{x \mid x = 0 \vee (\text{FLT}\ \&\ x = 1)\},$$

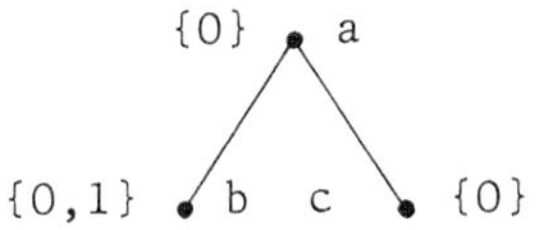

where 'FLT' abbreviates the statement of Fermat's Last Theorem. Since Fermat's Last Theorem has not yet been decided, we are currently in the state of knowledge represented by the node a: we know that $0 \in \underline{R}$, but we do not know whether 1 belongs to $\underline{R}$ or not. If Fermat's Last Theorem is ever proved, we shall attain the state of knowledge represented by b: we shall know that $\underline{R} = \{0,1\}$. If, on the other hand, Fermat's Last Theorem comes to be disproved, we shall then have advanced to the state of knowledge represented by c, and shall know that $\underline{R} = \{0\}$. Or, again, consider a Kripke tree with only two nodes, a and b, with $b < a$, $\underline{R}_a = \{0\}$ and $\underline{R}_b = \{0, 1\}$. This represents the situation in which we take as our domain a species as

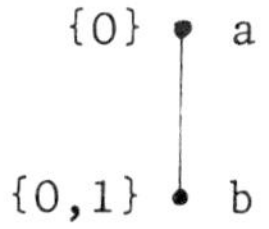

$$\underline{Q} = \{x \mid x = 0 \vee [(FLT \vee \neg FLT) \;\&\; x = 1]\} .$$

At present, we know that $0 \in \underline{Q}$, but do not know that $1 \in \underline{Q}$, and hence are at the node a. We shall never be in a position to say that $1 \notin \underline{Q}$; but, if Fermat's Last Theorem is ever decided, we shall advance to the node b, and shall know that $\underline{Q} = \{0, 1\}$. We are, of course, unlikely in practice ever to construct a mathematical theory in which the variables range over such species as $\underline{R}$ and $\underline{Q}$; nevertheless, since undecidable species are perfectly legitimate mathematical entities from an intuitionistic standpoint, there is no objection in principle to quantifying over an undecidable species (e.g. the positive real numbers), and hence no objection to a semantics which allows for this possibility.

For Kripke trees with variable domains, the specification of the condition for $\exists x\, B(x)$ to be true at a node becomes:

$\exists x\, B(x)$ is true at a under ϕ iff, for some $n \in \underline{R}_a$, $B(\bar{n})$ is true at a under ϕ.

The condition for $\forall x\ B(x)$ is more extensively modified: it is

$\forall x\ B(x)$ is true at a under ϕ iff, for every $b \leq a$, and every $n \in \underline{R}_b$, $B(\bar{n})$ is true at b under ϕ .

Intuitively, it is not enough, in order to be able, at a given stage, to assert $\forall x\ B(x)$, that we should be able to verify $B(\bar{n})$ for each n of which we already know, at that stage, that it belongs to the domain; it is necessary also that, for each n of which, at some later stage, we may come to recognize it as belonging to the domain, we should, at that stage, also be able to verify that $B(x)$ is true of it.

With this machinery, we can easily construct a counter-example to the formula B:

$$\forall x\ (p \vee Fx) \rightarrow (p \vee \forall x\ Fx).$$

Take, as above, the Kripke tree with two nodes, a and b, with

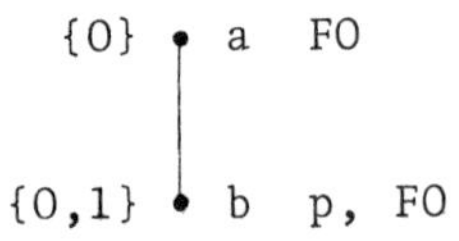

$b < a$, $\underline{R}_a = \{0\}$, and $\underline{R}_b = \{0,1\}$, and put $\phi(p) = \{b\}$, $\nu_\phi(F0) = \{a,b\}$, and $\nu_\phi(F1) = \emptyset$. Then $\nu_\phi(\forall x\ (p \vee Fx)) = \{a, b\}$, since $p \vee F0$ is true at a under ϕ, and $p \vee F0$ and $p \vee F1$ are both true at b under ϕ. On the other hand, $\nu_\phi(\forall x\ Fx) = \emptyset$, and hence $\nu_\phi(p \vee \forall x\ Fx) = \{b\}$: accordingly B is not true at a under ϕ. If the domain of the variables is taken as the species $\underline{Q}$ above, then p, under the interpretation ϕ, may be taken as the proposition FLT $\vee\ \neg$FLT, while Fx may be construed as the predicate $x = 0$.

It can in fact be shown that the family of Q-valuation systems generated by Kripke trees with the *fixed* domain $\underline{N}$ is strictly characteristic for that extension IC^+ of IC obtained by adding the rule (*):

$$\frac{\Gamma : \forall x(A \vee B(x))}{\Gamma : A \vee \forall x\ B(x)}$$

(where x does not occur free in A). If we have a formulation of the sequent calculus which admits sequents with more than one formula in the succedent, the same effect may be obtained by allowing the :∀ rule to assume its classical form:

$$\frac{\Gamma : \Delta, B(y)}{\Gamma : \Delta, \forall x\, B(x)}$$

(where y does not occur free anywhere in the conclusion of the rule). It follows from the fact that IC^+ is complete with respect to Kripke trees with a fixed domain that IC^+ and IC have their sentential fragment in common.

The logic IC^+ is, of course, useless for intuitionistic purposes: we have already observed that, within it, the Continuity Principle would be inconsistent. It may, nevertheless, appear to answer to an intuitively clear, though non-intuitionistic, interpretation of the logical constants, both because the intuitionists' rejection of the rule (*) is harder to swallow than their rejection of any other classical law, and seems to be based on more radical principles than need be appealed to in order to justify the rejection of the other laws, and also because the Kripke trees provide IC^+ with such a readily intelligible semantics. IC^+ has, however, one great disadvantage in lacking a property, which we may call the *relativization property*, which is possessed by most logics and appears desirable. Where A(x) is any formula with a single free variable x, the relativization to A(x) of a closed formula B is the formula $A(c_1)$ & ... & $A(c_n)$ & B*, where B* is obtained from B by replacing each part of the form ∀y C(y) by ∀y(A(y) → C(y)) and each part of the form ∃y C(y) by ∃y (A(y) & C(y)), and where $c_1,\ldots,c_n$ are all the individual constants occurring in B. (This is stated for the case in which B does not contain function-symbols.) A logic $\mathcal{L}$ possesses the relativization property just in case, whenever $\Gamma \vdash_{\mathcal{L}} B$, where B is a closed formula and Γ a set of closed formulas, then also $\exists x\, A(x), \Gamma' \vdash_{\mathcal{L}} B'$, where A(x) is any formula with the single free variable x, B' is the relativization of B to A(x), and Γ' is the set of relativizations to A(x) of formulas in Γ. Both IC and classical logic PC can be shown to have the relativization property. Intuitively, relativization of a formula to a predicate has the effect of restricting the domain of the variables to the extension of that predicate (under a given interpretation); hence the relativization property guarantees that the logical laws which hold good whenever the individual variables are taken as ranging over any admissible domain also hold good when they are confined to some inhabited subdomain which can be characterized by a predicate of the language. It can easily be seen that IC^+ does not have the relativization property, since the counter-example to the sequent

$$\forall x\ (p \vee Fx) : p \vee \forall x\ Fx$$

provided by Kripke trees with variable domains can very readily be adapted to give a counter-example, with a fixed domain, to the relativization of this sequent to a predicate Qx, that is, a counter-example to the sequent

$$\exists x\ Qx,\ \forall x(Qx \to p \vee Fx) : p \vee \forall x(Qx \to Fx).$$

We again consider a Kripke tree with two nodes a and b, with a > b, this time with the set $\underline{N}$ of natural numbers as the fixed domain of the variables. As before, we set $\phi(p) = \{b\}$, $\nu_\phi(F0) = \{a, b\}$ and $\nu_\phi(F1) = \emptyset$. We also set $\nu_\phi(Q0) = \{a, b\}$, $\nu_\phi(Q1) = \{b\}$ and $\nu_\phi(F\bar{n}) = \nu_\phi(Q\bar{n}) = \emptyset$ for $n > 1$:

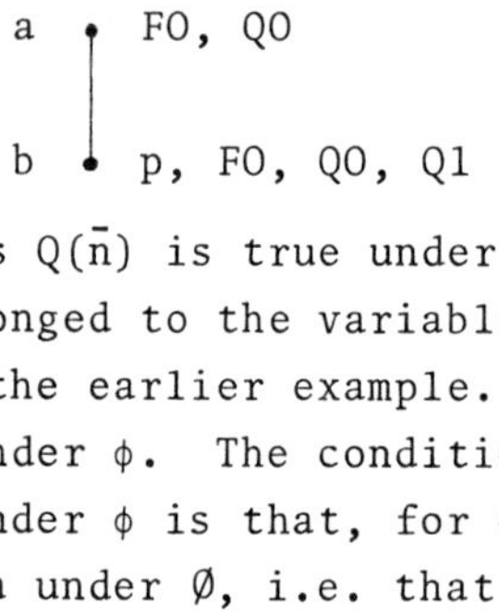

thus $Q(\bar{n})$ is true under ϕ at either node just in case n belonged to the variable domain associated with that node in the earlier example. Then, evidently, $\exists x\ Qx$ is true at a under ϕ. The condition for $\forall x(Qx \to p \vee Fx)$ to be true at a under ϕ is that, for every n, $Q\bar{n} \to p \vee F\bar{n}$ should be true at a under $\emptyset$, i.e. that, for each n for which $Q\bar{n}$ is true at either node under ϕ, $p \vee F\bar{n}$ should be true at that node. $Q\bar{n}$ is true at a only for $n = 0$, and $p \vee F0$ is true at a, since F0 is; since p is true at b, $p \vee F\bar{n}$ is true at b for every n. Thus $\forall x(Qx \to p \vee Fx)$ is true at a under ϕ. The condition for $p \vee \forall x(Qx \to Fx)$ to be true at a under ϕ is, of course, that either p or $\forall x(Qx \to Fx)$ should be. p is not true at a. The condition for $\forall x(Qx \to Fx)$ to be true at a is that, for each n, if $Q\bar{n}$ is true at either node, $F\bar{n}$ should be true at that node. This condition is not, however, satisfied: Q1 is true at b, but F1 is not. Thus $p \vee \forall x(Qx \rightarrow Fx)$ is not true at a under ϕ, and hence the sequent fails under ϕ on this tree.

It can be argued as a justification for the failure of the relativization property for IC^+ that the logic is intended for a decidable domain, whereas a subdomain determined by an arbitrary predicate may well not be decidable.

A similar difficulty does not arise for the Beth trees. When these are viewed as topological spaces whose points are the paths of the trees, it is again true that in these spaces, any intersection of open sets is open, and hence that, in the Q-valuation system generated by a Beth tree, $\bigwedge$ is simply set-theoretic intersection. The condition for the truth of a formula $\forall x\, B(x)$ at a node can therefore be simply stated as:

$\forall x\, B(x)$ is true at a iff, for every n, $B(\bar{n})$ is true at a.

As always, $\bigvee$ is infinite union. Just as in the case of disjunction, this has the effect of yielding the stipulation:

$\exists x\, B(x)$ is true at a iff, for some species $\underline{S}$ of nodes, $\underline{S}$ bars a and, for each $b \in \underline{S}$, there is an n such that $B(\bar{n})$ is true at b.

It is now quite simple to construct a counter-example to

$$\forall x(p \vee Fx) \rightarrow (p \vee \forall x\, Fx).$$

Consider a Beth tree in which each non-terminal node stands immediately above two other nodes, one terminal and the other

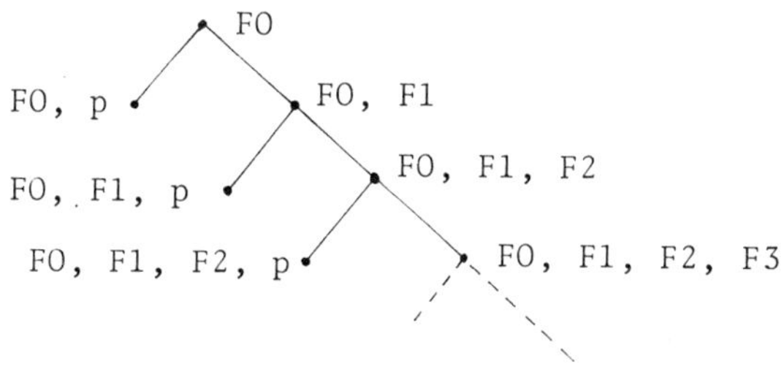

non-terminal: there is therefore just one infinite path in the tree. p is taken to be true at a node if and only if that node is terminal. $F\bar{n}$ is true at a node a on the infinite path, and at the terminal node immediately below a, just in case a is of depth $\geq$ n. In order to show that $\forall x(p \vee Fx)$ is true at the vertex, we have to show that $p \vee F\bar{n}$ is true at the vertex for each n; and, in order to do this, we have, for each given n, to find a species $\underline{S}$ which bars the vertex and is such that, for each $b \in \underline{S}$, either p is true at b or $F\bar{n}$ is. This is easy, however: we simply choose $\underline{S}$ to consist of the node of depth n on the infinite path, together with the finitely many terminal nodes which do not lie below it. Were $p \vee \forall x\, Fx$ to be true at the vertex, there would have to be a species $\underline{S}'$ barring the vertex such that, at each $b \in S'$. either p was true or $\forall x\, Fx$ was. This is impossible, however, since any species that bars the vertex must contain a node on the infinite path, whereas at no node on the infinite path is either p or $\forall x\, Fx$ true.

Because it is possible to give, by means of Beth trees, a counter-example to the above invalid formula without having recourse to variable domains, the advantage, in supplying a representation of the intended meanings of the intuitionistic logical constants, lies heavily with the Beth trees as against the Kripke trees. It is true enough that the use of variable domains can be given a sound intuitionistic sense as a representation of quantification over an undecidable species; but the fact that, with the Kripke trees, it is essential to use variable domains in order to falsify our formula makes it appear that this formula is invalid only in view of the possibility of quantifying over an undecidable domain; from an intuitionistic viewpoint, however, this is not so at all -- the formula remains just as invalid when we take the variables to be ranging over the natural numbers. (We shall inquire in the last chapter how close the Beth trees come to representing the intuitionistic meanings of the constants.)

It is possible, by various means, to give a classical proof of the completeness of intuitionistic predicate logic with respect to Beth trees, with the domain taken as $\underline{N}$; but, as already remarked, such proofs have only an oblique interest. In the next section we investigate how far completeness may be established by intuitionistically acceptable reasoning.

We conclude this section by citing an argument of Kripke's to show that intuitionistic monadic predicate logic (predicate logic with only one-place predicate letters) is recursively undecidable. This illustrates once again the great difference between classical and intuitionistic predicate logic, a difference already exemplified by the decidability of prenex formulas in IC, since it is well known that classical monadic predicate logic is decidable. Kripke's argument is non-constructive, since it requires us to assume the completeness of classical predicate logic with respect to denumerable models; it does not, however, require us to assume a completeness result for intuitionistic logic, but only to assume that each Q-valuation system generated by a Kripke tree, even if it is non-finitary, with fixed domain $\underline{N}$ is faithful to IC.

Let A be a closed formula, containing no individual constant or function-symbol, and only one predicate-letter, a two-place predicate-letter H. We suppose that A is a prenex formula, and that its quantifier-free part is in disjunctive normal form (i.e. is a disjunction of conjunctions of atomic formulas and their negations). It is known that there is no effective procedure for deciding the classical provability of such formulas, or of their negations. We now take C as $\forall x\ \forall y\ (Hxy \vee \neg Hxy)$, D as $\exists x\ \exists y\ (Hxy\ \&\ \neg Hxy)$, and B as $(A\ \&\ C) \rightarrow D$. Then plainly, where PC is classical logic, $\vdash_{PC} B \leftrightarrow \neg A$. In A, C, D and B, $\neg$ occurs only in front of atomic formulas, and we let A_1, C_1, D_1 and B_1 be formed from them, respectively, by replacing each part of the form $\neg$ Huv by Kuv, where K is a new two-place predicate-letter (u and v being any individual variables, not necessarily distinct). If $\vdash_{PC} B_1$, then also $\vdash_{PC} B$, whence $\vdash_{PC} \neg A$. On the other hand, we evidently have

$$\vdash_{PC} \forall x\ \forall y\ (Kxy \leftrightarrow \neg Hxy) \leftrightarrow \forall x\ \forall y\ (Hxy \vee Kxy)\ \&$$

$$\neg \exists x\ \exists y\ (Hxy\ \&\ Kxy),$$

i.e. $\vdash_{PC} \forall x\ \forall y\ (Kxy \leftrightarrow \neg Hxy) \leftrightarrow (C_1\ \&\ \neg D_1)$.

Now $\vdash_{PC} (A_1\ \&\ \neg A) \rightarrow \neg \forall x\ \forall y\ (Kxy \leftrightarrow \neg Hxy)$

whence $\vdash_{PC} \neg A \rightarrow [(A_1\ \&\ C_1) \rightarrow D_1]$,

i.e. $\vdash_{PC} \neg A \rightarrow B_1$.

It follows that $\vdash_{PC} B_1$ if and only if $\vdash_{PC} \neg A$.

$\vdash_{PC} B_1$ if and only if the sequent

(1) $$A_1, C_1 : D_1$$

is derivable in the classical sequent calculus LK. A_1, C_1 and D_1 are all prenex formulas. By an argument based on the cut-elimination theorem for LK, Gentzen showed (extended Hauptsatz) that, if a sequent all of whose formulas are in prenex form is derivable, then it is possible to find a cut-free proof of it in which the sentential rules are used only in the top half of the proof and the quantifier rules only in the bottom half. The two halves of such a proof are separated by a quantifier-free 'mid-sequent', derived by sentential rules alone, and from which the conclusion can itself be derived by quantifier rules alone. In our case, this mid-sequent will take the form:

(2) $$A_1^{(1)}, A_1^{(2)}, \ldots, A_1^{(k)}, C_1^{(1)}, \ldots, C_1^{(m)} : D_1^{(1)}, \ldots, D_1^{(n)} ,$$

where each $A_1^{(i)}$ $(1 \leq i \leq k)$ is obtained from the quantifier-free part of A_1 by replacing variables by variables, and similarly for the $C_1^{(i)}$ and $D_1^{(i)}$. In our case $A_1^{(i)}$, $C_1^{(i)}$ and $D_1^{(i)}$ contain only the operators & and $\vee$, and it is easy to show that the fragment of intuitionistic sentential logic that uses only & and $\vee$ coincides with the fragment of classical logic that uses only these operators. It follows that (2) is also derivable in the version L' of the intuitionistic sequent calculus which admits succedents with more than one formula. Furthermore, since there is no occurrence of ∀ in the succedent of (1), the only rules used in deriving (1) from (2) will be :∃, ∃:, and ∀:, which are the same classically as intuitionistically. Hence, if (1) is provable classically, it is provable intuitionistically, and we have:

$$\vdash_{PC} B_1 \text{ if and only if } \vdash_{IC} B_1$$

and hence:

$$\vdash_{PC} \neg A \text{ if and only if } \vdash_{IC} B_1.$$

We now form A_2, C_2, D_2, and B_2 from A_1, C_1, D_1 and B_1, respectively, by replacing each part Huv by (Fu → Gv), and each part Kuv by (Fu → ¬Gv), where F and G are one-place predicate-letters. It is evident that if $\vdash_{IC} B_1$, then also $\vdash_{IC} B_2$. We have thus established that if $\vdash_{PC} \neg A$, then $\vdash_{IC} B_2$.

Now suppose, conversely, that $\vdash_{IC} B_2$. Then, by assumption, B_2 will be true at the vertex of every Kripke tree with fixed domain N̲, under every interpretation. We consider in particular the tree with denumerably many nodes a, b_o, b_1,..., with $b_i < a$ for each i, and b_i, b_j incomparable for $i \neq j$. We confine ourselves to those interpretations ϕ for which

$F\bar{n}$ is not true at a under ϕ
$G\bar{n}$ is not true at a under ϕ
$F\bar{n}$ is true at b_i under ϕ iff $i = n$.

By assumption, B_2, which is $(A_2 \,\&\, C_2) \rightarrow D_2$, is true at a under any such ϕ. D_2 is

$$\exists x\, \exists y\, [(Fx \rightarrow Gy) \,\&\, (Fx \rightarrow \neg Gy)],$$

which is equivalent to $\exists x\, \neg Fx$. This, however, cannot be true at a, since, for each n, $F\bar{n}$ is true at b_n. C_2 is

$$\forall x\, \forall y\, [(Fx \rightarrow Gy) \vee (Fx \rightarrow \neg Gy)].$$

For each n, $F\bar{n}$ is true only at b_n, and, since b_n is terminal, for each m, either $G\bar{m}$ is true at b_n or $\neg G\bar{m}$ is: hence C_2 is true at a. It follows that, for B_2 to be true at a, A_2 cannot be true at a.

For each interpretation ϕ satisfying the above conditions, we can take $\underline{R}_\phi$ as the binary relation on $\underline{N}$ such that

$$m \; \underline{R}_\phi \; n \text{ iff } G\bar{n} \text{ is true at } a_m \text{ under } \phi.$$

Classically speaking, we can say that, for each binary relation $\underline{R}$ on $\underline{N}$, there is an interpretation ϕ such that $\underline{R} = \underline{R}_\phi$ (intuitionistically, this can be asserted only for decidable relations $\underline{R}$, since, for undecidable $\underline{R}$, the required ϕ would not be well-defined). Further, for given ϕ, it is plain that

$$(F\bar{m} \to G\bar{n}) \text{ is true at a under } \phi \text{ iff } m \; \underline{R}_\phi \; n, \text{ and}$$

$$(F\bar{m} \to \neg G\bar{n}) \text{ is true at a under } \phi \text{ iff not } m \; \underline{R}_\phi \; n.$$

A_2 is built up out of formulas of the forms $(Fu \to Gv)$ and $(Fu \to \neg Gv)$ solely by means of &, $\vee$, $\forall$ and $\exists$. On a Kripke tree with $\underline{N}$ as fixed domain, the truth-value at a given node of a formula so built up depends only upon the truth-values of the numerical instances of the constituent formulas at that node (and not at nodes below it), i.e. it is classically determined from the truth-values of those instances at that node. It follows that A itself will be classically true, with respect to the domain $\underline{N}$, when the predicate-letter H is interpreted as denoting a binary relation $\underline{R}$ on $\underline{N}$ just in case A_2 comes out true at the node a under that interpretation ϕ, satisfying the above conditions, for which $\underline{R} = \underline{R}_\phi$, an interpretation which, as we have seen, must, from a classical standpoint, exist. But we have already seen that A_2 cannot be true at a under any such ϕ. It follows that A does not come out true, classically, under any interpretation with respect to $\underline{N}$, and hence that $\neg A$ comes out true under every such interpretation; therefore, by the completeness theorem for classical predicate logic, $\vdash_{PC} \neg A$. We have thus established, non-constructively, that, if $\vdash_{IC} B_2$, then $\vdash_{PC} \neg A$, the converse having already been established constructively. Hence we have shown, as a classical result, that, since there is no decision procedure for the classical provability of negations of prenex formulas with one two-place predicate-letter, there is no decision procedure for the intuitionistic provability of formulas with at least two one-place predicate-letters.

Historical Note. The use of finite distributive lattices to characterize intuitionistic sentential logic was initiated by Jaśkowski in 1936. This was generalized further by McKinsey and Tarski, who in 1946 introduced the notion of a Heyting lattice (or rather its dual, called by them a Brouwerian algebra). They also showed how any topological space will generate a Brouwerian algebra (which they took as consisting of the closed, rather than of the open, sets), and proved the corresponding representation theorem. They further proved, by the means used in section 5.2, that any unprovable formula of IC is falsified by some finite Brouwerian algebra. They did not, however, investigate the finite topological spaces generating such finite Brouwerian algebras (i.e. finite distributive lattices), but concentrated on seeking to identify a single topological space the Brouwerian algebra generated by which should constitute a weakly characteristic valuation system for IC, and showed that the usual topology on the real line, or on Euclidean n-space, would serve this purpose. As a result, they did not connect up this topological interpretation with anything that could be regarded as a semantics for intuitionistic logic, i.e. which provided a semantic interpretation of the sentential operators. This work was continued by Rasiowa and Sikorski, who extended it to apply to first-order predicate logic. The first use of a particular type of topological space, the Beth trees, to give a semantics for intuitionistic predicate logic was made by Beth as early as 1947, and in 1956 he offered a completeness proof relative to it. In so far as this proof was claimed to rest only on intuitionistic reasoning, it was criticized by Kleene and by Kreisel; Kreisel and Dyson subsequently showed that an appeal was made in the proof to a form of Markov's principle, and Kreisel also gave a proof, originating with Gödel, that this was a best possible result. In 1959, what later became known as Kripke trees were introduced, for sentential logic only, by Lemmon and myself, as also the more general notion of a PO-space. In 1965 Kripke showed how Kripke trees could be used for predicate logic as well, stated their intuitive interpretation, and demonstrated their relation to Beth trees. In 1968 Scott extended the topological interpretation to intuitionistic analysis; and in 1973 Smorynski obtained metamathematical results for intuitionistic arithmetic by applications of Kripke trees.

5.6. THE COMPLETENESS OF INTUITIONISTIC PREDICATE LOGIC

We have shown, by intuitionistically acceptable arguments, that intuitionistic sentential logic (IC) is complete with respect to Beth trees and to Kripke trees. It is not difficult, by the use of reasoning which is classically but not intuitionistically acceptable, to extend this result to intuitionistic first-order predicate logic (ICP); but it is quite obscure what interest, from an intuitionistic standpoint, such an extension would have. Before inquiring how far we can get, using only intuitionistically acceptable reasoning, towards a completeness proof for ICP, it is best to take stock of the general question of the semantics of intuitionistic logic.

In treating of classical logic, we are in no uncertainty about which is the intended semantics, namely the standard two-valued one. For this reason, the statement that some branch of classical logic is complete, not relative to any specified semantics, but absolutely, has an unambiguous sense. In the case of intuitionistic logic, the intended meanings of the logical constants, and the intended notions of the meanings and of the truth of a sentence, have indeed been made intuitively clear, in terms of the notions of a mathematical construction and of a construction's being a proof of a statement. The difficulty lies in the fact that these notions have not yet been formulated in a manner that makes them amenable to mathematical treatment, such as is required in order to give a proof of completeness. Kreisel and Goodman have, indeed, made strenuous efforts to devise a mathematical theory of constructions which should serve precisely this purpose; and when this work is completed, it will supply a mathematically usable semantics that all can recognize as that intended for intuitionistic logic, as the two-valued semantics is so recognized for classical logic. Despite the progress that has been made, however, the theory of constructions has not yet attained a satisfactory state, and, until it does, completeness results for intuitionistic logic have to be stated relative to some specified semantics such as the Beth trees. It will be left until Chapter 7 to discuss how close the Beth trees come to capturing the intended meanings of the logical constants.

It is, however, possible to discuss the completeness of ICP without appealing either to a semantics of the sort provided by the Beth trees or to a mathematically rigorous version of the intuitive explanations in terms of the notion of a construction. We can give an interpretation of one or more formulas of ICP by specifying some inhabited species $\underline{D}$ (a species which we can show to have at least one element) as the domain of the individual variables, and assigning to each individual constant an element of $\underline{D}$ and to each n-place predicate-letter a subspecies of $\underline{D}^n$. Without attempting to give any non-circular explanations of the logical constants, but simply taking their intuitionistic meanings for granted, we are then entitled to assume that, from an intuitionistic standpoint, to say that a formula comes out true under an interpretation of this kind has a perfectly determinate content. Let us call an interpretation of this sort an *internal* interpretation. We may then discuss the validity or satisfiability of a given formula with respect to internal interpretations, i.e. its truth under all or some internal interpretations, and hence may also discuss the completeness of ICP, or any fragment of it, with respect to internal interpretations. With internal interpretations, no question can arise over whether we are attaching to the logical constants their intended meanings, since we are simply assuming that these are taken as carrying whatever meanings they are intended to carry. Hence completeness results with respect to internal interpretations may legitimately be stated as absolute completeness results.

What status does an internal interpretation, or the notion of validity with respect to internal interpretations, have? It bears an obvious resemblance to the standard semantic notion of an interpretation of a formula of classical predicate logic (PCP), which consists in specifying a non-empty set $\underline{D}$ as the domain of the individual variables, and in assigning to each individual constant an element of $\underline{D}$ and to each n-place predicate-letter a subset of $\underline{D}^n$. If, then, an internal interpretation is the analogue, for formulas of ICP, of a standard semantic interpretation of formulas of PCP, to what, in the classical case, are we to compare an interpretation of formulas of ICP in terms of Beth trees (or, equally, in terms of a theory of constructions)? The natural answer is that there is no analogue. An account of the meanings of the intuitionistic logical constants in terms of Beth trees, or of constructions, is an attempt to explain those meanings without presupposing them as already understood, that is, in terms which are, as far as possible, neutral between the classical and the intuitionistic ways of construing them (for instance, by using the sentential operators, in the explanations, only as applied to decidable statements). No such attempt is made in classical logic: the explanations of the logical constants given in standard classical semantics presuppose their classical meanings, because those constants are used, in their full classical strength, in framing the explanations. The standard two-valued classical semantics may therefore, it seems, rightly also be described as internal.

This way of looking at the matter is rather compelling; that it is, nevertheless, mistaken may be seen if we suppose that the formula of ICP of which we have to give an internal interpretation contains some sentence-letters, and consider how we are to fill in the missing entry in the following table:

	intuitionistic interpretation	classical interpretation
individual constant	element of D	element of D
n-place predicate-letter	subspecies of D^n	subset of D^n
sentence-letter	?	truth-value

Without resorting to some external semantics, such as that of Beth trees or the theory of constructions, the only possible entry we can put in the empty space is 'proposition'; and a proposition does not appear to be the intuitionistic analogue of a truth-value in the way that a species appears, at first sight, simply to be the intuitionistic analogue of a set. A proposition is, rather, the meaning of a sentence, and is a notion as much in place in a classical context as in an intuitionistic one, the whole point of classical semantics being that it does not purport, at least in any direct way, to give an account of meaning. What, in classical semantics, is assigned by any one interpretation to a sentence-letter or a predicate-letter is not supposed to be sufficient to determine the meaning of any actual sentence or predicate, but only that feature of a sentence or predicate which is necessary to determine the truth or otherwise of a complex sentence of which it is a component, and hence the truth or falsity, under the interpretation, of the formula in which the schematic letter occurs. To say that, in specifying an internal interpretation of a formula of ICP, we assign a proposition to each sentence-letter is just to say that we interpret each sentence-letter as a specific sentence of intuitionistic mathematics. But, in just the same way, the intuitionistic notion of a species is just that of the meaning of some predicate; hence to say that, in specifying an internal interpretation, we assign a species to each predicate-letter is just to say that we interpret each predicate-letter as a specific predicate. It is true that we cannot in the same way equate a mathematical object (element of the domain) with the meaning of a term; we have to allow that terms with distinct meanings (e.g. a numeral and a complex numerical term) may denote the same object. But since, in intuitionistic mathematics, identity, strictly so called, is required to be decidable (this being, in effect, a requirement on the way the meanings of terms are given), and since the principle of the substitutivity of identicals in all contexts necessarily obtains, it makes no effective difference whether we take an interpretation as assigning, to each individual constant, an element of the domain or an actual term.

An internal interpretation of formulas of ICP is, thus, to be regarded, not as any sort of semantic interpretation, but simply as an interpretation by replacement. It is, indeed, one of the salient differences between a classical and an intuitionistic theory of meaning that the former imposes on us a distinction, for expressions of all categories, between sense and reference, while the latter leaves no place for such a distinction except as applied to terms: in any genuine semantics for intuitionistic logic, whatever is assigned to a sentence-letter or predicate-letter must be an abstract representation of the whole sense of a sentence or predicate, not, as in classical semantics, something that could be a common feature of non-synonymous sentences or predicates. But, for a type of interpretation to be a genuinely semantic notion, it must display the mechanism whereby the condition for the truth of each sentence is determined in accordance with its composition, as the standard semantics for PCP does and as an internal interpretation for ICP does not. The idea that the standard classical notion of an interpretation is an internal, i.e. non-explanatory, one contains only a grain of truth; we need to distinguish between circularity and triviality. A trivial explanation tells us nothing that we did not already explicitly know; a circular one merely presumes an implicit understanding of what it describes explicitly. Classical semantics is circular, in that the specific explanations which it offers of the logical constants require, for their understanding, that those constants, as used in the explanations, should be taken in their full classical senses; but it is not trivial, because it provides an account of the way in which a sentence may be determined as true in accordance with its composition. An explicit grasp of this account, which involves (for a language containing a term denoting each element of the domain) that the truth-value of any sentence depends ultimately only upon the truth-values of atomic sentences, is by no means required for an implicit understanding of the classical logical constants. By contrast, the notion of an internal interpretation of ICP *is* trivial in this sense. The difference arises because of a disanalogy between the intuitionistic notion of a species and the classical notion of a set. It is constitutive of the latter that a set both determines and is uniquely determined by which elements of the domain belong to it; hence to lay down that a predicate-letter is to be interpreted by assigning a set to it is tantamount to saying that the contribution of a predicate to determining the truth-condition of a sentence in which it occurs depends solely upon which objects it is true of. A species, on the other hand, possesses all those properties, whatever they may be, which its defining predicate has in virtue of its meaning; to lay down that a predicate-letter is to be interpreted by assigning a species to it therefore tells us nothing at all about how a predicate contributes to the truth-condition of a sentence. A species thus, after all, stands to a set exactly as a proposition stands to a truth-value. It is thus equally misleading to regard the standard classical notion of an interpretation as internal and to regard an internal interpretation of ICP as in any sense a semantic interpretation; it is nothing but an interpretation by replacement.

The use of schematic letters in logic is, of course, as old as Aristotle, whereas the idea of a semantics is an invention of the modern era. Before that, the only available notion of an interpretation was that of an interpretation by replacement of the schematic letters by specific expressions of the appropriate logical category, the way in which any sentence resulting from such a replacement is determined as true being left unanalysed; and the only available notion of validity was that of truth under all such replacements. Just because the semantics of classical logic is, from a classical standpoint, so well under control, an appeal to interpretations by replacement no longer has any utility in a classical context; but, in the case of intuitionistic logic, certain valuable results can as yet be obtained only by appeal to them.

However, once it is recognized that the notion of an internal interpretation is not a semantic one, but, rather, represents the pre-semantic idea of an interpretation by replacement, it is in one respect preferable to regard such an interpretation as assigning objects, rather than terms, to the individual constants, and species, rather than predicates, to the predicate-letters (although in practice it is usually indifferent in which of these two ways we choose to describe it). The reason is that it is essential that we do not view the range of internal interpretations as subject to restrictions arising from the limitations of language: in particular, we assume that, for any one particular mathematical object (including a choice sequence), we are able to refer to it; hence any one-place predicate maps every element of the domain over which it is defined on to a proposition, and every two-place predicate maps every element of the domain of one of its variables on to a subspecies of the domain of the other. In consequence, if a formula is to be valid with respect to internal interpretations, it is necessary that, for any replacements of its schematic letters by expressions containing one or more extraneous parameters, the universal closure of the resulting open sentence should be true. A formula may thus be shown to be invalid with respect to internal interpretations, not merely by demonstrating the falsity of a specific sentence obtained from it by replacement, but also by showing false the closure of an open sentence obtained by replacements involving parameters.

We may thus say that a formula A is *valid under replacements* just in case, where $P_1,\ldots,P_k$ are all the schematic letters (sentence-letters, predicate-letters, and individual constants) occurring in A, for every inhabited species $\underline{D}$ and all specific expressions $P_1^*,\ldots,P_k^*$ of the appropriate logical types, the sentence A* which results from restricting the individual variables of A to $\underline{D}$ and replacing each P_i by $P_i{}^*$ is true. Providing that we interpret this definition in a generous sense, so as to allow that, for any finite number of particular mathematical objects, the $P_i{}^*$ may be taken from a language permitting reference to those objects, this is equivalent to saying that, for every inhabited species $\underline{D}$ and all specific expressions $P_1{}^*,\ldots,P_k{}^*$ involving parameters, the universal closure of A* is true. It is also equivalent to saying that A is *internally valid*, viz. that, for every inhabited species $\underline{D}$ and all mathematical entities $P_1{}^+,\ldots,P_k^+$ (propositions, species, and objects) of the appropriate categories, the proposition A^+ which results from interpreting the individual variables of A as ranging over $\underline{D}$ and each P_i as meaning $P_i{}^+$ is true. (These are, of course, just two ways of formulating the same notion.) In any case, completeness with respect to replacements must be distinguished from what Kreisel has called completeness by substitution. A fragment of ICP is, obviously, *complete with respect to replacements* if every formula A belonging to the fragment which is valid under replacements is provable; we may correspondingly define *internal completeness*. The fragment is *complete by substitution*, on the other hand, if, for each formula A belonging to the fragment, there exist an inhabited species $\underline{D}$ and specific expressions $P_1{}^*,\ldots,P_k{}^*$ such that, if A* is true, then A is provable. If we admit classical reasoning (in the metalanguage), these two notions of completeness are indeed equivalent, but not if we confine ourselves to intuitionistic reasoning. Kreisel points out that even IC is definitely not complete by substitution; for, if it were, then, since the formula $P \vee \neg P$ is not provable, there would have to be a sentence P* such that $P^* \vee \neg P^*$ is not true, and hence (since the word 'true' is being so used in this context that 'C is true' is equivalent to C) $\neg(P^* \vee \neg P^*)$ would be true, which is impossible. It by no means follows, from an intuitionistic standpoint, that IC is incomplete with respect to replacements. Indeed, we can show it complete with respect to replacements for the particular formula $P \vee \neg P$ by instancing an open sentence $P^*(\alpha)$ containing a parameter for a choice sequence for which $\neg \forall \alpha(P^*(\alpha) \vee \neg P^*(\alpha))$ is true; we reason, as above, that if $P \vee \neg P$ were true under all replacements, it would, in particular, be true when P was replaced by a sentence $P^*(\alpha^*)$, where α^* is any specific value of α, and hence $\forall\alpha(P^*(\alpha) \vee \neg P^*(\alpha))$ would be true. To show by such means that IC was complete by substitution for $P \vee \neg P$, it would be necessary to show the truth of $\exists\alpha \neg (P^*(\alpha) \vee \neg P^*(\alpha))$, which is of course impossible. (Throughout this section, we shall be concerned only with completeness for single formulas, or, equivalently, for finite sets -- the propositions, respectively, that, for every A, if A is valid, then $\vdash A$, and that, for every finite Γ and every A, if A is entailed by Γ, then $\Gamma \vdash A$.)

We could say of a formula valid under replacements, or internally valid, that it was absolutely valid, or valid simpliciter, and of a fragment of ICP complete with respect to replacements, or internally complete, that it was absolutely complete, or complete simpliciter, since the internal validity of a formula guarantees its validity under the intended semantics, independently of any formulation of the latter. The first use we make of internal interpretations is to show that, if a formula is internally valid, then it is valid on Beth trees; accordingly, positive completeness results with respect to Beth trees imply corresponding results for internal completeness. For simplicity, we consider formulas of a first-order language which has no free individual variables, but only bound ones, and which contains no function-symbols, but contains, for each natural number n, a numeral $\bar{n}$ as an individual constant; we shall restrict ourselves to interpretations under which, for each n, $\bar{n}$ denotes n. Let $\underline{T}$ be a Beth tree with respect to which an interpretation of this language is given. We first transform $\underline{T}$ into a tree $\underline{T}'$ in which every path is infinite by replacing each terminal node a by an infinite chain a_o, a_1, a_2,... such that an (atomic) formula is true at each a_i just in case it was true, in $\underline{T}$, at a. We can then represent $\underline{T}'$ by a spread s, so that each node of $\underline{T}'$ corresponds to a finite sequence admissible in s, and conversely. For any formula A of our language and any finite sequence $\vec{u}$ admissible in s, let us write 'Tr(A, $\vec{u}$)' to mean that A is true at the node of $\underline{T}$' which corresponds to $\vec{u}$. Let α be any *lawless* sequence in s. A lawless element α of s is, of course, one upon the choice of whose terms no restriction is imposed at any stage save the initial requirement confining it to s, and is distinguished by the fact that any statement B(α) can be recognized as true of it only on the basis of some initial segment of it, together with its membership of s and (possibly) the fact that it *is* lawless, and therefore implies

$$\exists n\ \forall\beta_{\beta \in s,\ \beta \in \bar{\alpha}(n)}\ B(\beta),$$

at least where the variable β ranges only over lawless sequences. We assume that, for any admissible finite sequence $\vec{u}$, there is a lawless sequence of which $\vec{u}$ is an initial segment. We now specify an internal interpretation which takes every formula A of our first-order language into a proposition A*(α) about α. The domain of the individual variables is to be the species $\underline{N}$ of natural numbers. Since the language contains a term for each element of the domain, all we need to do in addition is to define the interpretation for atomic formulas: where Q is any atomic formula, Q*(α) is to be the proposition that $\exists$n Tr(Q, $\bar{\alpha}$(n)). This stipulation determines the interpretations of the predicate-letters occurring in A, and hence A*(α) is well-defined. We now state a lemma, in the proof of which the variables β and γ are again to be taken as ranging over lawless sequences.

Lemma. A*(α) iff $\exists$n Tr(A, $\bar{\alpha}$(n)).

Proof. The proof is by induction on the complexity of A.

(i) If A is atomic, the lemma is immediate from the specification of the meaning of A*(α).

(ii) If A is B & C, A*(α) is B*(α) & C*(α). By the induction hypothesis, this is equivalent to $\exists$n Tr(B, $\bar{\alpha}$(n)) & $\exists$m Tr(C, $\bar{\alpha}$(m)). Since a formula true at any node is also true at any lower node, it is plain, by taking the maximum of the required n and m, that this is in turn equivalent to

$\exists n\ Tr(A, \bar{\alpha}(n))$.

(iii) If A is B $\vee$ C, $A^*(\alpha)$ is, by the induction hypothesis, equivalent to

(a) $$\exists m\ (Tr(B, \bar{\alpha}(m)) \vee Tr(C, \bar{\alpha}(m)).$$

(a) obviously implies $\exists n\ Tr(A, \bar{\alpha}(n))$. Conversely, by the definition of truth at a node on a Beth tree, $\exists n\ Tr(A, \bar{\alpha}(n))$ is equivalent to

(b) $$\exists n\ \forall\beta_{\beta\in s, \beta\in\bar{\alpha}(n)}\ \exists m(Tr(B, \bar{\beta}(m)) \vee Tr(C, \bar{\beta}(m))),$$

which, taking β as α, implies (a).

(iv) If A is B $\rightarrow$ C, $A^*(\alpha)$ is by the induction hypothesis equivalent to

(c) $$\exists m\ Tr(B, \bar{\alpha}(m)) \rightarrow \exists n\ Tr(C, \bar{\alpha}(n)).$$

By the definition of truth at a node on a Beth tree, $\exists n\ Tr(A, \bar{\alpha}(n))$ is equivalent to

(d) $$\exists n\ \forall\beta_{\beta\in s, \beta\in\bar{\alpha}(n)}\ \forall m_{m\geq n}\ (Tr(B, \bar{\beta}(m)) \rightarrow Tr(C, \bar{\beta}(m))).$$

If (d) holds, we obtain

$$\exists n\ \forall m_{m\geq n}\ (Tr(B, \bar{\alpha}(m)) \rightarrow Tr(C, \bar{\alpha}(m))),$$

which, in view of the fact that a formula true at a node is true at all lower nodes, implies (c). Conversely, suppose that (c) holds. Now, since α is a lawless element of s, we must know the truth of (c) on the strength of an initial segment $\bar{\alpha}(k)$ of α, i.e. we have

$$\forall\beta_{\beta\in s, \beta\in\bar{\alpha}(k)}\ (\exists m\ Tr(B, \bar{\beta}(m)) \rightarrow \exists n\ Tr(C, \bar{\beta}(n)))$$

for some k. (Note that this is the first time we have appealed to the fact that α is a lawless sequence.) Now suppose that $\beta \in$ s and $\beta \in \bar{\alpha}(k)$ and $Tr(B, \bar{\beta}(m))$ and $m \geq k$. Then

$$\forall\gamma_{\gamma\in s, \gamma\in\bar{\beta}(m)}\ \exists n\ Tr(C, \bar{\gamma}(n)).$$

But then the node a of <u>T</u>' corresponding to $\bar{\beta}(m)$ is barred by a species of nodes at which C is true, and C is therefore true at a, i.e. we have $Tr(C, \bar{\beta}(m))$. We have thus shown that

$$\forall\beta_{\beta\in s,\beta\in\bar{\alpha}(k)}\ \forall m_{m\geq k}\ (Tr(B,\ \bar{\beta}(m)) \rightarrow Tr(C,\ \bar{\beta}(m))),$$

and thus have shown (d) to be true.

(v) If A is $\neg B$, $A^*(\alpha)$ is by the induction hypothesis equivalent to $\neg\exists m\ Tr(B,\ \bar{\alpha}(m))$. Since α is a lawless element of s, this is equivalent to

$$\exists n\ \forall\beta_{\beta\in s,\beta\in\bar{\alpha}(n)}\ \neg\exists m\ Tr(B,\ \bar{\beta}(m)).$$

This says, in effect, that, for some n, B is not true at any node in <u>T</u>' below the node a which corresponds to $\bar{\alpha}(n)$; it follows that A is true at a, i.e. that $\exists n\ Tr(A,\ \bar{\alpha}(n))$.

(vi) If A is $\forall x\ B(x)$, $A^*(\alpha)$ is by the induction hypothesis equivalent to $\forall m\ \exists n\ Tr(B(\bar{m}),\ \bar{\alpha}(n))$. Since α is a lawless sequence, we must have, for some k,

$$\forall\beta_{\beta\in s,\beta\in\bar{\alpha}(k)}\ \forall m\ \exists n\ Tr(B(\bar{m}),\bar{\beta}(n)).$$

For each m, therefore, the node a in <u>T</u>' corresponding to $\bar{\alpha}(k)$ is barred by a species of nodes at which $B(\bar{m})$ is true, and hence $B(\bar{m})$ is true at a for each m. $\forall x\ B(x)$ is therefore true at a, and we have $Tr(A,\ \bar{\alpha}(k))$.

(vii) If A is $\exists x\ B(x)$, $A^*(\alpha)$ is by the induction hypothesis equivalent to

(e) $$\exists m\ \exists n\ Tr(B(\bar{m}),\ \bar{\alpha}(n)).$$

$\exists n\ Tr(A,\ \bar{\alpha}(n))$ is, by the definition of truth at a node on a Beth tree, equivalent to

(f) $$\exists n\ \forall\beta_{\beta\in s,\beta\in\bar{\alpha}(n)}\ \exists m\ \exists k\ Tr(B(\bar{m}),\ \bar{\beta}(k)).$$

As under (iii), (f) follows easily from (e), but also implies (e), by taking β as α.

The upshot of this lemma is that the Beth trees may be viewed as giving a representation of the logic of statements involving a parameter for a lawless sequence: if <u>T</u> is any Beth tree, then the formula A will be true at its vertex just in case the statement $\forall\alpha_{\alpha\in s,\ \alpha\ \text{lawless}}A^*(\alpha)$ is true, where s is the spread which represents $\underline{T}'$. As an illustration, consider the following Beth tree which provides a refutation of the invalid formula

$$\forall x\ (Fx \vee \neg Fx)\ \&\ \neg\neg\exists x\ Fx \rightarrow \exists x\ Fx.$$

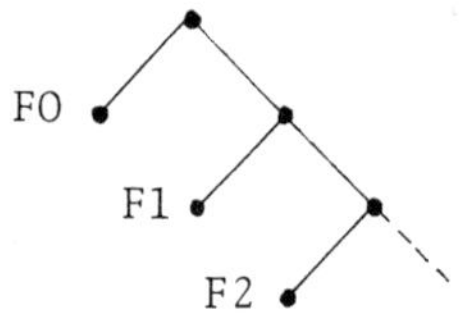

Only the rightmost path is infinite; each node on this path has just two nodes immediately below it, of which the left-hand one is terminal. $F\bar{n}$ is true at a node a just in case a is a terminal node of level n + 1. For each n, the vertex is barred by the species consisting of the terminal nodes of level $\leq n + 1$ and the node on the rightmost path of level n + 1; of these, $F\bar{n}$ is true at the terminal node of level n + 1, and $\neg F\bar{n}$ is true at the others. Hence $F\bar{n} \vee \neg F\bar{n}$ is true at the vertex for every n, and therefore $\forall x\ (Fx \vee \neg Fx)$ is true at the vertex. At every terminal node $\exists x\ Fx$ is true, and every node on the rightmost path has a terminal node below it; it follows that $\neg\exists x\ Fx$ is not true at any node, and hence that $\neg\neg\exists x\ Fx$ is true at the vertex. On the other hand, for $\exists x\ Fx$ to be true at the vertex, there would have to be a species $\underline{S}$ of nodes barring the vertex such that, for every $a \in \underline{S}$, $F\bar{n}$ was true at a for some n; but there can be no such species, since for no n is $F\bar{n}$ true at any node on the rightmost path. Thus the cited formula is not true at the vertex.

The spread s which corresponds to this tree is that which contains just those choice sequences α such that $\forall n\ (\alpha(n) = 0 \vee \forall m_{m\leq n}\ \alpha(m) = 1)$; by applying the internal interpretation to which the lemma relates, we make $F\bar{n}$ come out as meaning, in effect that $\alpha(n) = 0\ \&\ \forall m_{m<n}\ \alpha(m) = 1$. The fact that $\forall x(Fx \vee \neg Fx)$ was true at the vertex represents the fact that $\forall\alpha_{\alpha\in s}\ \forall n(\alpha(n) = 0 \vee \alpha(n) \neq 0)$. The fact that $\neg\neg\exists x\ Fx$ was true at the vertex represents the fact that $\forall\alpha_{\alpha\in s}\neg\neg\exists n\ \alpha(n) = 0$ (where the variable α is to be taken as ranging over lawless sequences only); this holds because if $\neg\exists n\ \alpha(n) = 0$ for some

lawless $\alpha \in s$, then there would exist a number k such that (where β is again taken as ranging over lawless sequences) $\forall\beta_{\beta \in s, \beta \in \bar{\alpha}(k)} \neg \exists n\ \beta(n) = 0$, and this is absurd for $\beta \in \bar{\alpha}(k)^\frown 0$. On the other hand, the failure of $\exists x\ Fx$ at the vertex reflects the fact that $\neg \forall\alpha_{\alpha \in s}\ \exists n\ \alpha(n) = 0$; for, since s is a fan, if $\forall\alpha_{\alpha \in s}\ \exists n\ \alpha(n) = 0$, then by the Fan Theorem there would exist m such that $\forall\alpha_{\alpha \in s}\ \exists n_{n \le m}\ \alpha(n) = 0$, which is obviously false. A refutation of any formula by means of a Beth tree can be similarly handled.

Provided that the notion of a lawless sequence is accepted as an intuitionistically meaningful one, the internal validity of a formula of ICP must entail its truth under an interpretation involving a parameter for a lawless sequence. Hence we have, as an immediate consequence of the lemma,

Theorem 1. Every internally valid formula of ICP is valid on Beth trees.

We now investigate whether the completeness theorem for IC with respect to Beth trees can be extended to ICP. Since there is no decision procedure for ICP, we cannot make a straightforward extension of the methods we used for IC, if we wish our argument to be itself intuitionistically valid, but must approach the matter from a slightly different direction. This is the method of *semantic tableaux*, introduced by Beth for both intuitionistic and classical logic. The idea is to use a suitable formulation of a cut-free sequent calculus in such a way that a systematic search for a proof of a given sequent, such as we described in giving the decision procedure for IC, can be simultaneously construed as an attempt to construct a counter-example, i.e. an interpretation under which the sequent does not hold; in our case, a Beth tree. If the sequent is not provable, we cannot hope that the construction will in all cases terminate, since, if it did, we should have a decision procedure for ICP; but our hope will be that, by indefinitely continuing the construction, we shall obtain an infinite Beth tree on which the sequent does not hold. If the sequent is valid, then the construction cannot yield such a counter-example; if we could show that whenever it does not yield a counter-example, it will yield a proof of the sequent, we should have a completeness proof.

We again consider the same first-order language as above, containing all the numerals but no free variables; all formulas are, therefore, closed. We use a version of that form of the sequent calculus in which the succedent, as well as the antecedent, of a sequent may contain any finite number of formulas (positive or zero): to say that such a sequent $\Gamma : \Delta$ holds at a node a of a Beth tree is to say that, if every formula of Γ is true at a, then at least one formula of Δ is true at a. We take as basic sequents all those of the form $\Gamma, A : A, \Delta$. The rules are then as follows:

$$\&: \frac{\Gamma, A, B : \Delta}{\Gamma, A \& B : \Delta} \qquad :\& \frac{\Gamma : A, \Delta \quad \Gamma : B, \Delta}{\Gamma : A \& B, \Delta}$$

$$\vee: \frac{\Gamma, A : \Delta \quad \Gamma, B : \Delta}{\Gamma, A \vee B : \Delta} \qquad :\vee \frac{\Gamma : A, B, \Delta}{\Gamma : A \vee B, \Delta}$$

$$\rightarrow: \frac{\Gamma, A \rightarrow B : A, \Delta \quad \Gamma, B : \Delta}{\Gamma, A \rightarrow B : \Delta} \qquad :\rightarrow \frac{\Gamma, A : B}{\Gamma : A \rightarrow B, \Delta}$$

$$\neg: \frac{\Gamma, \neg A : A, \Delta}{\Gamma, \neg A : \Delta} \qquad :\neg \frac{\Gamma, A :}{\Gamma : \neg A, \Delta}$$

$$\forall: \frac{\Gamma, \forall x\, A(x), A(\overline{n_1}), \ldots, A(\overline{n_r}) : \Delta}{\Gamma, \forall x\, A(x) : \Delta} \qquad :\forall \frac{\Gamma : A(\bar{n})}{\Gamma : \forall x\, A(x), \Delta}$$

$$\exists: \frac{\Gamma, A(\bar{n}) : \Delta}{\Gamma, \exists x\, A(x) : \Delta} \qquad :\exists \frac{\Gamma : A(\overline{n_1}), \ldots, A(\overline{n_r}), \exists x\, A(x), \Delta}{\Gamma : \exists x\, A(x), \Delta}$$

In the quantifier rules x is any bound variable, and, in $\forall$: and $:\exists$, $\overline{n_1}, \ldots, \overline{n_r}$ are any numerals. In $\exists$: it is required that the numeral $\bar{n}$ should not occur anywhere in the conclusion of the inference, and in $:\forall$ that it should not occur in Γ or in

∀x A(x). As far as possible, the rules have been formulated in such a way that, not only must the conclusion hold if the premiss or premisses hold, but, conversely, if the conclusion holds, then so does each of the premisses. This applies to all the left-introduction rules and to :&, :∨, and :∃ (in the case of ∃:, we can say only that, if there is an interpretation under which the conclusion holds, then there is one under which the premiss holds). In particular, this provides a motive for the inclusion of the introduced formula in the premiss, or one of the premisses, of →:, ¬:, ∀:, and :∃. The purpose of this is to minimize the number of alternative proof tree-trunks we have to construct at any given stage in the systematic search for a proof of a given sequent: for the rules which have this property, we do not have to consider the possibility that the rule which, in searching for a proof, we apply in reverse was not the right one to have appealed to at that particular stage, since, by appealing to it, we shall not have destroyed the opportunity to find a proof if there is one. (The inclusion of the introduced formula in the premisses of the rules mentioned will also be seen to be of importance in the construction of a counter-example.) However, because of the presence in the intuitionistic system of three rules, :→ , :¬, and :∀ , which do not admit the presence of more than one formula in the succedent of the premiss, it is impossible to carry through this plan completely. If, for instance, we arrive, at any stage, at a sequent of the form Γ : A → B, Δ, with Δ non-empty, we are bound to consider separately the possibilities that it was ultimately derived by →-introduction on the right, Δ then being brought in by thinning, and that it was derived in some other way. Hence, even though we have adopted the most general form of basic sequent, we cannot effectively dispense with a rule of thinning on the right, although thinning on the left is superfluous. In the above formulation, thinning on the right has not been presented as a separate step, but has been incorporated into the rules for introducing →, ¬, and ∀ on the right.

Actually, the procedure for constructing a counter-example for a sequent $\Gamma : \Delta$ cannot, strictly speaking, be identified with that of searching for a proof of $\Gamma : \Delta$ in the sequent calculus, but is, rather, dual to the latter procedure, and may be regarded as going on simultaneously with it. For this reason, we shall describe whatever is generated at any stage in the process of attempting to construct a counter-example as a dual tree-trunk for $\Gamma : \Delta$. The duality consists in the fact that wherever, in the course of attempting to construct a proof, we put two sequents, as premisses for a rule of inference, above a given sequent, we choose one or other of these two sequents when we are trying to construct a counter-example; furthermore, wherever, in the course of trying to construct a proof, we have a choice of putting any one of several sequents above a given one, we put all of these other sequents above it when we are trying to construct a counter-example. (If, for the given sequent to be valid, it is necessary that both of two other sequents should be valid, then it is sufficient, for the first sequent to fail, that either one of the other two sequents should fail; conversely, if, for a given sequent to be valid, it is sufficient that any one of several other sequents should be valid, then it is necessary, if the given sequent is to fail, that all the others should also fail.) The process of attempting to construct a counter-example will be shown to be related to the search for a proof in this way, that if, at any stage, the attempt to construct a counter-example is found to break down, then, by that stage, the other process must have yielded a proof. Because of this relationship, it is sufficient to give a formal description only of the procedure for constructing dual tree-trunks; the systematic procedure for constructing proof tree-trunks can be left to take care of itself, without our having to provide a formal description of it. We can accordingly be content to define a *proof tree-trunk of level k for* $\Gamma : \Delta$ without reference to a specific process of construction, but simply as being a finite tree with each node a of which is associated a sequent $\Gamma_a : \Delta_a$, such that (i) the sequent associated with the vertex is $\Gamma : \Delta$, (ii) for each non-terminal node a, the sequents associated with the nodes immediately above a are premisses, under one of the rules of inference, for $\Gamma_a : \Delta_a$ and (iii) each path in the tree is of length $\leq k$, and is of length k unless, for the terminal node a on the path, $\Gamma_a : \Delta_a$ is a basic sequent. Note that we are here visualizing a proof tree-trunk as having its vertex at the bottom, unlike most (mathematical) trees; throughout this section, we shall take all proof tree-trunks and dual tree-trunks as so oriented. Further, we are taking a path of length k to be one on which there are k + 1 nodes, so that a proof tree-trunk of level 0 for $\Gamma : \Delta$ consists of a single node with which $\Gamma : \Delta$ is associated. A *proof-tree of level k for* $\Gamma : \Delta$ is a proof tree-trunk of level k for $\Gamma : \Delta$ which satisfies the further condition (iv) that, for each terminal node a, $\Gamma_a : \Delta_a$ is a basic sequent.

Dual tree-trunks will be defined by reference to a particular process of construction, and, in order to define this, we need an auxiliary definition. Where $\underline{T}$ is any tree with each node a of which two sets of formulas, Γ_a and Δ_a are associated, we shall say that a formula A is *fulfilled in* $\underline{T}$ *at* a node a *with respect to* a number m provided that:

(i) if A is B & C and $A \in \Gamma_a$, then $B \in \Gamma_a$ and $C \in \Gamma_a$;

(ii) if A is B & C and $A \in \Delta_a$, then either $B \in \Delta_a$ or $C \in \Delta_a$;

(iii) if A is $B \vee C$ and $A \in \Gamma_a$, then either $B \in \Gamma_a$ or $C \in \Gamma_a$;

(iv) if A is $B \vee C$ and $A \in \Delta_a$, then $B \in \Delta_a$ and $C \in \Delta_a$;

(v) if A is $B \rightarrow C$ and $A \in \Gamma_a$, then either $B \in \Delta_a$ or $C \in \Gamma_a$;

(vi) if A is $B \rightarrow C$ and $A \in \Delta_a$, then, for some node $b \geq a$, $B \in \Gamma_b$ and $C \in \Delta_b$;

(vii) if A is $\neg B$ and $A \in \Gamma_a$, then $B \in \Delta_a$;

(viii) if A is $\neg B$ and $A \in \Delta_a$, then, for some $b \geq a$, $B \in \Gamma_b$;

(ix) if A is $\forall x\, B(x)$ and $A \in \Gamma_a$, then $B(\bar{m}) \in \Gamma_a$;

(x) if A is $\forall x\, B(x)$ and $A \in \Delta_a$, then, for some number n and for some $b \geq a$, $B(\bar{n}) \in \Delta_b$;

(xi) if A is $\exists x\, B(x)$ and $A \in \Gamma_a$, then, for some n, $B(\bar{n}) \in \Gamma_a$;

(xii) if A is $\exists x\, B(x)$ and $A \in \Delta_a$, then $B(\bar{m}) \in \Delta_a$.

'$b \geq a$' means, of course, that b either coincides with a or lies above a on some path through a; we are, as before, visualizing $\underline{T}$ as having its vertex at the bottom.

In order to make the process of constructing dual tree-trunks for any given sequent as specific as possible, we assume given some fixed enumeration of all formulas. The notion we shall now define is that of an *infinite sequence* $\underline{T}_0, \underline{T}_1, \underline{T}_2, \ldots$ *of dual tree-trunks for* $\Gamma : \Delta$. In order to do this, we need one more auxiliary notion, an arithmetical function ψ depending on $\Gamma : \Delta$. We first put

$$\phi(r, d, n) = \begin{cases} r(2^{d+1} - 1) & \text{if } n = 0 \\ \frac{r}{n}((n + 1)^{d+1} - 1) & \text{if } n \geq 1. \end{cases}$$

To define ψ from ϕ, we appeal to the standard notion of the *degree* of a formula: the degree of an atomic formula is 0, the degree of $\neg B$ is d + 1, where d is the degree of B, the degree of B & C, of $B \vee C$, and of $B \rightarrow C$ is max(d,e) + 1, where d and e are the degrees of B and C, and the degree of $\forall x\, B(x)$ and of $\exists x\, B(x)$ is d + 1, where d is the degree of B(0). Then, where r is the number of formulas in $\Gamma : \Delta$ and d is the maximum degree of any formula in $\Gamma : \Delta$, we put:

$$\psi(m) = \sum_{i=0}^{m} \phi(r, d, i).$$

Each member $\underline{T}_k$ of any infinite sequence of dual tree-trunks for $\Gamma : \Delta$ is a finite tree with each node a of which is associated a sequent $\Gamma_{a,k} : \Delta_{a,k}$. The first member $\underline{T}_0$ of such a sequence is always the tree consisting of a single node with which is associated $\Gamma : \Delta$. In order to define such a sequence, it therefore remains only to specify how, from any member $\underline{T}_k$ of the sequence, the next member $\underline{T}_{k+1}$ is to be obtained. This we do with the help of a partial function $\lambda a.\ A_{a,k}$ from nodes of $\underline{T}_k$ to formulas; $A_{a,k}$, when it exists, is to be a formula which occurs in $\Gamma_{a,k} : \Delta_{a,k}$. We specify $A_{a,k}$ by induction on the level of the node a. First, let m be the smallest number such that $k \leq \psi(m)$. If $\Gamma_{a,k} : \Delta_{a,k}$ is a basic sequent, $A_{a,k}$ is undefined. If $\Gamma_{a,k} : \Delta_{a,k}$ is not a basic sequent, and, for every $b < a$, either $A_{b,k}$ does not exist or $A_{b,k} \in \Delta_{b,k}$, we let $A_{a,k}$ be, with respect to our fixed enumeration, the earliest formula, if any, which is not fulfilled in $\underline{T}_k$ at a with respect to every number $\leq m$; if there is no such formula, $A_{a,k}$ is again undefined. Finally, if $\Gamma_{a,k} : \Delta_{a,k}$ is not a basic sequent, $b < a$, $A_{b,k} \in \Gamma_{b,k}$ and, for every $c < b$, either $A_{c,k}$ does not exist or $A_{c,k} \in \Delta_{c,k}$, we let $A_{a,k}$ be $A_{b,k}$.

In order to form $\underline{T}_{k+1}$ from $\underline{T}_k$, we consider in turn each node a of $\underline{T}_k$, and either replace $\Gamma_{a,k} : \Delta_{a,k}$ by a sequent $\Gamma_{a,k+1} : \Delta_{a,k+1}$, or, leaving $\Gamma_{a,k+1} : \Delta_{a,k+1}$ identical with $\Gamma_{a,k} : \Delta_{a,k}$, add a new node b and associate with it a sequent $\Gamma_{b,k+1} : \Delta_{b,k+1}$, according to the following rules:

if $A_{a,k}$ does not exist, we take $\Gamma_{a,k+1} : \Delta_{a,k+1}$ as

$$\Gamma_{a,k} : \Delta_{a,k};$$

if $A_{a,k}$ is B & C and $A_{a,k} \in \Gamma_{a,k}$, we take $\Gamma_{a,k+1} : \Delta_{a,k+1}$ as

$$\Gamma_{a,k}, B, C : \Delta_{a,k};$$

if $A_{a,k}$ is B & C and $A_{a,k} \in \Delta_{a,k}$, we take $\Gamma_{a,k+1} : \Delta_{a,k+1}$ either as $\Gamma_{a,k} : B, \Delta_{a,k}$ or as $\Gamma_{a,k} : C, \Delta_{a,k}$;

if $A_{a,k}$ is B $\vee$ C and $A_{a,k} \in \Gamma_{a,k}$, we take $\Gamma_{a,k+1} : \Delta_{a,k+1}$ either as $\Gamma_{a,k}, B : \Delta_{a,k}$ or as $\Gamma_{a,k}, C : \Delta_{a,k}$;

if $A_{a,k}$ is B $\vee$ C and $A_{a,k} \in \Delta_{a,k}$, we take $\Gamma_{a,k+1} : \Delta_{a,k+1}$ as $\Gamma_{a,k} : B, C, \Delta_{a,k}$;

if $A_{a,k}$ is B $\rightarrow$ C and $A_{a,k} \in \Gamma_{a,k}$, we take $\Gamma_{a,k+1} : \Delta_{a,k+1}$ either as $\Gamma_{a,k} : B, \Delta_{a,k}$ or as $\Gamma_{a,k}, C : \Delta_{a,k}$;

if $A_{a,k}$ is B $\rightarrow$ C and $A_{a,k} \in \Delta_{a,k}$, we leave $\Gamma_{a,k+1} : \Delta_{a,k+1}$ as $\Gamma_{a,k} : \Delta_{a,k}$, and put a new node b immediately above a (on a path distinct from any existing path through a), taking $\Gamma_{b,k+1} : \Delta_{b,k+1}$ as $\Gamma_{a,k}, B : C$; further, if there is no node c > a in $\underline{T}_k$, we also put another new node c, distinct from b, immediately above a, and take $\Gamma_{c,k+1} : \Delta_{c,k+1}$ as $\Gamma_{a,k} : \Delta_{a,k}$;

if $A_{a,k}$ is $\neg$B and $A_{a,k} \in \Gamma_{a,k}$, we take $\Gamma_{a,k+1} : \Delta_{a,k+1}$ as $\Gamma_{a,k} : B, \Delta_{a,k}$;

if $A_{a,k}$ is $\neg$B and $A_{a,k} \in \Delta_{a,k}$, we leave $\Gamma_{a,k+1} : \Delta_{a,k+1}$ as $\Gamma_{a,k} : \Delta_{a,k}$, and put a new node b immediately above a (on a path distinct from any existing path through a), and take $\Gamma_{b,k+1} : \Delta_{b,k+1}$ as $\Gamma_{a,k}, B :$; further, if there is no node c > a in $\underline{T}_k$, we also put another new node c, distinct from b, immediately above a, and take $\Gamma_{c,k+1} : \Delta_{c,k+1}$ as $\Gamma_{a,k} : \Delta_{a,k}$;

if $A_{a,k}$ is $\forall x\, B(x)$ and $A_{a,k} \in \Gamma_{a,k}$, we take $\Gamma_{a,k+1} : \Delta_{a,k+1}$ as $\Gamma_{a,k}, B(0),\ldots,B(\bar{m}) : \Delta_{a,k}$;

if $A_{a,k}$ is $\forall x\, B(x)$ and $A_{a,k} \in \Delta_{a,k}$, we leave $\Gamma_{a,k+1} : \Delta_{a,k+1}$ as $\Gamma_{a,k} : \Delta_{a,k}$, and put a new node b immediately above a (on a path distinct from any existing path through a), taking $\Gamma_{b,k+1} : \Delta_{b,k+1}$ as $\Gamma_{a,k} : B(\bar{n})$, where n is the smallest number such that $\bar{n}$ does not occur in $\Gamma_{a,k} : \Delta_{a,k}$; further, if there is no node $c > a$ in $\underline{T}_k$, we also put another new node c, distinct from b, immediately above a, and take $\Gamma_{c,k+1} : \Delta_{c,k+1}$ as $\Gamma_{a,k} : \Delta_{a,k}$;

if $A_{a,k}$ is $\exists x\, B(x)$ and $A_{a,k} \in \Gamma_{a,k}$, we take $\Gamma_{a,k+1} : \Delta_{a,k+1}$ as $\Gamma_{a,k}, B(\bar{n}) : \Delta_{a,k}$, where n is the smallest number such that $\bar{n}$ does not occur in $\Gamma_{b,k} : \Delta_{b,k}$ for any b in $\underline{T}_k$ which is on a path through a;

if $A_{a,k}$ is $\exists x\, B(x)$ and $A_{a,k} \in \Delta_{a,k}$, we take $\Gamma_{a,k+1} : \Delta_{a,k+1}$ as $\Gamma_{a,k} : B(0),\ldots,B(\bar{m}), \Delta_{a,k}$.

The dual tree-trunk $\underline{T}_{k+1}$ is the tree which results from applying the above operations to every node of $\underline{T}_k$. As before, the number m is the smallest one such that $k \leq \psi(m)$. For convenience of future reference, in the three cases in which we introduce a new node, the node c, if it is to be added, is to be taken as on the leftmost path through a in $\underline{T}_{k+1}$, and, in any case, the node b is not to be on the leftmost path. Also for convenience of future reference, in the three cases in which there is a choice as to which of two sequents we are to take as being $\Gamma_{a,k+1} : \Delta_{a,k+1}$, we shall refer to the first of those mentioned in the above statement of the rules as $\Gamma'_{a,k+1} : \Delta'_{a,k+1}$ and to the second as $\Gamma''_{a,k+1} : \Delta''_{a,k+1}$. An infinite sequence of dual tree-trunks for $\Gamma : \Delta$ will, for brevity, also be called a *dual sequence for* $\Gamma : \Delta$, and only a member of such a dual sequence will be recognized as being a *dual tree-trunk for* $\Gamma : \Delta$. It should be noted that, in any dual sequence, if a node a belongs to $\underline{T}_k$, then it also belongs to $\underline{T}_m$ for every m > k; if a belongs to $\underline{T}_k$ but not to $\underline{T}_j$ for any j < k, we say that a is *introduced at stage* k. It should also be noted (a) that if a belongs to $\underline{T}_j$, and j < k, then $\Gamma_{a,j} \subseteq \Gamma_{a,k}$ and $\Delta_{a,j} \subseteq \Delta_{a,k}$; (b) that if a and b belong to $\underline{T}_k$, and a > b, then $\Gamma_{b,k} \subseteq \Gamma_{a,k}$, and hence that, as claimed, $A_{a,k}$, when it exists, always occurs in $\Gamma_{a,k} : \Delta_{a,k}$; and (c) that if a > b, and a lies on the leftmost path through b in $\underline{T}_k$, then $\Gamma_{a,k} : \Delta_{a,k}$ is identical with $\Gamma_{b,k} : \Delta_{b,k}$.

By a *dual tree for* $\Gamma : \Delta$ is meant a tree $\underline{T}$ with each of whose nodes a is associated an ordered pair $\Gamma_a : \Delta_a$ of sets of formulas, either or both of which may be infinite, such that, for some dual sequence $\underline{T}_0, \underline{T}_1, \ldots$ for $\Gamma : \Delta$, each node a of $\underline{T}$ belongs to some $\underline{T}_k$, and, if a was introduced at stage k, then $\Gamma_a = \sum_{i=k}^{\infty} \Gamma_{a,i}$ and $\Delta_a = \sum_{i=k}^{\infty} \Delta_{a,i}$. A dual tree-trunk $\underline{T}_k$ is a *refutation tree-trunk* if $\Gamma_{a,k} : \Delta_{a,k}$ is not a basic sequent for any node a in $\underline{T}_k$; and a dual sequence is a *refutation sequence*, and its associated dual tree a *refutation tree*, if all the members of the sequence are refutation tree-trunks.

We can now state:

Theorem 2. If there is no proof-tree for $\Gamma : \Delta$ of level $\leq$ k, then there is a dual sequence $\underline{T}_0, \underline{T}_1, \ldots$ for $\Gamma : \Delta$ such that $\underline{T}_k$ is a refutation tree-trunk.

Proof. Assume that there is no proof-tree for $\Gamma : \Delta$ of level $\leq$ k.

We specify a particular dual sequence $\underline{T}_0, \underline{T}_1, \ldots$ for

$\Gamma : \Delta$, and claim that, for each i, $0 \leq i \leq k$, and each node a belonging to $\underline{T}_i$, there is no proof-tree for $\Gamma_{a,i} : \Delta_{a,i}$ of level $\leq k - i$. In order to determine the dual sequence, it is necessary only to lay down, for any case in which, in constructing $\underline{T}_{i+1}$ from $\underline{T}_i$, we have, for some node a in $\underline{T}_i$, a choice between taking $\Gamma_{a,i+1} : \Delta_{a,i+1}$ as $\Gamma'_{a,i+1} : \Delta'_{a,i+1}$ and taking it as $\Gamma''_{a,i+1} : \Delta''_{a,i+1}$, which we are to choose. The rule we are to follow in such a case is this: if there is no proof-tree of level $\leq k - i - 1$ for $\Gamma'_{a,i+1} : \Delta'_{a,i+1}$, we shall choose it; otherwise, we shall choose $\Gamma''_{a,i+1} : \Delta''_{a,i+1}$. Note that, since there are only finitely many proof tree-trunks of a given level for a given sequent, this is an effective rule.

We now prove by induction on i that, if $0 \leq i \leq k$ and a belongs to $\underline{T}_i$, there is no proof-tree for $\Gamma_{a,i} : \Delta_{a,i}$ of level $\leq k - i$. If $i = 0$, $\Gamma_{a,i} : \Delta_{a,i}$ is $\Gamma : \Delta$, and the statement is the hypothesis of the theorem. For the induction step, there are three cases:

(i) a was introduced at stage i. Then a stands immediately above a node b such that $\Gamma_{b,i-1} : \Delta_{b,i-1}$ is either identical with $\Gamma_{a,i} : \Delta_{a,i}$ or can be derived from it by $: \rightarrow$, $: \neg$, or $: \forall$. By the induction hypothesis, there is no proof-tree for $\Gamma_{b,i-1} : \Delta_{b,i-1}$ of level $\leq k - i + 1$; hence there is no proof-tree for $\Gamma_{a,i} : \Delta_{a,i}$ of level $\leq k - i$.

(ii) a was introduced at a stage earlier than i, and we had no choice in forming $\Gamma_{a,i} : \Delta_{a,i}$ from $\Gamma_{a,i-1} : \Delta_{a,i-1}$. Then $\Gamma_{a,i-1} : \Delta_{a,i-1}$ is either identical with $\Gamma_{a,i} : \Delta_{a,i}$ or can be derived from it by a single-premiss inference. As under (i), it follows from the induction hypothesis that there is no proof-tree for $\Gamma_{a,i} : \Delta_{a,i}$ of level $\leq k - i$.

(iii) a was introduced at a stage earlier than i, but the general rules for constructing a dual sequence gave us an option, in forming $\Gamma_{a,i} : \Delta_{a,i}$, between taking it as $\Gamma'_{a,i} : \Delta'_{a,i}$ and taking it as $\Gamma''_{a,i} : \Delta''_{a,i}$. If we chose the former, then according to the rule for constructing the particular dual sequence $\underline{T}_0$, $\underline{T}_1, \ldots$, this means that there is no proof-tree for $\Gamma_{a,i} : \Delta_{a,i}$ of level $\leq k - i$. If, on the other hand, we chose $\Gamma''_{a,i} : \Delta''_{a,i}$ to be $\Gamma_{a,i} : \Delta_{a,i}$, this means that there is a proof-tree for $\Gamma'_{a,i} : \Delta'_{a,i}$ of level $\leq k - i$. Suppose, in this case, that there is also a proof-tree for $\Gamma''_{a,i} : \Delta''_{a,i}$ of level $\leq k - i$. Then, since $\Gamma_{a,i-1} : \Delta_{a,i-1}$ can be derived from $\Gamma'_{a,i} : \Delta'_{a,i}$ and $\Gamma''_{a,i} : \Delta''_{a.i}$ by a two-premiss inference, it follows that there is a proof-tree for $\Gamma_{a,i-1} : \Delta_{a,i-1}$ of level $\leq k - i + 1$. This, however, is contrary to the induction hypothesis, and therefore there is not, in this case, a proof-tree for $\Gamma''_{a,i} : \Delta''_{a,i}$, that is, for $\Gamma_{a,i} : \Delta_{a,i}$, of level $\leq k - i$.

It follows, in particular, that, if a belongs to $\underline{T}_k$, there is no proof-tree of level 0 for $\Gamma_{a,k} : \Delta_{a,k}$, i.e. that $\Gamma_{a,k} : \Delta_{a,k}$ is not a basic sequent. $\underline{T}_k$ is thus a refutation tree-trunk.

<u>Theorem 3</u>. If every attempt to construct a refutation tree for $\Gamma : \Delta$ fails, there is a proof of $\Gamma : \Delta$.

Proof. We can represent the dual sequences for $\Gamma : \Delta$ as elements of a dressed fan $\langle s, h \rangle$. We take $h(\langle\ \rangle)$ as the dual tree-trunk which figures as $\underline{T}_0$ in all dual sequences for $\Gamma : \Delta$. Suppose $\vec{u}$ is admissible in s and $h(\vec{u}) = \underline{T}_k$, and let $\underline{T}^{(0)}_{k+1}, \ldots, \underline{T}^{(r)}_{k+1}$ be all the dual tree-trunks which can occur in a dual sequence as the successor of $\underline{T}_k$. Then we take $\vec{u} \frown i$ as admissible just in case $i \leq r$, and put $h(\vec{u} \frown i) = \underline{T}^{(i)}_{k+1}$.

We interpret the statement that every attempt to

construct a refutation tree for $\Gamma : \Delta$ fails as meaning that, for every $\alpha \in s$, there exists a number m such that $h(\bar{\alpha}(m))$ contains a basic sequent. It follows by the Fan Theorem that there exists a bound k such that, in every dual sequence $\underline{T}_0, \underline{T}_1, \ldots$ for $\Gamma : \Delta$, $\Gamma_{a,m} : \Delta_{a,m}$ is a basic sequent for some $m \leq k$ and some a in $\underline{T}_m$, and hence that $\Gamma_{a,k} : \Delta_{a,k}$ is also a basic sequent and $\underline{T}_k$ therefore not a refutation tree-trunk. It follows, by Theorem 2, that there cannot fail to be a proof-tree for $\Gamma : \Delta$ of level $\leq k$, and hence, since it is decidable whether or not there is such a proof-tree, that there is one.

If we could establish that, whenever a formula A is valid on Beth trees, then every attempt to construct a refutation tree for the sequent :A will fail, we should have proved a completeness theorem for single formulas with respect to Beth trees. In order to make any approach to such a result, we must settle on a way of treating a dual tree as a Beth tree, i.e. we must specify a particular interpretation of our first-order language relative to any given dual tree, considered as a Beth tree. We shall take our domain to be the species $\underline{N}$ of natural numbers. It then remains only to lay down at which nodes the atomic formulas are true: if A is an atomic formula, then A is to be true at a node a of a dual tree $\underline{T}$ just in case a is barred in $\underline{T}$ by the species of nodes b such that $A \in \Gamma_b$. We next state a lemma.

Lemma. If $\underline{T}$ is a refutation tree, then every formula is fulfilled in $\underline{T}$ at every node and with respect to every number.

Proof. Let $\underline{T}$ be a refutation tree for $\Gamma : \Delta$, formed from a refutation sequence $\underline{T}_0, \underline{T}_1, \ldots$ for $\Gamma : \Delta$. Let r be the number of formulas occurring in $\Gamma : \Delta$, and let d be the maximum degree of any formula occurring in $\Gamma : \Delta$.

Now, for any fixed node a and number m, consider any formula B which occurs in $\Gamma_{a,i} : \Delta_{a,i}$ for some $i \leq \psi(m)$. B must have been generated by some formula A occurring in $\Gamma : \Delta$, in the following sense. In applying the rule for constructing $\underline{T}_{j+1}$ from $\underline{T}_j$ for any $j < i$, each $A_{b,j}$, for b a node in $\underline{T}_j$, will cause us to introduce at most $m + 1$ (or 2 if $m = 0$) new subformulas of $A_{b,j}$, which did not occur in $\Gamma_{b,j} : \Delta_{b,j}$, either into $\Gamma_{b,j+1} : \Delta_{b,j+1}$ or into $\Gamma_{b',j+1} : \Delta_{b',j+1}$ for some new node b': let us say that $A_{b,j}$ *immediately generates* these new formulas. We shall then say that a formula A *generates* a formula B iff either A is identical with B or A immediately generates a formula which generates B. Let us suppose that a is introduced at stage j. Then, where $m' = \max(m,1)$, any one formula of degree $e \leq d$ can generate at most $1 + (m' + 1) + \ldots + (m' + 1)^e = \frac{1}{m'}((m' + 1)^{e+1} - 1)$ formulas in $\Theta_{a,m} = \bigcup_{i=j}^{\psi(m)} (\Gamma_{a,i} \cup \Delta_{a,i})$. Since every formula in $\Theta_{a,m}$ is generated by some formula occurring in $\Gamma : \Delta$, it follows that there are at most $\phi(r, d, m)$ formulas in $\Theta_{a,m}$.

Now suppose that $B \in \Gamma_a \cup \Delta_a$; we wish to show that B is fulfilled in $\underline{T}$ at a with respect to m. Let n be such that $n \geq m$ and B occurs in $\Gamma_{a,k} : \Delta_{a,k}$, where $k = 1$ if $n = 0$ and $k = \psi(n - 1) + 1$ if $n > 0$. Suppose that B is not fulfilled in $\underline{T}_{\psi(n)+1}$ at a with respect to m. Then B is also not fulfilled in $\underline{T}_i$ at a with respect to m for any i such that $k \leq i \leq \psi(n)$; it follows that $A_{a,i}$ must be defined for each such i. Moreover, by the rules of construction, for all j such that $i < j \leq \psi(n)$, $A_{a,i}$ is fulfilled in $\underline{T}_j$, with respect to every number $\leq n$, at a and at each $b < a$ such that $A_{a,i} \in \Gamma_{b,j}$; hence $A_{a,i}$ and $A_{a,j}$ are distinct whenever $k \leq i < j \leq \psi(n)$. Furthermore, each $A_{a,1} \in \Gamma_{a,i} \cup \Delta_{a,i} \subseteq \Theta_{a,n}$, and $\Theta_{a,n}$ has at most $\phi(r, d, n)$ members. By the definition of ψ, however, there are just $\phi(r, d, n)$ numbers i such that $k \leq i \leq \psi(n)$. It follows that B must be $A_{a,i}$ for some such i, and therefore that B is fulfilled in $\underline{T}_{\psi(n)+1}$ at a with respect to m, contrary to hypothesis. Since the hypothesis was a decidable one, we have shown that B is fulfilled in $\underline{T}_{\psi(n)+1}$, and hence in $\underline{T}$, at a with respect to m.

It is now very easy to prove the next theorem.

Theorem 4. If $\underline{T}$ is a refutation tree, and a a node in $\underline{T}$, then every formula in Γ_a is true at a and no formula in Δ_a is true at a.

Proof. Let $\underline{T}$ be formed from the refutation sequence $\underline{T}_0, \underline{T}_1, \ldots$, and let a be a node in $\underline{T}$ and A a formula in $\Gamma_a \cup \Delta_a$. The proof is by induction on the degree of A.

For the induction basis, let A be atomic. If $A \in \Gamma_a$, it is immediate that A is true at a. If $A \in \Delta_a$, consider any node b on the leftmost path through a. If $b < a$, $\Gamma_b \subseteq \Gamma_a$; hence $a \notin \Gamma_b$, since otherwise we should have $A \in \Gamma_a \cup \Delta_a$, and hence $A \in \Gamma_{a,k} \cup \Delta_{a,k}$ for some k, and so $\underline{T}$ would not be a refutation tree. If $b \geq a$, $\Gamma_b : \Delta_b$ is identical with $\Gamma_a : \Delta_a$, whence, for the same reason, $A \notin \Gamma_b$. Hence a cannot be barred by $\{b | A \in \Gamma_b\}$, and so A is not true at a.

For the induction step, there are twelve cases.

(i) A is B & C and $A \in \Gamma_a$. By the lemma, $B \in \Gamma_a$ and $C \in \Gamma_a$. By the induction hypothesis, B and C are true at a, whence A is true at a.

(ii) A is B & C and $A \in \Delta_a$. By the lemma, either $B \in \Delta_a$ or $C \in \Delta_a$. By the induction hypothesis, B and C are not both true at a, whence A is not true at a.

(iii) A is $B \vee C$ and $A \in \Gamma_a$. By the lemma, either $B \in \Gamma_a$ or $C \in \Gamma_a$. By the induction hypothesis, either B or C is true at a, whence A is true at a.

(iv) A is $B \vee C$ and $A \in \Delta_a$. By the lemma, $B \in \Delta_a$ and $C \in \Delta_a$. Hence, for any $b \geq a$ on the leftmost path through a, $B \in \Delta_b$ and $C \in \Delta_b$. By the induction hypothesis, neither B nor C is true at any such b. Hence a is not barred by $\{b \mid B$ is true at b or C is true at b$\}$, and so A is not true at a.

(v) A is $B \rightarrow C$ and $A \in \Gamma_a$. Then for any $b \geq a$, $A \in \Gamma_b$. Hence, by the lemma, for any $b \geq a$, either $B \in \Delta_b$ or $C \in \Gamma_b$. Therefore, by the induction hypothesis, for any $b \geq a$ at which B is true, C is true. Thus A is true at a.

(vi) A is $B \rightarrow C$ and $A \in \Delta_a$. By the lemma, there exists $b \geq a$ such that $B \in \Gamma_b$ and $C \in \Delta_b$. By the induction hypothesis, B is true at b and C is not true at b. Hence A is not true at a.

(vii) A is $\neg B$ and $A \in \Gamma_a$. Then for any $b \geq a$, $A \in \Gamma_b$. Hence by the lemma, for any $b \geq a$, $B \in \Delta_b$. Therefore, by the induction hypothesis, B is not true at any $b \geq a$. Thus A is true at a.

(viii) A is $\neg B$ and $A \in \Delta_a$. By the lemma, there exists $b \geq a$ such that $B \in \Gamma_b$. By the induction hypothesis, B is true at b, and so A is not true at a.

(ix) A is $\forall x\, B(x)$ and $A \in \Gamma_a$. By the lemma, $B(\bar{m}) \in \Gamma_a$ for every m. By the induction hypothesis, $B(\bar{m})$ is true at a for every m. Hence A is true at a.

(x) A is $\forall x\, B(x)$ and $A \in \Delta_a$. By the lemma, there exists $b \geq a$ such that $B(\bar{n}) \in \Delta_b$ for some n. By the induction hypothesis, $B(\bar{n})$ is not true at b, nor, accordingly, at a. Hence A is not true at a.

(xi) A is $\exists x\, B(x)$ and $A \in \Gamma_a$. By the lemma, $B(\bar{n}) \in \Gamma_a$ for some n. By the induction hypothesis, $B(\bar{n})$ is true at a. Hence A is true at a.

(xii) A is $\exists x\, B(x)$ and $A \in \Delta_a$. By the lemma, $B(\bar{m}) \in \Delta_a$ for every m. Hence, for every $b \geq a$ on the leftmost path through a, $B(\bar{m}) \in \Delta_b$, and so, by the induction hypothesis, $B(\bar{m})$ is not true at b, for every m. Thus a is not barred by $\{b|$ for some m, $B(\bar{m})$ is true at $b\}$, and so A is not true at a.

This completes the proof of the theorem.

It now appears at first sight that we have successfully accomplished our strategy for a completeness proof. For suppose that A is valid with respect to Beth trees, i.e. true at the vertex of every Beth tree. Theorem 4 entails that A cannot be true at the vertex of a refutation tree for the sequent :A, from which it follows that there cannot be a refutation tree for :A. This would appear to imply that every attempt to construct a refutation tree for :A must fail, whence we could conclude, by Theorem 3, that there is a proof of :A. The flaw in this plausible reasoning lies in the step from saying that there cannot be a refutation tree for the sequent to saying that every attempt to construct a refutation tree for it must fail: when we scrutinize the forms of the statements involved, we see that this step conceals a hidden shift of the negation sign.

Completeness requires that, for every formula A, the proposition

(1) A is valid

should imply

(2) $\vdash A$.

(For the time being, validity is understood as being with respect to Beth trees.) By Theorem 4, we have an implication from (1) to

(3) There is no refutation tree for :A.

(3) is to be understood as meaning

(3a) There is no dual sequence for :A every member of which is a refutation tree-trunk.

To check this analysis of the logical structure of (3), we recall the dressed fan ⟨s,h⟩ mentioned in the proof of Theorem 3 as containing, as its elements, all the dual sequences for some fixed sequent, which we here take as the sequent :A, for some given formula A. Let us take $\underline{R}$ as the decidable species $\{\vec{u} \mid h(\vec{u})$ is a refutation tree-trunk$\}$. Theorem 4 tells us that, if $\alpha \in s$, if h correlates α with a refutation sequence $\underline{T}_0, \underline{T}_1, \ldots$ for :A, and if $\underline{T}$ is the corresponding refutation tree for :A, then A does not hold at the vertex of $\underline{T}$. To say that h correlates α with a refutation sequence for :A is to say that

$$(4) \qquad \alpha \in s \ \& \ \forall n\ \bar{\alpha}(n) \in \underline{R}.$$

Let us write '$T_\alpha(A)$' to mean that A is true at the vertex of the dual tree corresponding to the dual sequence which h correlates with α. Then, by Theorem 4, we have an implication from (4) to

$$(5) \qquad \neg T_\alpha(A).$$

Contraposing and quantifying, we obtain an implication from

$$(6) \qquad \forall\alpha\ T_\alpha(A)$$

to

$$(3b) \qquad \forall\alpha_{\alpha \in s}\ \neg\forall n\ \bar{\alpha}(n) \in \underline{R},$$

or, equivalently, to

$$(3c) \qquad \neg\exists\alpha_{\alpha \in s}\ \forall n\ \bar{\alpha}(n) \in \underline{R},$$

which is simply the formalization of (3a). Since, if A is valid, it will, a fortiori, be true at the vertex of every dual tree, we have an implication from (1) to (6), and hence from (1) to (3b).

Now Theorem 3 asserts an implication from

(7) Every attempt to construct a refutation tree for : A fails

to (2). However, in the proof of Theorem 3 it was noted that we need to interpret (7) as meaning

(7a) In every dual sequence for : A we can find a member which is not a refutation tree-trunk,

which can be formalized as

(7b) $\forall\alpha_{\alpha \in S} \exists n\, \bar{\alpha}(n) \notin \underline{R}$.

This was essential, if the crucial application of the Fan Theorem was to be possible. Hence, in order by these means to prove completeness, we should have to establish an implication from (3a) to (7a), i.e. to prove

(8) $\forall\alpha_{\alpha \in S} \neg \forall n\, \bar{\alpha}(n) \in \underline{R} \rightarrow \forall\alpha_{\alpha \in S} \exists n\, \bar{\alpha}(n) \notin \underline{R}$.

Since $\underline{R}$ is decidable, (3b) is also equivalent to

(3d) $\forall\alpha_{\alpha \in S} \neg\neg \exists n\, \bar{\alpha}(n) \notin \underline{R}$,

so that (8) may equivalently be written

(8a) $\forall\alpha_{\alpha \in S} \neg\neg \exists n\, \bar{\alpha}(n) \notin \underline{R} \rightarrow \forall\alpha_{\alpha \in S} \exists n\, \bar{\alpha}(n) \notin \underline{R}$.

In view of the decidability of $\underline{R}$, this is a special case of the schema

(9) $\forall\vec{u}(A(\vec{u}) \vee \neg A(\vec{u}))\ \&\ \forall\alpha \neg\neg \exists n\, A(\bar{\alpha}(n)) \rightarrow \forall\alpha\, \exists n\, A(\bar{\alpha}(n))$.

(If we use a suitable coding of finite sequences as natural numbers, our species $\underline{R}$ can be seen to be primitive recursive.) We can thus assert

Theorem 5. If schema (9) holds, ICP is complete, for single formulas, with respect to Beth trees.

There is, unfortunately, good reason to think (9) intuitionistically invalid. It is, in fact, equivalent to Markov's principle:

(10) $\forall n\, (P(n) \vee \neg P(n))\ \&\ \neg\neg \exists n\, P(n) \rightarrow \exists n\, P(n)$.

From (10), (9) follows by taking $P(n)$ as $A(\bar{\alpha}(n))$ for any given α; from (9), (10) follows by taking $A(\vec{u})$ as $P(lh(\vec{u}))$. (But note that, while (9) restricted to primitive recursive $A(\vec{u})$ implies (10) restricted to primitive recursive $P(n)$, the converse does not hold: for $A(\bar{\alpha}(n))$ will not, in general, be a primitive recursive predicate of n whenever $A(\vec{u})$ is a primitive recursive predicate of $\vec{u}$, but only one primitive recursive in α.)

Markov's principle expresses a perfectly clear platonistic notion of a constructive proof or effective procedure. Classically, if P(n) is an effectively decidable predicate, and ∃n P(n) is true, then there is an effective procedure for finding a number k such that $P(\bar{k})$ is true: we simply run through all the numbers 0, 1, 2, ... in turn until we find such a k, which we must eventually do. Hence, classically, *any* proof of ∃n P(n) will be a constructive proof; i.e. the distinction between constructive and non-constructive proofs simply does not arise at this level. In the mixed notation we used before, where the ordinary symbols are to be understood as expressing the classical logical constants, but ∪ and Ǝ have a (classically) constructive meaning, we may write:

$$(10^C) \quad \forall n\ (P(n) \cup \neg P(n))\ \&\ \exists n\ P(n) \to Ǝn\ P(n).$$

And this is precisely the principle we intuitively need for passing from (3a) to (7a): if no dual tree for :A can be a refutation tree, then, since it is decidable whether a given dual tree-trunk is a refutation tree-trunk, all we need to do, for each dual sequence $\underline{T}_0$, $\underline{T}_1$, $\underline{T}_2$,... for :A, is to try out, for k = 0, 1, 2, ... in turn, whether or not $\underline{T}_k$ is a refutation tree-trunk, and we must eventually find a k for which it is not.

Intuitionistically, this reasoning is invalid; or rather, it is not so much invalid as unintelligible, since, in the general case, it starts from the proposition, platonistically understood, that there exists an n such that P(n), or that it is not the case that ¬P(n) for all n; in our example, that, for a given dual sequence, it is not the case that every member of it is a refutation tree-trunk. Given this platonistic proposition, it must indeed follow that, by trying out each n in turn, we shall eventually find one that satisfies P(n); for, if it were not so, we should have ¬P(n) for all n. This is, however, classical reasoning: the intuitionistic statement that ¬¬∃n P(n), or that ¬∀n ¬P(n), does not express the classical proposition that there exists an n such that P(n), or that it will not happen that we check each n in turn, and find, in every case, that ¬P(n). The intuitionistic statement merely expresses that we shall never be able to prove that ∀n ¬P(n); i.e. that, for however large a number m we may have verified that $\forall n_{n \leq m} \neg P(n)$, the possibility will remain open that we may find an n > m for which P(n); and, from this proposition, $\exists n\, P(n)$ does not follow, even in its classical sense. It may be objected that, in our particular case, the proof which we gave of Theorem 4 shows more than merely that, where A is valid, we shall never be able, for a given dual sequence for :A, to *prove* that each of its members if a refutation tree-trunk: it shows that it could not *be the case* that each of its members was a refutation tree-trunk. This latter proposition is, however, intelligible only from a platonistic standpoint. If we take it as intelligible, then we shall indeed take the proof of Theorem 4 as demonstrating its truth, and hence, by Markov's principle in the form (10^{C}), shall conclude that we can find a member of any given dual sequence for :A which is not a refutation tree-trunk. Intuitionistically, however, the proposition cannot even be understood, and we can therefore take Theorem 4 as asserting no more than that it is contradictory to suppose that we could prove that every member of the dual sequence was a refutation tree-trunk.

We have, in fact, already refuted Markov's principle (10) by means of a Beth tree, namely the Beth tree cited in illustration of the lemma to Theorem 1. Hence, if Markov's principle holds generally, under the intended meanings of the logical constants, ICP is complete, for single formulas, with respect to Beth trees, but is certainly incomplete with respect to the intended semantics, since Markov's principle itself is not provable in it. However, in view of Theorem 1, Markov's principle is definitely incorrect intuitionistically, since it is inconsistent with the theory of lawless sequences; it can also be shown to be inconsistent with the theory of the creative subject. (Less compellingly, it is demonstrably underivable either in the system HA of intuitionistic arithmetic or in the standard systems of intuitionistic analysis; its restriction to primitive recursive predicates can be expressed by a single arithmetical formula, whose negation may be consistently added to HA.)

We may also consider the weaker schema

(11) $\forall\vec{u}(A(\vec{u}) \vee \neg A(\vec{u})) \ \& \ \forall\alpha \neg\neg\exists n\, A(\bar{\alpha}(n)) \rightarrow$

$$\neg\neg\forall\alpha\, \exists n\, A(\bar{\alpha}(n)).$$

A special case will be

(12) $\forall\alpha_{\alpha \in S} \neg\neg\exists n\, \bar{\alpha}(n) \notin \underline{R} \rightarrow \neg\neg\forall\alpha_{\alpha \in S}\exists n\, \bar{\alpha}(n) \notin \underline{R}.$

Since, by Theorem 3, we have an implication from (7b) to (2), we also have an implication from

(13) $$\neg\neg\forall\alpha_{\alpha \in S}\exists n\, \bar{\alpha}(n) \notin \underline{R}$$

to

(14) $$\neg\neg\vdash A.$$

If we appeal to (12), therefore, we shall obtain an implication from (3d) to (14), and hence from (1) to (14). Let us speak of a fragment of ICP as being *quasi-complete*, relative to any given notion of an interpretation, if, for every valid formula A in the fragment, A is not unprovable, or, equivalently, if, for every unprovable formula A, A is invalid (where 'A is invalid' is taken to mean that A does not hold under every interpretation, rather than that there is an interpretation under which A does not hold). (I use the terms 'complete' and 'quasi-complete' in preference to 'strongly complete' and 'weakly complete', since the latter pair is often used to mean 'complete for infinite sets of formulas' and 'complete for finite sets (equivalently, for single formulas)' respectively.) We can then assert

Theorem 6. If schema (11) holds, ICP is quasi-complete, for single formulas, with respect to Beth trees.
Unfortunately, there is no particular reason for supposing schema (11) to be intuitionistically valid; it can again be shown to be underivable in the usual systems of intuitionistic analysis, although there is not the same positive reason to suppose it invalid as there was in the case of (9).

From Theorem 1, 5, and 6 we immediately obtain:

Theorem 7. ICP is internally complete, for single formulas, if schema (9) holds, and internally quasi-complete, for single formulas, if schema (11) holds.

Now for all that we have seen so far, the notion of internal validity might be more restrictive than that of validity on Beth trees, and, accordingly, starting from the strengthened hypothesis that A was internally valid, we might be able to derive the conclusion that A was provable, or that it was not unprovable, without needing to appeal to schema (9) or to schema (11). Disappointingly, this hope is dashed by a result of Gödel's, expounded by Kreisel, which, by establishing a near-converse of Theorem 7, shows that it is virtually a best possible result.

In order to state Gödel's result, we consider the following variants on schemata (9) and (11):

$$(9')\quad \forall\alpha_{\alpha \in b} \neg\neg \exists n\, A(\alpha, n) \rightarrow \forall\alpha_{\alpha \in b}\, \exists n\, A(\alpha, n)$$

and

$$(11')\quad \forall\alpha_{\alpha \in b} \neg\neg \exists n\, A(\alpha, n) \rightarrow \neg\neg \forall\alpha_{\alpha \in b}\, \exists n\, A(\alpha, n),$$

where, in both cases, $A(\alpha, n)$ expresses a primitive recursive relation, and b is the full binary spread (i.e. $\alpha \in b$ iff $\forall n\, \alpha(n) \leq 1$). The restriction to the full binary spread is a matter of convenience only, since we can code every sequence α as a binary sequence α^*; for example, by putting

$$\alpha^*(i) = 0 \text{ if } 0 \leq i < \alpha(0) + 1 \text{ or}$$

$$\sum_{j=0}^{2n+1} (\alpha(j) + 1) \leq i < \sum_{j=0}^{2n+2} (\alpha(j) + 1),$$

$$\alpha^*(i) = 1 \text{ if } \sum_{j=0}^{2n} (\alpha(j) + 1) \leq i < \sum_{j=0}^{2n+1} (\alpha(j) + 1).$$

Furthermore, if $A(\alpha, n)$ is primitive recursive, then, for some primitive recursive predicate $B(\vec{u}, n)$, $A(\alpha, n)$ holds iff $\exists m\, B(\bar{\alpha}(m), n)$. Hence, if we put $C(\bar{\beta}(k))$ iff k is of the form $2^m.3^n$, where $B(\bar{\alpha}(m), n)$, $C(\vec{u})$ is primitive recursive, and $\exists n\ A(\alpha, n)$ iff $\exists n\ C(\bar{\alpha}(n))$. (9') and (11') are therefore equivalent to the restrictions of (9) and (11) to primitive recursive $A(\vec{u})$.

Gödel's result is, then, the following.

Theorem 8. If ICP is internally complete for single formulas, then schema (9') holds, and if it is internally quasi-complete for single formulas, schema (11') holds.

Proof. The proof proceeds by constructing, for any primitive recursive predicate $A(\alpha, n)$, a closed formula B such that, if ICP is internally complete for B, then (9') holds for the given $A(\alpha, n)$, and, if it is internally quasi-complete for B, then (11') holds for that predicate. This is done by showing that if, for all α in the full binary spread, $\neg\neg\exists n\ A(\alpha, n)$, then B is internally valid, and that, if B is provable, then, for all α in the full binary spread, $\exists n\ A(\alpha, n)$. Although we have hitherto been considering a first-order language with infinitely many terms (numerals), we shall, for greater generality, set out the construction in a language without individual constants or function-symbols.

We begin by constructing a closed formula P which axiomatizes the theory of the successor relation. P contains a one-place predicate-letter Z (where Zx means intuitively 'x = 0') and two two-place predicate-symbols = and S (where Sxy means intuitively 'y is the successor of x' and = represents equality), and is the conjunction of the universal closures of the formulas:

$$x = x$$
$$x = y \ \&\ x = z \rightarrow y = z$$
$$x = y \ \&\ Zx \rightarrow Zy$$
$$x = y \ \&\ Sxz \rightarrow Syz$$
$$x = y \ \&\ Szx \rightarrow Szy$$
$$Zx \ \&\ Zy \rightarrow x = y$$
$$Sxy \ \&\ Sxz \rightarrow y = z$$
$$Sxz \ \&\ Syz \rightarrow x = y$$
$$Sxy \rightarrow \neg Zy.$$

We then take G as the formula $\exists x\ Zx \ \&\ \forall x\ \exists y\ Sxy$, and H as

P & G.

Now let $A(\alpha, n)$ be some given predicate expressing a primitive recursive relation between choice sequences in the full binary spread and natural numbers, and let $\Phi(\alpha, n)$ be its characteristic functional, i.e.

$$\Phi(\alpha, n) = \begin{cases} 0 & \text{if } A(\alpha, n) \\ 1 & \text{if } \neg A(\alpha, n). \end{cases}$$

Then there is a set $\underline{E}$ of equations containing function-symbols $f_0, f_1, \ldots, f_k$ such that, for each n and each $\alpha \in b$, if we add to $\underline{E}$ sufficiently many equations of the form $f_0(\bar{m}) = \bar{i}$, where $\alpha(m) = i$, we can derive the equation $f_k(\bar{n}) = \bar{j}$, where $j = \Phi(\alpha, n)$. Furthermore, for each i, $1 \leq i \leq k$, $\underline{E}$ will contain an equation or pair of equations of one of the following forms (where m and n, with or without a subscript, are free variables):

(i) $f_i(n) = 0$

(ii) $f_i(n) = n'$

(iii) $f_i(n_1, \ldots, n_r) = n_j$ for some j, $1 \leq j \leq r$.

(iv) $f_i(n_1, \ldots, n_r) = f_{s_0}(f_{s_1}(n_1, \ldots, n_r), \ldots,$

$f_{s_q}(n_1, \ldots, n_r))$ where $0 \leq s_j < i$ for $0 \leq j \leq q$.

(v) $f_i(0, m) = f_r(m)$ and

$f_i(n', m) = f_s(n, m, f_i(n, m))$

for some r and s, $0 \leq r < i$, $0 \leq s < i$.

We construct a formula E which formalizes this set $\underline{E}$ of equations; E contains, besides Z, S, and =, a one-place predicate-letter Q (where Qx means intuitively '$\alpha(x) = 0$'), and, for each i, $0 \leq i \leq k$, an $(r + 1)$-place predicate-letter F_i, where f_i is an r-ary function-symbol (and where $F_i x_1 \ldots x_r x$ means intuitively '$f_i(x_1, \ldots, x_r) = x$'). E is the conjunction of the universal closures of all formulas of the following forms:

$x = y \;\&\; Qx \rightarrow Qy$

$y = z \;\&\; F_i x_1 \ldots x_{j-1} y x_{j+1} \ldots x_r x_{r+1} \rightarrow F_i x_1 \ldots x_{j-1} z x_{j+1} \ldots x_r x_{r+1}$

for each i, $0 \leq i \leq k$, and each j, $1 \leq j \leq r + 1$.

$F_i x_1 \ldots x_r y \;\&\; F_i x_1 \ldots x_r z \rightarrow y = z$

for each i, $0 \leq i \leq k$.

$Qx \;\&\; Zy \rightarrow F_0 xy$

$\neg Qx \;\&\; Zy \;\&\; Syz \rightarrow F_0 xz$

$Zy \rightarrow F_i xy$ for each i such that $\underline{E}$ contains an equation of the form (i) governing f_i.

$Sxy \rightarrow F_i xy$ for each i such that $\underline{E}$ contains an equation of the form (ii) governing f_i.

$F_i x_1 \ldots x_r x_j$ for each i such that $\underline{E}$ contains an equation of the form (iii) governing f_i.

$F_{s_1} x_1 \ldots x_r y_1 \;\&\ldots\&\; F_{s_q} x_1 \ldots x_r y_q \;\&\; F_{s_0} y_1 \ldots y_q z \rightarrow F_i x_1 \ldots x_r z$

for each i such that $\underline{E}$ contains an equation of the form (iv) governing f_i.

$Zx \;\&\; F_r yz \rightarrow F_i xyz$ and

$F_i xyz \;\&\; F_s xyzu \;\&\; Sxv \rightarrow F_i vyu$

for each i such that $\underline{E}$ contains a pair of equations of the form (v) governing f_i.

We now take C as the formula $H \;\&\; E \;\&\; \forall x\,(Qx \vee \neg Qx)$; we take D as $\exists x\, \exists y\, (Zy \;\&\; F_k xy)$; and, finally, we take B as $\neg(C \;\&\; \neg D)$.

Lemma 1. If $\forall\alpha_{\alpha \in b} \neg\neg\exists n\, A(\alpha, n)$, then B is internally valid.

Proof. Assume that $\forall\alpha_{\alpha \in b} \neg\neg\exists n\, A(\alpha, n)$.

We wish to show that B holds under any internal interpretation. An internal interpretation of B is a structure $\langle \underline{M}, \underline{Z}^*, \underline{S}^*, \underline{I}^*, \underline{Q}^*, \underline{F}_0^*, \underline{F}_1^*, \ldots, \underline{F}_k^* \rangle$, where $\underline{M}$ is an inhabited species taken as the domain of the variables, and the rest are subspecies of $\underline{M}^n$, for varying values of n, taken as interpreting the several predicate-symbols of B ($\underline{I}^*$ being the interpretation of =). Consider any one such interpretation, and suppose that C & $\neg$D holds under it.

Since G holds under the interpretation, there is, for every element $d \in \underline{M}$, an element $d' \in \underline{M}$ such that $\underline{S}^*(d,d')$; hence, by the Axiom of Choice, there is a function g defined over $\underline{M}$ such that $\underline{S}^*(d,g(d))$ for every $d \in \underline{M}$. Further, there is an element $d \in \underline{M}$ satisfying $\underline{Z}^*(d)$. We define a mapping of the natural numbers on to elements $n^* \in \underline{M}$ by letting 0^* be any one element such that $\underline{Z}^*(0^*)$, and setting $(n + 1)^* = g(n^*)$ for every n. From the fact that P holds under the interpretation, it is easy to see that the species of elements $n^* \in \underline{M}$ is isomorphic to the species $\underline{N}$ of natural numbers with respect to the successor relation.

Since $\underline{Q}^*(d) \vee \neg\underline{Q}^*(d)$ for each $d \in \underline{M}$, there is, again by the Axiom of Choice, a function α^* defined over $\underline{M}$ and satisfying

$$\alpha^*(d) = \begin{cases} 0^* & \text{if } \underline{Q}^*(d) \\ 1^* & \text{if } \neg\underline{Q}^*(d). \end{cases}$$

Let α be that element of the full binary spread such that, for each n, $\alpha(n) = 0$ if $\alpha^*(n^*) = 0^*$ and $\alpha(n) = 1$ if $\alpha^*(n^*) = 1^*$. Then $\alpha(n) = 0$ iff $\underline{Q}^*(n^*)$, and therefore iff $\underline{F}_0^*(n^*,0^*)$.

For each i, $1 \leq i \leq k$, let h_i be the function such that, for each n and m, $h_i(n) = m$ iff the equation $f_i(\bar{n}) = \bar{m}$ is derivable from the set $\underline{E}$ supplemented by sufficiently many equations of the form $f_0(\bar{r}) = \bar{j}$ where $\alpha(r) = j$. We argue by induction on i that, for each $n_1,\ldots,n_r$ and m, $h_i(n_1,\ldots,n_r) = m$ iff $\underline{F}_i^*(n_1^*,\ldots,n_r^*,m^*)$. Hence, in particular, $h_k(n) = \Phi(\alpha, n) = 0$ iff $\underline{F}_k^*(n^*, 0^*)$. Now by assumption $\neg D$ holds under the interpretation. It follows that, for every d and d' in $\underline{M}$, $\neg(\underline{Z}^*(d') \;\&\; \underline{F}_k^*(d,d'))$, and hence that, for every n, $\neg \underline{F}_k^*(n^*,0^*)$. Accordingly, $\Phi(\alpha, n) = 1$ for every n, and therefore $\neg \exists n\, A(\alpha, n)$. This, however, contradicts our assumption that $\neg\neg \exists n\, A(\alpha, n)$ for every α in the full binary spread. We have thus shown that C & $\neg$D cannot hold under any interpretation and hence that B is valid.

(By considering that interpretation with the species of natural numbers as domain which is obtained by giving the predicate-symbols of B their intuitive meaning, relative to any given α, we can easily establish the converse of Lemma 1; and, by precisely parallel arguments, we can show that C → D is valid iff $\forall\alpha_{\alpha \in b}\, \exists n\, A(\alpha, n)$. However, we do not need these results for the proof of the theorem.)

We now observe that C & $\neg$D is equivalent to a prenex formula, say U. For C is a conjunction of prenex formulas, and $\neg$D is equivalent to $\forall x\, \forall y\, \neg(Zx \;\&\; F_k xy)$, and a conjunction of prenex formulas can readily be brought to prenex form by first making all the variables distinct. In particular, U may be seen to have the form

$$\exists x\, \forall y\, \exists z\, \forall u_1 \ldots \forall u_s\, (Zx \;\&\; Syz \;\&\; W(u_1,\ldots,u_s)).$$

B is equivalent to $\neg$U. On the strength of this we assert

Lemma 2. If $\vdash$B, then $\forall\alpha_{\alpha \in b}\, \exists n\, A(\alpha, n)$.

Proof. Assume $\vdash$B. Then there is a proof of the sequent U: .

Since U is a prenex formula, we can, by Gentzen's extended Hauptsatz, find a cut-free proof of U: which consists of a proof by purely sentential rules of a quantifier-free 'mid-sequent' U': , followed by a derivation of U: from U': by quantifier rules alone. U': will have the form

$U_1, \ldots U_q :$, where each U_j, $1 \leq j \leq q$, is a substitution instance, obtained by replacing free variables by free variables, of the quantifier-free part of U. These free variables may be taken as drawn from $b_1, \ldots, b_q$, $c_1, \ldots, c_q$, where each U_j takes the form $Zb_j \;\&\; St_jc_j \;\&\; W_j$, where t_j is either b_i for some $i \leq j$ or c_i for some $i < j$, and W_j is a substitution instance of $W(u_1, \ldots, u_s)$. Since the sequent U' : is provable, we have a proof of the formula $\neg U'$, i.e. $\neg(U_1 \;\&\; \ldots \;\&\; U_q)$.

We assume the soundness of IC, so that $\neg U'$ is valid. We now interpret $\neg U'$ over the natural numbers, with a particular assignment to its free variables $b_1, \ldots, b_q$, $c_1, \ldots, c_q$: we assign 0 to each b_i, and, where n is assigned to t_i, we assign $n + 1$ to c_i (t_1 can only be b_1). We select some one α in the full binary spread, and interpret the predicate-symbols according to their intuitive meanings, relative to that α. Under this interpretation and assignment, each substitution instance C_j of the quantifier-free part of C, contained as a conjunct in U_j, comes out true; hence, since $\neg U'$ is true under this interpretation and assignment, we conclude that a formula of the form

$$\neg[\neg(Zs_1 \;\&\; F_kr_1s_1) \;\&\; \ldots \;\&\; \neg(Zs_q \;\&\; F_kr_qs_q)]$$

is true, where the r_i and s_i are drawn from $b_1, \ldots, b_q$, $c_1, \ldots, c_q$. Where n_i is, for each i, the number assigned to the variable r_i, this comes out as holding under the interpretation provided that it is not the case that $h_k(n_i) \neq 0$ for each i, $1 \leq i \leq q$. Since $h_k(n) = 0 \vee h_k(n) \neq 0$ for each n, it follows that $h_k(n_i) = 0$ for some i, and therefore, since $h_k(n) = \Phi(\alpha, n)$, that $\exists n\, A(\alpha, n)$. Since α was any element of the full binary spread, we have shown that $\forall\alpha_{\alpha \in b}\, \exists n\, A(\alpha, n)$.

The theorem now follows easily. If $\forall\alpha_{\alpha \in b} \neg\neg\exists n\, A(\alpha, n)$, then B is internally valid by Lemma 1. If ICP is internally complete for B, B is therefore provable, whence, by Lemma 2, $\forall\alpha_{\alpha \in b} \exists n\, A(\alpha, n)$. If ICP is internally quasi-complete for B, then B is not unprovable, whence, by contraposing Lemma 2 twice, $\neg\neg\forall\alpha_{\alpha \in b} \exists n\, A(\alpha, n)$.

It is an immediate corollary of Theorem 8 that, for each primitive recursive predicate P(n), there is a formula B such that, if ICP is internally complete for B, then Markov's principle (10) holds for that P(n). In fact, this result can be strengthened by taking B as a so-called *negative* formula, that is, one built up from negations of atomic formulas without the use of $\vee$ or $\exists$ (in the case of a formula of some actual theory, it is sufficient that it be so built up from stable atomic formulas). We sketch the proof of this result.

Theorem 9. If the negative fragment of ICP is internally complete, for single formulas, then schema (10) holds for each primitive recursive predicate P(n).

Proof. As before, we find, for each primitive recursive predicate P(n), a negative formula B' such that, if $\neg\neg\exists n\, P(n)$, B' is internally valid, and, if $\vdash B'$, $\exists n\, P(n)$. We rely on the soundness of ICP. ($\vdash$, without subscript, means 'is provable in ICP'; $\vdash_C$ means 'is provable in PCP'.)

Let $\phi(n)$ be the characteristic function for P(n), and let $\underline{E}'$ be the set of defining equations for ϕ ($\underline{E}'$ contains only the function-symbols $f_1,\ldots,f_k$). Let P, G, H, and D be as before, and let E' be formed from $\underline{E}'$ as E was formed from $\underline{E}$, but omitting formulas in which Q or F_0 appear; let C' be H & E'.

Suppose, for given n, that P(n), i.e. $\phi(n) = 0$. Then the equation $f_k(\bar{n}) = 0$ can be derived from $\underline{E}'$ by substitution and replacement of equals by equals. It follows that, where $\bar{m}$ is the largest numeral used in the derivation, and G_m is $\exists x_o \exists x_1 \ldots \exists x_m (Zx_o \,\&\, Sx_ox_1 \,\&\, \ldots \,\&\, Sx_{m-1}x_m)$,

$\vdash P \,\&\, G_m \,\&\, E' \to D$, and hence, since $\vdash G \to G_m$, $\vdash C' \to D$. By contraposition and quantification, if not $\vdash C' \to D$, then $\forall n \neg P(n)$.

Now, for any formula K, let K^- be the result of prefixing $\neg\neg$ to each atomic formula, and let K^o be the result of replacing each part of the form $\exists x\, R(x)$ by $\neg \forall x \neg R(x)$, and each part of the form $R \vee Q$ by $\neg(\neg R \,\&\, \neg Q)$. If $\vdash K$, then $\vdash_C K$ and hence $\vdash_C K^{-o}$, since $\vdash_C K \leftrightarrow K^{-o}$. But K^{-o} is a negative formula, and it can be shown that if L is negative and $\vdash_C L$, then $\vdash L$. We thus have that if $\vdash K$, then $\vdash K^{-o}$. It follows that, if not $\vdash (C' \to D)^{-o}$, then not $\vdash C' \to D$, and hence $\forall n \neg P(n)$.

Since ICP is sound, if $(C' \,\&\, \neg D)^{-o}$ comes out true under some interpretation, then not $\vdash (C' \to D)^{-o}$, and hence $\forall n \neg P(n)$. Therefore if $\neg\neg \exists n\, P(n)$, $(C' \,\&\, \neg D)^{-o}$ is not true under any interpretation, whence $\neg(C' \,\&\, \neg D)^{-o}$ is valid. Hence, taking B' as $\neg(C' \,\&\, \neg D)^{-o}$, we have that, if ICP is internally complete for B', and $\neg\neg\exists n\, P(n)$, then $\vdash B'$. Now, as previously noted, $C' \,\&\, \neg D$ is equivalent to a prenex formula. For any quantifier-free formula K, we can show that $\vdash K \to K^{-o}$. Since $\vdash \exists x\, R(x) \to \neg\forall x \neg R(x)$, it follows by repeated quantification that, for any prenex formula K, $\vdash K \to K^{-o}$, and hence $\vdash \neg K^{-o} \to \neg K$. Thus if $\vdash B'$, i.e. $\vdash \neg(C' \,\&\, \neg D)^{-o}$, then also $\vdash \neg(C' \,\&\, \neg D)$; from this, as a special case of Lemma 2 of Theorem 8, we have $\exists n\, P(n)$. We have therefore shown that, if ICP is internally complete for B', $\neg\neg\exists n\, P(n) \to \exists n\, P(n)$.

Theorem 8 itself cannot be strengthened by taking B negative. In fact, Kreisel has proved:

Theorem 10. The negative fragment of ICP is internally quasi-complete for single formulas.

Sketch of proof. Let B be a negative formula containing predicate-letters $F_1,\ldots,F_m$. Let Pf(x,y) be a formula of intuitionistic arithmetic HA expressing the proof-predicate for ICP, and let $Pf_C(x,y)$ be one expressing the proof-predicate for PCP; since HA contains a function-symbol for every primitive recursive function, these may be taken as atomic formulas. Let b be the Gödel number of B. Then, by arithmetizing the proof that, since B is negative, if $\vdash_C B$, then $\vdash B$, and contraposing, we obtain

$$\vdash_{HA} \forall x \neg Pf(x,\bar{b}) \rightarrow \forall x \neg Pf_C(x,\bar{b}).$$

By the completeness theorem for PCP, we can find arithmetical predicates $P_1,\ldots,P_m$, which can of course be taken as written without $\vee$ and $\exists$, such that, where B* is obtained from B by replacing F_i by P_i for each i, $1 \leq i \leq m$, and where PA is classical arithmetic enriched by the necessary function-symbols,

$$\vdash_{PA} \forall x \neg Pf_C(x,\bar{b}) \rightarrow \neg B^*$$

and hence

$$\vdash_{PA} \forall x \neg Pf(x,\bar{b}) \rightarrow \neg B^*.$$

Since this formula is itself negative, we also have

$$\vdash_{HA} \forall x \neg Pf(x,\bar{b}) \rightarrow \neg B^*.$$

This is the formal version of the statement that, if not $\vdash B$, then B is not true when interpreted over the natural numbers, with each F_i interpreted as the relation expressed by P_i; hence, if B is valid, $\neg\neg\vdash B$.

A quasi-completeness proof of this kind can plainly be given only for a fragment of predicate logic within which the intuitionistically and classically provable formulas coincide (and not, as Kreisel points out, for every such fragment). As for the general case, it is evident from Theorem 8 that, unless we are prepared to accept schema (11) for primitive recursive predicates, we have no hope of proving even the quasi-completeness of any formalization of intuitionistic logic for which the extended Hauptsatz, which is a version of Herbrand's Theorem, holds. Yet another result of Kreisel's shows that the assumption that there is *any* formal system which is complete with respect to constructive interpretations is in conflict with Church's Thesis, i.e. that Church's Thesis implies that the set of constructively valid formulas is not recursively enumerable. (A formula is constructively valid if it comes out true under every internal interpretation given in terms of constructive functions and completely defined species, i.e. species into whose definition there enters no parameter for a choice sequence.)

Theorem 11. If Church's Thesis holds, the species of constructively valid formulas of first-order logic is not recursively enumerable.

Proof. A predicate $B(\vec{u})$ represents a binary tree if (i) $B(\langle\ \rangle)$, (ii) $\forall\vec{u}\ \forall k\ (B(\vec{u}^\frown k) \to B(\vec{u})$, and (iii) $\forall\vec{u}\ (B(\vec{u}) \to b(\vec{u}) = 0)$, where b is the full binary spread. We shall speak of such a predicate as being (primitive) recursive if, under some fixed effective coding of finite sequences as natural numbers, the corresponding number-theoretic predicate is (primitive) recursive, and similarly for other predicates or functions one or more of whose arguments is a finite sequence- and, where $B(\vec{u})$ is recursive, and e is the index of the characteristic function for the corresponding number-theoretic predicate, we shall write $B(\vec{u})$ as $B_e(\vec{u})$, and speak of e as the index of $B(\vec{u})$, and of the tree represented by $B(\vec{u})$ as being the tree with index e.

If $B(\vec{u})$ represents a binary tree, and $\alpha \in b$, then α is an infinite path in that tree iff $\forall n\ B(\bar{\alpha}(n))$. We set:

$\underline{I}$ = {e | the tree with index e has an infinite primitive recursive path}

$\underline{F}$ = {e | every recursive path in the tree with index e is finite} .

As is usual, we write ω_e for the r. e. set with index e. Then we have the following

Classical Lemma. The sets $\underline{I}$ and $\underline{F}$ are effectively not separable by an r. e. set; i.e. there is a recursive function p such that, for all e, $p(e) \in (\omega_e \cap \underline{I}) \cup (\underline{F} - \omega_e)$.

Proof of Lemma. (A reader who is prepared to take for granted this purely classical proof of a result in recursive function theory may prefer to skip at once to the proof of the theorem from the lemma.)

We put:

$$R(e_o, e_1, \vec{u}) \leftrightarrow b(\vec{u}) = 0 \ \& \ {}^{i}{}_{i} < \ell h(\vec{u}) \ \forall^{j} j < \ell h(\vec{u})$$

$$[(T_1(e_o, i, j) \rightarrow a_i = 0) \ \& \ (T_1(e_1, i, j) \rightarrow a_i = 1)].$$

Then R is primitive recursive and, for fixed e_o and e_1, $R(e_o, e_1, \vec{u})$ represents a binary tree. For $\alpha \in b$, α is an infinite path in that tree, i.e. $\forall n\ R(e_o, e_1, \bar{\alpha}(n))$, iff $\omega_{e_o} \subseteq \{n \mid \alpha(n) = 0\} \ \& \ \omega_{e_1} \cap \{n \mid \alpha(n) = 0\} = \emptyset$, i.e. iff $\{n \mid \alpha(n) = 0\}$ separates ω_{e_o} and ω_{e_1}. We can find a primitive recursive function h such that for all a

$$\{h(e_o, e_1)\}(\vec{u}) = \begin{cases} 0 & \text{if } R(e_o, e_1, \vec{u}) \\ 1 & \text{otherwise.} \end{cases}$$

Then, for each e_o and e_1, $h(e_o, e_1)$ is the index of a binary tree such that, for each α, α is an infinite path in the tree iff α is the characteristic function of a set separating ω_{e_o} and ω_{e_1}.

Take k as a primitive recursive function such that, for all e, $\omega_{k(e)}$ is finite if ω_e is finite, and $\omega_{k(e)} = \underline{N}$ if ω_e is infinite. Consider any pair of disjoint recursively inseparable r. e. sets $\underline{A}$ and $\underline{B}$, and let g_o, g_1 be primitive recursive functions such that, for every e, $\omega_{g_o}(e) = \underline{A} \cap \omega_{k(e)}$ and $\omega_{g_1(e)} = \underline{B} \cap \omega_{k(e)}$. If ω_e is finite, then $\omega_{g_o(e)}$ and $\omega_{g_1(e)}$ are finite and disjoint, and hence can be separated by a primitive recursive set. If, on the other hand, ω_e is infinite, $\omega_{g_o(e)} = \underline{A}$ and $\omega_{g_1(e)} = \underline{B}$, and so $\omega_{g_o(e)}$ and

$\omega_{g_1(e)}$ are recursively inseparable. It follows that, where $f(e) = h(g_0(e), g_1(e))$, if ω_e is finite, the tree with index $f(e)$ contains an infinite primitive recursive path, i.e. $f(e) \in \underline{I}$, and, if ω_e is infinite, every recursive path in the tree with index $f(e)$ is finite, i.e. $f(e) \in \underline{F}$.

Now take q as a primitive recursive function such that, for each e, $\omega_{q(e)} = \{n \mid f(n) \in \omega_e\}$; and let r be the recursive production function for the productive set $\underline{K} = \{e \mid \omega_e \text{ is infinite}\}$. Then, for every e,

$$r(e) \in (\omega_e - \underline{K}) \cup (\underline{K} - \omega_e)$$

and so

$$r(q(e)) \in (\omega_{q(e)} - \underline{K}) \cup (\underline{K} - \omega_{q(e)}).$$

Finally, we take $p(e) = f(r(q(e)))$. By the definition of q, we have

$$r(q(e)) \in \omega_{q(e)} \leftrightarrow p(e) \in \omega_e .$$

By the definition of f, we have

if $r(q(e)) \notin \underline{K}$, $\omega_{r(q(e))}$ is finite, and hence $p(e) \in \underline{I}$, and

if $r(q(e)) \in \underline{K}$, $\omega_{r(q(e))}$ is infinite, and hence $p(e) \in \underline{F}$.

It follows that, for all e,

$$p(e) \in (\omega_e \cap \underline{I}) \cup (\underline{F} - \omega_e).$$

Proof of Theorem. Put:

$\underline{Q} = \{e \mid$ no recursive path in the tree with index e is infinite$\}$.

By the lemma, we have, classically:

(1) $\forall e\ (p(e) \in \omega_e \rightarrow p(e) \in \underline{I})$

(2) $\forall e\ (p(e) \notin \omega_e \rightarrow p(e) \in \underline{F})$.

(2) can be expressed purely arithmetically, by quantification only over natural numbers (and finite sequences), as follows:

$$\forall e\{\forall k\ \neg T_1(e,\ p(e),\ k) \rightarrow$$
$$\forall m[\forall n\ \exists k\ (T_1(m,\ n,\ k)\ \&\ U(k) \leq 1) \rightarrow$$
$$\exists \vec{u}\ (\neg B_{p(e)}(\vec{u})\ \&$$
$$\forall i_{i < \ell h(\vec{u})}\ \forall k\ (T_1(m,\ i,\ k) \rightarrow U(k) = a_i))]\}\ .$$

This statement is classically provable, and hence, by Godel's translation from classical into intuitionistic arithmetic, we can prove intuitionistically:

$$\forall e\{\forall k\ \neg T_1(e,\ p(e),\ k) \rightarrow$$
$$\forall m[\forall n\ \neg\neg\ \exists k\ (T_1(m,\ n,\ k)\ \&\ U(k) \leq 1) \rightarrow$$
$$\neg\neg\ \exists \vec{u}\ (\neg B_{p(e)}(\vec{u})\ \&$$
$$\forall i_{\ i < \ell h(\vec{u})}\ \forall k\ (T_1(m,\ i,\ k) \rightarrow U(k) = a_i))]\}\ ,$$

and hence, a fortiori,

$$\forall e\{\forall k\ \neg T_1(e,\ p(e),\ k) \rightarrow$$
$$\forall m[\forall n\ \exists k\ (T_1(m,\ n,\ k)\ \&\ U(k) \leq 1) \rightarrow$$
$$\neg\neg\exists \vec{u}\ (\neg B_{p(e)}(\vec{u})\ \&$$
$$\forall i_{\ i < \ell h(\vec{u})}\ \forall k\ (T_1(m,\ i,\ k) \rightarrow U(k) = a_i))]\}\ .$$

This latter statement serves to express:

$$(3)\quad \forall e\ (p(e) \notin \omega_e \rightarrow p(e) \in \underset{\sim}{Q})\ ,$$

which is thus intuitionistically true. Similarly, where f_o, f_1, ... is an enumeration of all unary primitive recursive functions, (1) can be expressed as:

$$\forall e\ \forall k\ (T_1(e,\ p(e),\ k) \rightarrow \exists i\ \forall n\ B_{p(e)}(\overline{f_i}(n)))$$

and we can therefore prove intuitionistically:

$$\forall e\ \forall k\ (T_1(e,\ p(e),\ k) \rightarrow \neg\neg\ \exists i\ \forall n\ B_{p(e)}(\overline{f_i}(n))),$$

and hence also:

$$\forall e\ (\neg \exists i\ \forall n\ B_{p(e)}(\overline{f_i}(n)) \rightarrow \forall k\ \neg T_1(e, p(e), k)).$$

This latter statement expresses:

$$(4)\quad \forall e\ (p(e) \notin \underline{I} \rightarrow p(e) \notin \omega_e).$$

Now plainly we have intuitionistically:

$$(5)\quad \forall e\ (p(e) \in \underline{Q} \rightarrow p(e) \notin \underline{I}),$$

and hence:

$$(6)\quad \forall e\ (p(e) \in \underline{Q} \leftrightarrow p(e) \notin \omega_e).$$

From (6) it follows that $\underline{Q}$ cannot be r. e., since, if $\underline{Q}$ were ω_m, we should have $p(m) \in \omega_m \leftrightarrow p(m) \notin \omega_m$.

Now if, for given e, $B_e(\vec{u})$ is a primitive recursive predicate, we may take $\neg B_e(\bar{\alpha}(n))$ as the $A(\alpha, n)$ of Theorem 8, and hence, as in the proof of that theorem, construct the corresponding closed first-order formula B (call it $B^{(e)}$). By Lemma 1 of Theorem 8, and the remark in parentheses immediately following its proof, $B^{(e)}$ is internally valid iff $\forall\alpha_{\alpha \in b}\ \neg\neg \exists n\ \neg B_e(\bar{\alpha}(n))$, i.e. iff no path in the tree with index e is infinite. The proof of Lemma 1 and of its converse will still hold good if we restrict ourselves to constructive interpretations, so that $B^{(e)}$ is constructively valid iff no constructive path in the tree with index e is infinite. If Church's Thesis (CT) be assumed, this amounts to saying that no recursive path in the tree with index e is infinite, i.e. that $e \in \underline{Q}$. We thus have: $CT \rightarrow \{e \mid B^{(e)}$ is constructively valid$\} = \underline{Q}$. But if the set of all constructively valid formulas is r. e., then also the set of all e such that $B^{(e)}$ is constructively valid will be r. e., that is (given CT), $\underline{Q}$ will be r. e., which as we have seen, is impossible.

We can escape the conclusion that every formal system for intuitionistic first-order logic is internally incomplete only if we are prepared to reject either Church's Thesis that every constructive function is recursive or the proposition that every constructively valid first-order formula is internally valid. No proof exists of the latter proposition; although no first-order formula is known which is constructively valid but not true under all more general internal interpretations, it is not obviously impossible that one might be found. As for Church's Thesis, this is not particularly plausible from an intuitionistic standpoint. The assumption that we can effectively recognize a proof of a given statement of some mathematical theory, say elementary number theory, lies at the basis of all intuitionistic mathematics; but to hold that there is any recursive procedure for recognizing proofs of arithmetical statements would be to run foul of Gödel's Incompleteness Theorem. This is not to maintain that the set of arithmetical statements provable by some intuitionistically correct means is not recursively enumerable, but only to deny that the totality of intuitionistically correct proofs of such statements can be represented by a formal system (with a recursive proof-predicate). It is, indeed, true that Church's Thesis is, when expressed in a suitable form, demonstrably consistent with most intuitionistic formal systems; but that is not in itself surprising, because formal systems, as we presently understand them, have recursive proof-predicates. Most intuitionistic formal systems have the existential definability property, namely (for quantification over the natural numbers) that if $\exists x\, A(x)$ is a closed formula, and $\vdash \exists x\, A(x)$, then $\vdash A(\bar{m})$ for some m. It follows, provided that the proof-predicate is recursive, that if $\forall x\, \exists y\, B(x, y)$ is a closed formula, and $\vdash \forall x\, \exists y\, B(x, y)$, then we can find a recursive function f such that $\vdash B(\bar{n}, f(\bar{n}))$ for every n, namely by putting $f(n)$ = the smallest m such that $\vdash B(\bar{n}, \bar{m})$. A system with this property is unlikely to be inconsistent with Church's Thesis. Thus Theorem 11 is not necessarily an insuperable bar to proving the completeness of intuitionistic logic; and, even if it is taken as showing that intuitionistic logic is incomplete -- that not every valid formula is provable -- that does not of itself rule out the possibility of showing that intuitionistic logic is quasi-complete -- that no valid formula is unprovable. Theorem 8 continues to provide the graver obstacle.

5.7. GENERALIZED BETH TREES

Despite the dismayingly negative results set out at the end of the last section, an attempt has recently been made by Veldman, de Swart, and others in Nijmegen to rectify the situation by giving an unconditional proof of completeness with respect to a modified type of Beth trees. We may approach the matter via what is easily seen to be an unworkable suggestion, namely to strengthen Theorem 4 of section 5.6 to:

> If $\underline{T}$ is a dual tree, and a a node in $\underline{T}$, then every formula in Γ_a is true at a, and, if Δ_a contains a formula true at a, then, for some b and some C, $C \in \Gamma_b \cap \Delta_b$.

From Theorem 4 itself we merely derived the corollary that, if $\underline{T}_0$, $\underline{T}_1$,... is a dual sequence for :A, and A is valid on Beth trees, then not every $\underline{T}_i$ is a refutation tree-trunk. Suppose, however, that we could prove the strengthened version, that A is valid on Beth trees, that $\underline{T}_0$, $\underline{T}_1$, ... is a dual sequence for :A, and $\underline{T}$ the corresponding dual tree. Then we could, by the strengthened theorem, find a formula C and a node b such that $C \in \Gamma_b$ and $C \in \Delta_b$; from this it would be possible to find a number i such that $C \in \Gamma_{b,i} \cap \Delta_{b,i}$, so that $\Gamma_{b,i} : \Delta_{b,i}$ would be a basic sequent. We could therefore find a particular $\underline{T}_i$ which was not a refutation tree-trunk: and this is what is needed to make our completeness proof independent of the validity of schema (9).

As things stand, there is no chance of proving this strengthened form of Theorem 4, since it relates to dual trees generally (rather than just to refutation trees, as Theorem 4 itself does), and, where a is a node of a dual tree, there is no guarantee that Γ_a will not be a set of formulas which are inconsistent with respect to Beth trees, that is, a set not every member of which can be true at a node of a Beth tree (as, for example, if Γ_a contains both B and $\neg$B, or the single formula $\neg(B \vee \neg B)$). This prompts us to ask whether it is possible so to modify the notion of a Beth tree that no set of formulas is inconsistent with respect to Beth trees.

It is in fact quite straightforward to do so. For present purposes, it is convenient to take the sentential constant $\perp$ as primitive, and to treat $\neg A$ as an abbreviation of $A \rightarrow \perp$; for the rest, we here revert to our previous form of first-order language, having no free variables, but a numeral $\bar{n}$ for every natural number n (so that all formulas are closed). We keep the same rules of inference as before, save that we omit $\neg$: and : $\neg$; as basic sequents we count all those of the form $\Gamma, A : A, \Delta$, as before, together with those of the form $\Gamma, \perp : \Delta$.

The notion of a Beth tree has three ingredients: the underlying abstract tree structure; the decidable relation of an atomic formula's being verified at a node of the tree; and the relation of a formula's being true at a node of the tree, defined inductively in terms, ultimately, of the relation of verification. We shall henceforward use the variable $\underline{T}$, not for the Beth tree as a whole, but for the underlying abstract tree; and, where A is an atomic formula, and a a node, we write '$\underline{V}$(A, a)' for 'A is verified at a'. We shall regard a Beth tree as determined by the first two ingredients, that is, as an ordered pair $\langle \underline{T}, \underline{V} \rangle$ consisting of an abstract tree $\underline{T}$ and a decidable relation $\underline{V}$ between atomic formulas and nodes of the tree. We can then give the following

Definition. If $\underline{T}$ is an abstract tree, and $\underline{V}$ a relation between atomic formulas and nodes of $\underline{T}$, then $\langle \underline{T}, \underline{V} \rangle$ is an *ordinary Beth tree* iff, for every atomic formula A and all nodes a, b of $\underline{T}$:

(i) $\underline{V}$(A, a) or not $\underline{V}$(A, a);

(ii) if $\underline{V}(A, a)$ and $b \leq a$, then $\underline{V}(A, b)$;

(iii) not $\underline{V}(\perp, a)$.

Applying our ordinary definition of truth at a node, and appealing to our present reading of $\neg B$ as $B \rightarrow \perp$, this yields just that notion of a Beth tree that we have hitherto been using; $\neg B$ will, by the clause for $\rightarrow$, be true at a just in case, for every $b \leq a$ at which B is true, $\perp$ is true, which, in view of condition (iii), is tantamount to saying that at no $b \leq a$ is B true.

In order to obtain a modified notion of Beth trees under which there is no bar to all the members of any set of formulas being true at some one node, it is obviously necessary, and, on reflection, also sufficient, to relax condition (iii) so as to allow $\perp$ to be true at a node. We accordingly give the

Definition. If $\underline{T}$ is an abstract tree, and $\underline{V}$ a relation between atomic formulas and nodes of $\underline{T}$, then $\langle \underline{T}, \underline{V} \rangle$ is a *generalized Beth tree* iff, for every atomic formula A and all nodes a, b of $\underline{T}$:

(i) $\underline{V}(A, a)$ or not $\underline{V}(A, a)$;

(ii) if $\underline{V}(A, a)$ and $b \leq a$, then $\underline{V}(A, b)$;

(iii') if $\underline{V}(\perp, a)$, then $\underline{V}(A, a)$.

(The members of the Nijmegen school merely drop condition (iii), without imposing condition (iii'); but it will readily be seen from what follows that imposing it involves no loss of generality, and simplifies the exposition.)

It is important to avoid thinking of the introduction of generalized Beth trees as, in itself, in any way counter-intuitive. It is, of course, essential to intuitionistic logic that a demonstration that A can never be proved should serve as a proof of $\neg A$; but the converse is by no means required for what may be termed the *elementary* properties of intuitionistic logic, that is to say, those embodied in the formulation of ICP. (It is, at least arguably, required for a full understanding of the notion of proof as employed in the standard intuitive explanations of the logical constants, but certainly not for a grasp of the most immediate features of their meanings, as given by those explanations.) We could, for example, take $\perp$ to be an absurd statement in some theory of whose consistency we are not assured -- if it is provable, then every statement is provable; when we go on to interpret $\neg B$ as $B \to \perp$, the validity of our logical laws will not depend upon the theory's being consistent. Or, if we had a stock of mutually compatible atomic statements, we might take $\perp$ as being, in effect, the conjunction of all of them; although this would yield a non-standard interpretation of negation, it would still be one under which all the usual logical laws held good for statements compounded only from atomic statements in that stock.

Since we have a slightly modified notion of a basic sequent, we need a minor revision of our procedure for generating a dual sequence for any given sequent. First, if $\underline{T}$ is any tree with each node a of which are associated two sets, Γ_a and Δ_a, of formulas, then we must take any formula as being fulfilled, with respect to any number m, at any node a at which either $\Gamma_a \cap \Delta_a$ is inhabited or $\perp \in \Gamma_a$. Secondly, as soon as, at a node a of a tree-trunk $\underline{T}_k$, $\Gamma_{a,k} : \Delta_{a,k}$ is found to be a basic sequent, we must add $\perp$ to $\Gamma_{a,k}$ (if it does not already contain it). Of course, the rules of construction relating to formulas of the form $\neg B$ no longer apply. Furthermore, in order to be able to construe a dual tree as a generalized Beth tree in accordance with condition (iii'), we must modify the definition of verification at a node of a dual tree: if A is atomic, A is now said to be verified at a node a of a dual tree iff either $A \in \Gamma_a$ or $\perp \in \Gamma_a$.

The question now arises whether the introduction of the generalized Beth tree is enough to enable us to prove the strengthened form of Theorem 4. For clarity, we here repeat the usual definition of truth at a node of a Beth tree, formulated now for generalized Beth trees. To say that A is true at a node a is, of course, relative to the generalized Beth tree $\langle \underline{T}, \underline{V} \rangle$ we are considering, so we shall abbreviate 'A is true at a in $\langle \underline{T}, \underline{V} \rangle$' as 'tr(A, a, $\underline{T}$, $\underline{V}$)'; but when no ambiguity is possible, we shall shorten this to 'tr(A, a)'.

Definition. If $\langle \underline{T}, \underline{V} \rangle$ is a generalized Beth tree, a a node of $\underline{T}$ and A a formula, we say that A is *true* at a in $\langle \underline{T}, \underline{V} \rangle$ (in symbols tr(A, a, $\underline{T}$, $\underline{V}$) or, for short, tr(A, a)) iff either:

(1) A is atomic and $\{b \mid \underline{V}(A, b)\}$ bars a;

(2) A is B & C and tr(B, a) and tr(C, a);

(3) A is $B \vee C$ and $\{b \mid$ tr(B, b) or tr(C, b)$\}$ bars a;

(4) A is $B \to C$ and, for every $b \leq a$, if tr(B, b), then tr(C, b);

(5) A is $\forall x\, B(x)$ and, for every n, tr($B(\bar{n})$, a);

(6) A is $\exists x\, B(x)$ and $\{b \mid$ for some n, tr($B(\bar{n})$, b)$\}$ bars a.

The notion of truth, thus defined, has four important general properties, as follows:

tr-(a): if tr($\bot$, a), then tr(A, a) for any A;

tr-(b): if tr(A, a) and $b \leq a$, then tr(A, b);

tr-(c): if $\underline{S}$ bars a and, for every $b \in \underline{S}$, tr(A, b), then tr(A, a);

tr-(d): tr(A, a, $\underline{T}$, $\underline{V}$) iff tr(A, a, $\underline{T}_a$, $\underline{V} \restriction \underline{T}_a$).

(In tr-(d), $\underline{T}_a$ is the subtree of $\underline{T}$ consisting of all nodes $b \leq a$, i.e. with a as vertex, and $V \restriction \underline{T}_a$ is simply the relation $\underline{V}$ confined to $\underline{T}_a$.)

When we scrutinize the proof of Theorem 4, we see that we cannot prove it in its strengthened form in relation to the dual trees, considered as generalized Beth trees, obtained by our modified procedure for constructing dual sequences. The difficulty lies in the induction step for the case when $A \in \Gamma_a$ and A is $B \to C$. In the actual proof of Theorem 4, we argued that, since $B \to C \in \Gamma_a$, $B \to C \in \Gamma_b$ for all $b \geq a$, and hence, for $b \geq a$, either $C \in \Gamma_b$ or $B \in \Delta_b$, since $B \to C$ is, by the lemma, fulfilled at every node. (Recall that, in dual trees, the vertex is at the *bottom*, not at the top.) Hence, by the induction hypothesis, for $b \geq a$ either tr(C,b) or not tr(B,b), and so, for all $b \geq a$, if tr(B, b), then tr(C,b), and thus tr($B \to C$,a).

Theorem 4, in its original form, related only to refutation trees. We are aiming to strengthen it to a theorem applying to all dual trees; but, in this more general form, the second half of the theorem will be weaker, and hence the induction hypothesis will be weaker. Adapted for our modified procedure for generating dual sequences, this generalized form of the theorem would naturally run:

> If a is a node of a dual tree, then, if $A \in \Gamma_a$, tr(A,a), and if $A \in \Delta_a$ and tr(A,a), then, for some $b \geq a$, tr($\bot$,b).

If, now, we attempt to prove this statement for the case in which $A \in \Gamma_a$, A is $B \to C$, we can argue as before that, for $b \geq a$, either $C \in \Gamma_b$ or $B \in \Delta_b$. However, when, at this stage, we apply the induction hypothesis, all that we obtain is that, for $b \geq a$, either tr(C,b) or, if tr(B,b), then tr($\bot$,c) for some $c \geq b$; and this is not enough to guarantee that, if tr(B,b), then tr(C,b).

We can in fact give a definite counter-example to the generalized form of Theorem 4 as formulated above. Suppose that we are constructing a dual sequence for the sequent Q, $(P \to Q) \to R$: . $\underline{T}_0$ is thus:

$$Q, (P \to Q) \to R : \bullet$$

Suppose $\underline{T}_1$ to be chosen as:

$$Q, (P \to Q) \to R : P \to Q \bullet$$

(Q, $(P \to Q) \to R$: $P \to Q$ is, of course, a valid sequent, so that this choice of $\underline{T}_1$ represents the wrong strategy for finding a refutation tree.) There is now only one possible choice for $\underline{T}_2$, namely

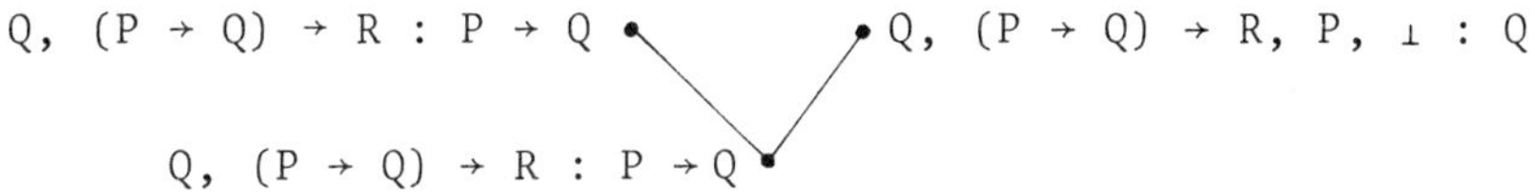

From this point on, $\underline{T}_3$, $\underline{T}_4$, ... are completely determined, by repetition of the operation that took us from $\underline{T}_1$ to $\underline{T}_2$. The dual tree formed from this dual sequence $\underline{T}_o$, $\underline{T}_1$, $\underline{T}_2$, ... therefore has the form:

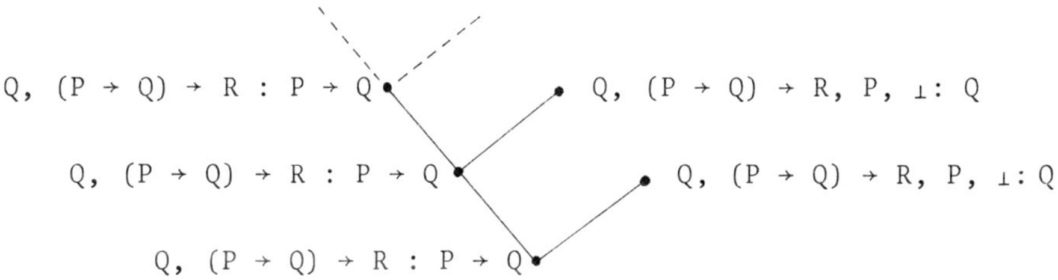

$(P \rightarrow Q) \rightarrow R \in \Gamma_a$ for every node a on the leftmost path in this dual tree, but at no such node is $(P \rightarrow Q) \rightarrow R$ true (since, at every such node, Q is verified, and so $P \rightarrow Q$ is true, but R is not verified at any such node).

It might be thought that, although this counter-example shows that the desired generalized version of Theorem 4 fails in respect of the particular search procedure (procedure for constructing dual sequences) that we have adopted, that was due only to the peculiarities of that search procedure. It is certainly true that the search procedure might have been varied in many different ways; but it seems highly unlikely that any such variation will allow us to prove the generalized form of Theorem 4. In order to be able to prove the induction step for the case when $A \in \Gamma_a$, A is $B \rightarrow C$, we should have to be able to show that if $B \in \Delta_b$, B is true at b and $B \rightarrow C \in \Gamma_b$, then C is true at b; it appears improbable that we can devise any search procedure which will guarantee this.

The only way round this difficulty is to adopt a modified notion of truth: the obvious modification to adopt is that given by the following

Definition. Where $\langle \underline{T}, \underline{V} \rangle$ is a generalized Beth tree, a a node of $\underline{T}$, and A a formula, A is *nearly true* at a in $\langle \underline{T}, \underline{V} \rangle$ (in symbols ntr(A, a, $\underline{T}$, $\underline{V}$) or, for short, ntr(A, a)) iff either:

(1) A is atomic and either $\{b \mid \underline{V}(A, b)\}$ bars a or, for some $b \leq a$, $\underline{V}(\bot, b)$:

(2) A is B & C and ntr(B, a) and ntr(C, a);

(3) A is $B \vee C$ and $\{b \mid$ ntr(B, b) or ntr(C, b)$\}$ bars a;

(4) A is $B \rightarrow C$ and, for every $b \leq a$, if ntr(B, b), then ntr(C, b);

(5) A is $\forall x\, B(x)$ and, for every n, ntr(B($\bar{n}$),a);

(6) A is $\exists x\, B(x)$ and $\{b \mid$ for some n, ntr(B($\bar{n}$), b)$\}$ bars a.

(For the purposes of this definition, we have reverted to the usual orientation of the trees, with the vertex at the top.) It will be noted that the inductive clauses (2) - (6) are exactly the same as the corresponding ones in the definition of *true*, save, of course, for having 'ntr(,)' in place of 'tr(,)'. It would have made no difference, however, if every clause had been modified in the same way as clause (1), since it is straightforward to show, by induction on the complexity of A:

For any formula A, if $b \leq a$ and $\underline{V}(\bot, b)$, then ntr(A,a), and hence, more generally:

ntr-(a'): if $b \leq a$ and ntr($\bot$,b), then ntr(A,a) for any A.

Armed with this definition, we can now indeed prove our generalized version of Theorem 4, in the form:

Theorem 4 (ntr). If a is a node of a dual tree, then if $A \in \Gamma_a$, ntr(A, a) and if $A \in \Delta_a$ and ntr(A, a), then ntr($\bot$, a).

Proof. In order to carry through the induction, we actually prove:

(i) if $A \in \Gamma_a$, then ntr(A, a), and

(ii) if ntr(A, a), $b \geq a$ and $A \in \Delta_b$, then ntr($\bot$, b).

We give two cases of the induction step.

(1) Assume that A is $B \rightarrow C$ and that $A \in \Gamma_a$. Then for every $b \geq a$, $A \in \Gamma_b$. So, by the lemma, $C \in \Gamma_b$ or $B \in \Delta_b$. Hence, by the induction hypothesis, if ntr(B, b), then either ntr(C, b) or ntr($\bot$, b), and so in either case ntr(C, b) (by ntr-(a')). Thus ntr(A, a).

(2) Assume that A is $\forall x\, B(x)$, that ntr(A, a), that $b \geq a$ and that $A \in \Delta_b$. By the lemma, for some n and for some $c \geq b$, $B(\bar{n}) \in \Delta_c$. Since ntr(A, a), ntr($B(\bar{n})$,a). Hence, by the induction hypothesis, ntr($\bot$, c), whence ntr($\bot$, b).

The notion of validity is, of course, relative not merely to whether we are considering ordinary or generalized Beth trees, but also to the notion of truth on a Beth tree that we are using. To avoid complexity of notation and terminology, however, we shall speak, not of a formula's being valid with respect to near truth, but of its being 'nearly valid': A is *nearly valid* on generalized Beth trees iff it is nearly true at the vertex of every generalized Beth tree. Our question now is whether we can use Theorem 4 (ntr) to give an unconditional proof of the completeness of IPC with respect to near validity, i.e. to show that, if A is nearly valid on generalized Beth trees, then $\vdash A$.

Theorems 2 and 3 present no difficulty: our search procedure has been very slightly modified, but not in such a way as to invalidate the proofs of the analogous theorems. In order to obtain our completeness result, however, we need also to derive from Theorem 4 (ntr) the following corollary:

Corollary (ntr). If $\underline{T}_0$, $\underline{T}_1$, ... is a dual sequence for :A, and A is nearly valid on generalized Beth trees, then, for some i, $\underline{T}_i$ is not a refutation tree-trunk.

The proof of this is not absolutely straightforward. Since $\bot$ is nearly true at a node a of a dual tree iff, for some $b \geq a$, $\bot \in \Gamma_b$, Theorem 4 (ntr) yields the result that, on the dual tree corresponding to $\underline{T}_o$, $\underline{T}_1$, ... , we can find a node b such that $\bot \in \Gamma_b$. What we need to do, therefore, is to find an i such that $\bot \in \Gamma_{b,i}$. Now, on our original search procedure, it was relatively easy, for any formula B, to find a bound on i such that, for a given node b on a dual tree for a given sequent $\Gamma : \Delta$, if $B \in \Gamma_b$, then $B \in \Gamma_{b,i}$. The reason is that any such formula B must be a subformula of some formula in Γ or in Δ, and so, by noting the greatest numeral occurring in B, the height of b and the maximum number of operations needed to extract B from some formula in $\Gamma \cup \Delta$, we can calculate the distance we need to go in the dual sequence before B appears at b. On the present search procedure, however, the matter is not so simple, since, from the fact that $\bot \in \Gamma_b$, it does not follow that $\bot$ was a subformula of any formula in $\Gamma : \Delta$ (although it may have been).

Fortunately, the difficulty is superficial: it is due only to our rule that $\bot$ must be added to the antecedent of any basic sequent that appears at some node of a tree-trunk; by so doing, we obliterate the trace of the reason for the appearance of $\bot$. We can overcome the difficulty by dropping this rule; having thus slightly modified our search procedure (for the purpose of the present corollary only), we must also redefine the notion of verification at a node of a dual tree. We shall now take an atomic formula Q to be verified at a node a just in case (i) $Q \in \Gamma_a$, or (ii) $\bot \in \Gamma_a$, or (iii) for some formula C, $C \in \Gamma_a \cap \Delta_a$. In particular, $\bot$ will be verified at a just in case either $\bot \in \Gamma_a$ or, for some C, $C \in \Gamma_a \cap \Delta_a$. It is evident that a formula A will be nearly true at just the same nodes of a dual tree under these stipulations as it was under the earlier ones. We shall, moreover, still be able to prove Theorem 4 (ntr). From it we can infer that if

A is nearly valid on generalized Beth trees, then on any dual tree for :A we can find a node b at which $\bot$ is verified; and this means that either $\bot \in \Gamma_b$ or we can find a formula B in $\Gamma_b \cap \Delta_b$. In the latter case, by the means indicated previously, we can find a bound on i such that $B \in \Gamma_{b,i} \cap \Delta_{b,i}$, and hence so that $\Gamma_{b,i} : \Delta_{b,i}$ is a basic sequent; but in the former case also we can find a bound on i such that $\bot \in \Gamma_{b,i}$, since, under our present modified search procedure, $\bot$ can appear at a node only by being generated in the usual way as a subformula of A.

Taken together with Theorem 3, the corollary to Theorem 4 (ntr) yields:

Theorem 5 (ntr). If A is nearly valid on generalized Beth trees, then $\vdash A$.

We have thus obtained an unconditional proof of the completeness of IPC, for single formulas, with respect to near validity. Of what significance is this result? Certainly the notion of the near truth of a formula at a node of a generalized Beth tree cannot be taken as an intuitively plausible representation of the notion of the truth of an intuitionistic statement. We can see this by asking whether the notion of near truth possesses the analogues of the four general properties of the notion of truth expressed by tr-(a) to tr-(d) above. The answer can be tabulated as follows:

ntr-(a), viz. 'If ntr($\bot$, a), then ntr(A, a)', HOLDS, as does the stronger statement ntr-(a'): if $b \leq a$ and ntr($\bot$, b), then ntr(A, a).

ntr-(b), viz. 'If ntr(A, a) and $b \leq a$, then ntr(A, b)', FAILS.

ntr-(c), viz. 'If $\underline{S}$ bars a, and, for every $b \in \underline{S}$, ntr(A, b), then ntr(A, a)', CANNOT BE PROVED.

ntr-(d), viz. 'ntr(A, a, $\underline{T}$, $\underline{V}$) iff ntr(A, a, $\underline{T}_a$, $\underline{V} \upharpoonright \underline{T}_a$)', HOLDS.

We have already noted the correctness of ntr-(a'), and that

of ntr-(d) can be seen from the obvious fact that the near truth of a formula at a node a depends only on what happens at nodes b ≤ a. A weak counter-example to ntr-(c) will be given later. A counter-example to ntr-(b) is extremely easy to construct; for instance, a tree with just two nodes, a and b, immediately below the vertex v, on which $\bot$ is verified at a but not at b or at any nodes below b. Then ntr($\bot$, v), but not ntr($\bot$, b). (It was in fact because of the failure of ntr-(b) that we set out the proof of Theorem 4 (ntr) as we did, taking as the proposition to be proved by induction a stronger statement than the theorem itself; if ntr-(b) had held, the proof of the theorem would have been quite plain sailing.) The failure of ntr-(b) is, by itself, enough to disqualify near truth from being regarded as corresponding to intuitive truth: under the intuitive notion, whatever has been established as true remains true thereafter. The notion of near truth can be regarded only in the light in which it first presents itself, namely as resulting from a deviant reinterpretation of the atomic statements. If we are viewing formulas in terms of their near truth at nodes, then we are in effect taking each atomic formula Q, not as representing the proposition Q*, but as representing the proposition 'Either Q* is true or there now exists a possibility that $\bot$ will be verified'.

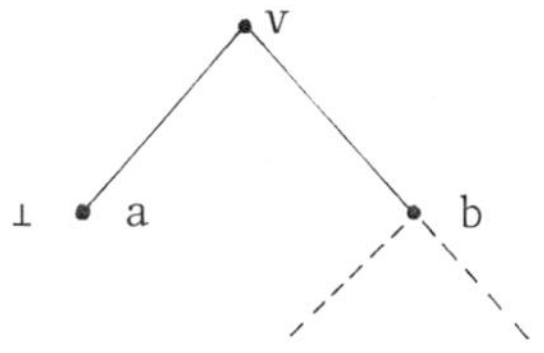

We cannot derive from Theorem 5 (ntr) an unconditional version of Theorem 7, that is, an outright proof of the internal completeness of ICP, since we cannot prove the analogue of Theorem 1, namely that every internally valid formula is nearly valid on generalized Beth trees. Of course, on ordinary Beth trees, near validity coincides with validity, and so every internally valid formula is nearly valid on all ordinary Beth trees. Further, where we define:

<u>Definition</u>. ⟨<u>T</u>, <u>V</u>⟩ *explodes* iff, for some a on <u>T</u>, <u>V</u>($\bot$, a),

every formula is nearly valid on those generalized Beth trees which explode. The difficulty arises over those generalized Beth trees of which we do not know whether or not they explode.

We cannot prove the analogue, for near truth, of Theorem 1 because we cannot prove the analogue of the lemma to Theorem 1. Suppose that the dressed spread $\langle s, h\rangle$ represents the generalized Beth tree $\langle \underline{T}, \underline{V}\rangle$ (the naked spread s represents $\underline{T}$, and, where $\vec{u}$ is a finite sequence admissible by s and representing the node a of $\underline{T}$, the correlation law h assigns to $\vec{u}$ the species $h(\vec{u}) = \{Q \mid \underline{V}(Q, a)\}$ of atomic formulas verified at a). Suppose further that γ is a lawless element of s, and that, for every atomic formula P, we set $P^*(\gamma)$ equivalent to $\exists n\ ntr(P, \bar{\gamma}(n))$. Then the analogue to the lemma would state that, for every formula A, $A^*(\gamma)$ holds iff $\exists n\ ntr(A, \bar{\gamma}(n))$. We can, however, give a weak counterexample to this proposition. Let 'beg(n)' express a decidable property of natural numbers such that we do not know that any natural number possesses it and we do not know that no natural number does. (Specifically, we may take 'beg(n)' to mean 'the n-th, (n + 1)-st, (n + 2)-nd, ... , (n + 9)-th digits in the decimal expansion of π are, respectively, 0,1,2,...,9'.) Let $\underline{T}$ be the full binary tree: let a be the left-hand of the two nodes immediately below the vertex v, and let $b_0, b_1, b_2, \ldots$ be the nodes (after v) on the rightmost path. Thus a is represented by $\langle 0\rangle$ and b_n by the finite sequence consisting of n + 1 1's. Let B be $\forall x\ Fx$ and let A be B & Q. For all nodes c, we set:

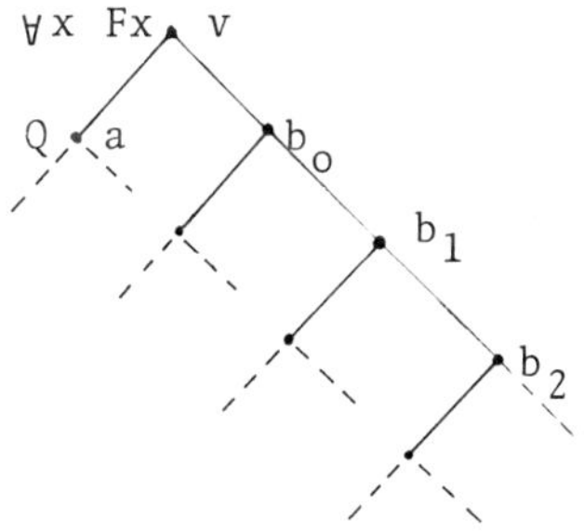

$$\underline{V}(\bot, c) \text{ iff for some } n, \text{beg}(n) \text{ and } c \le b_n$$

$$\underline{V}(Q, c) \text{ iff } \underline{V}(\bot, c) \text{ or } c \le a$$

$$\text{for each } n, \underline{V}(F\bar{n}, c) \text{ iff } \underline{V}(\bot, c) \text{ or } \neg\text{beg}(n).$$

Then, for each n, since beg(n) v $\neg$beg(n), ntr($F\bar{n}$, v), whence ntr(B, v) and so $B^*(\gamma)$. Also ntr(Q, a). We assume that $\gamma(0) = 0$, and so $Q^*(\gamma)$: thus $B^*(\gamma)$ & $Q^*(\gamma)$, i.e. $A^*(\gamma)$. To assert that $\exists n\ ntr(A, \bar{\gamma}(n))$, however, we must either know that ntr(B, a) or that ntr(Q, v). But we have:

$$\text{ntr}(Q, v) \text{ iff } \exists n\ \text{beg}(n)$$

$$\text{ntr}(B, a) \text{ iff } \forall n\ \neg\text{beg}(n),$$

and hence we are not in a position to assert either. This counter-example of course exploits the failure of ntr-(b).

The failure of the analogue, for near truth, of the lemma to Theorem 1 suggests that to view formulas in the light of the notion of near truth involves not merely a deviant reinterpretation of the atomic formulas, but non-standard readings of the logical constants. Completeness with respect to near validity is not, therefore, of any direct relevance to intuitionistic logic. In an attempt to remedy these defects of the notion of near truth, we may consider a further modification of the notion of truth, as follows:

Definition. Where $\langle \underline{T}, \underline{V} \rangle$ is a generalized Beth tree, a a node of $\underline{T}$, and A a formula, A is *largely true* at a in $\langle \underline{T}, \underline{V} \rangle$ (in symbols ltr(A, a, $\underline{T}$, $\underline{V}$) or, for short, ltr(A, a)) iff either:

(1) A is atomic and either $\{b \mid \underline{V}(A, b)\}$ bars a or $\langle \underline{T}, \underline{V} \rangle$ explodes;

(2) A is B & C and ltr(B, a) and ltr(C, a);

(3) A is $B \vee C$ and $\{b \mid \text{ltr}(B, b) \text{ or } \text{ltr}(C, b)\}$ bars a;

(4) A is $B \to C$ and, for every $b \leq a$, if ltr(B, b), then ltr(C, b);

(5) A is $\forall x\, B(x)$ and, for every n, ltr($B(\bar{n})$, a);

(6) A is $\exists x\, B(x)$ and $\{b \mid \text{for some } n, \text{ltr}(B(\bar{n}), b)\}$ bars a.

As with the definition of *nearly true*, the inductive clauses (2) - (6) are again the same as those in the definition of *true*, save for having 'ltr(,)' in place of 'tr(,)'. But, again, it would have made no difference if we had added to each of these clauses ' ... or $\langle \underline{T}, \underline{V} \rangle$ explodes', since we can show:

ltr-(a"): if ltr(⊥, b), then ltr(A, a), for all a, b, and A.

We can, more simply than before, prove the analogue for large truth of Theorem 4 (ntr), namely:

Theorem 4 (ltr). If a is a node of a dual tree, then if $A \in \Gamma_a$, ltr(A, a), and if $A \in \Delta_a$ and ltr(A, a), then for some node b, $\bot \in \Gamma_b$.

With the same slight awkwardness over the corollary to Theorem 4 (ltr), we can then derive:

Theorem 5 (ltr). If A is largely valid on generalized Beth trees, then $\vdash A$,

where 'largely valid' is to be understood by analogy with 'nearly valid'. We are thus as well off with the notion of large truth as with that of near truth; but are we any better off?

When we inquire whether the notion of large truth possesses the analogues of the four general properties of the notion of truth, we find that the situation has changed, but not for the better. The results can be tabulated as follows:

ltr-(a), viz. 'If ltr(⊥, a), then ltr(A, a)', HOLDS, as does the stronger statement ltr-(a"): if ltr(⊥, b), then ltr(A, a).

ltr-(b), viz. 'If ltr(A, a) and b ≤ a, then ltr(A, b)', HOLDS.

ltr-(c), viz. 'If S̲ bars a, and, for every $b \in$ S̲, ltr(A, b), then ltr(A, a)', CANNOT BE PROVED.

ltr-(d), viz. 'ltr(A, a, T̲, V̲) iff ltr(A, a, $\underline{T}_a$, $\underline{V} \upharpoonright \underline{T}_a$)', FAILS.

We have thus purchased ltr-(b) at the expense of ltr-(d). To give a counter-example to ltr-(d) we can in fact use just the same tree T̲ that we used to give a counter-example to ntr-(b): ⊥ is verified at a but not at b or at any node below b. Then since $\langle \underline{T}, \underline{V} \rangle$ explodes, $ltr(\bot, b, \underline{T}, \underline{V})$; but, since $\langle \underline{T}_b, \underline{V} \upharpoonright \underline{T}_b \rangle$ does not explode, not $ltr(\bot, b, \underline{T}_b, \underline{V} \upharpoonright \underline{T}_b)$.

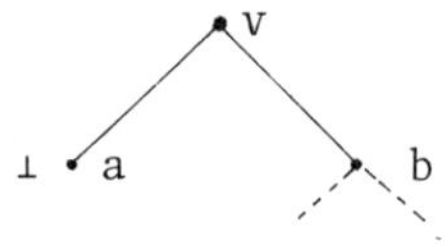

We can also give a weak counter-example to ltr-(c) for the case when A is atomic. Consider a tree $\underline{T}$ with denumerably many nodes b_o, b_1, b_2, ... immediately below the vertex v. For every node c, we set:

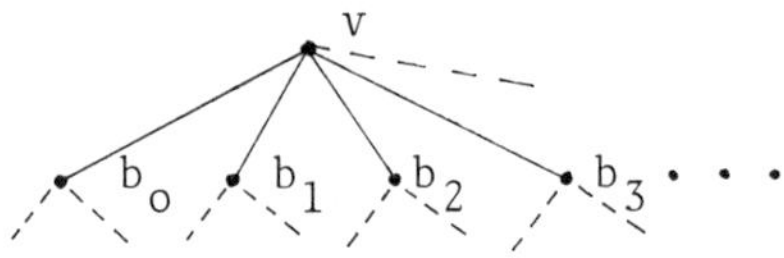

$$\underline{V}(\bot, c) \text{ iff for some } n,\ c \le b_n \text{ and } beg(n)\ \&\ \forall m_{m<n} \neg beg(m)$$

$$\underline{V}(P, c) \text{ iff for some } n,\ c \le b_n \text{ and } \forall m_{m<n} \neg beg(m).$$

Then $\langle \underline{T}, \underline{V} \rangle$ explodes iff $\exists n\ beg(n)$. For each n, either $\underline{V}(P, b_n)$ or $\langle \underline{T}, \underline{V} \rangle$ explodes, whence for each n $ltr(P, b_n)$. Thus $\{c \mid ltr(P, C)\}$ bars v. But ltr(P, v) iff $\forall n\ \neg beg(n) \vee \exists n\ beg(n)$, and hence we cannot assert that ltr(P, v).

To make this into a counter-example to ntr-(c), we suppose that each b_n has immediately below it two nodes a_{2n} and a_{2n+1}, and set:

$$\underline{V}'(\bot, c) \text{ iff for some } n,\ c \le a_{2n} \text{ and } beg(n)$$

$$\underline{V}'(P, c) \text{ iff } \underline{V}'(\bot, c) \text{ or, for some } n,\ c \le b_n \text{ and } \neg beg(n).$$

Then, for each n, $ntr(P, b_n, \underline{T}, \underline{V}')$, and so again $\{c \mid ntr(P, c, \underline{T}, \underline{V}')\}$ bars v. But to assert that $ntr(P, v, \underline{T}, \underline{V}')$ once more requires us to know that $\exists n\ beg(n) \vee \forall n\ \neg beg(n)$. ($\langle \underline{T}, \underline{V}' \rangle$ could also have been taken as a counter-example to ltr-(c).)

Thus the notion of large truth involves an even more deviant reinterpretation of the atomic formulas than did that of near truth: an atomic formula Q will, in effect, be taken, not as representing the proposition Q*, but as representing the proposition 'Either Q* is true or there is *or had been* a possibility of verifying ⊥'. While it is reasonable to count any statement as true whenever it is known that ⊥ will be verified, it is counter-intuitive to do so just on the score that it is possible, though not certain, that ⊥ will be verified, and even more so on the score that there was once such a possibility, though it no longer exists.

Just as with near truth, we cannot prove the analogue of Theorem 1 for large truth, and hence cannot derive from Theorem 5 (ltr) an unconditional version of Theorem 7. There is again an obstacle to proving the analogue of the lemma to Theorem 1, though, since ltr-(b) holds, not at the same place. It does not arise as a result of the failure of ltr-(d), because, while we used both tr-(b) and tr-(c) in proving the original lemma, we had no occasion to appeal to tr-(d): it has, therefore, to be connected with our inability to prove ltr-(c).

Suppose, as before, that the dressed spread ⟨s, h⟩ represents the generalized Beth tree $\langle \underline{T}, \underline{V} \rangle$, that γ is a lawless element of s, and that, for each atomic formula P, we set $P^*(\gamma)$ equivalent to $\exists n\ \mathrm{ltr}(P, \bar{\gamma}(n))$. In $\underline{T}$, let the nodes on the rightmost path be $a_0, a_1, a_2, \ldots$ (a_0 the vertex): each a_n has just two nodes, b_n and a_{n+1}, immediately below it, and each b_n has denumerably many nodes $d_{n0}, d_{n1}, d_{n2}, \ldots$ immediately below it. For each node c, we set:

$\underline{V}(P, c)$ iff $c \leq b_n$ for some n

$\underline{V}(\bot, c)$ iff for some k and n, $c \leq d_{nk}$ and

$\mathrm{beg}(k)\ \&\ \forall m_{m < k}\ \neg\,\mathrm{beg}(m)$

$\underline{V}(Q, c)$ iff for some k and n, $c \leq d_{nk}$ and

$\forall m_{m < k}\ \neg\,\mathrm{beg}(m)$.

Take A as $P \rightarrow Q$. We claim that $A^*(\gamma)$, but that we cannot show that $\exists n\ \mathrm{ltr}(A, \bar{\gamma}(n))$.

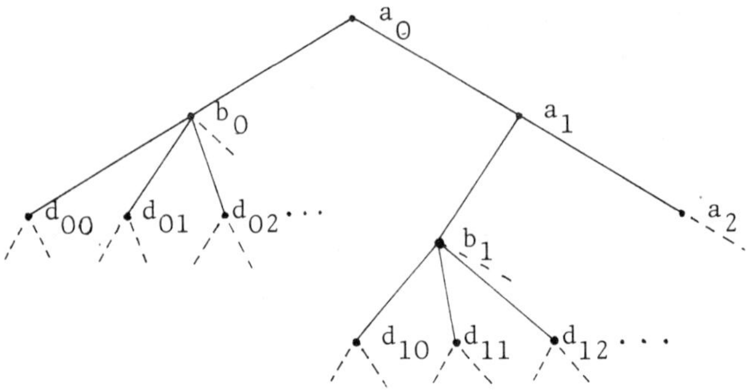

To show that $A^*(\gamma)$, suppose that $P^*(\gamma)$, say $1tr(P, \bar{\gamma}(n))$. If $\bar{\gamma}(n)$ corresponds to a_n, then, since a_n is not barred by $\{c \mid \underline{V}(P, c)\}$, it follows that $\langle \underline{T}, \underline{V} \rangle$ explodes, and hence that $1tr(Q, \bar{\gamma}(n))$ and so $Q^*(\gamma)$. If, on the other hand, $\bar{\gamma}(n)$ does not correspond to a_n, then, for some $m < n$, $\bar{\gamma}(m + 1)$ corresponds to b_m and $\bar{\gamma}(m + 2)$ to d_{mk} for some k. If $\forall j_{j < k} \neg beg(j)$, then $\underline{V}(Q, d_{mk})$, while if $\exists j_{j < k}\, beg(j)$, then $\langle \underline{T}, \underline{V} \rangle$ explodes. Hence in either case $1tr(Q, \bar{\gamma}(m + 2))$, and therefore $Q^*(\gamma)$. We have thus shown that $P^*(\gamma) \rightarrow Q^*(\gamma)$, i.e. that $A^*(\gamma)$.

Now suppose that we were in a position to assert that $1tr(A, \bar{\gamma}(n))$. $\bar{\gamma}(n)$ could not then correspond to a_n, because, since $1tr(P, b_n)$, in order to be able to assert that $1tr(A, a_n)$, we should have to be in a position to assert that $1tr(Q, b_n)$. This we are not in a position to do, since $1tr(Q, b_n)$ holds iff either $\{c \mid \underline{V}(Q, c)\}$ bars b_n or $\langle \underline{T}, \underline{V} \rangle$ explodes, that is, iff either $\forall n \neg beg(n)$ or $\exists n\, beg(n)$. It follows that, in order to be able to assert that $\exists n\, 1tr(A, \bar{\gamma}(n))$, we should have to be able to show that there exists an m such that $\bar{\gamma}(m + 1)$ corresponds to b_m. It is, however, absurd that we should be able to prove this for arbitrary γ.

It thus appears that we have reason to regard the notion

of large truth, too, as in effect imposing non-standard interpretations on the logical constants. Theorem 5 (ltr) thus appears to have as little intuitive significance as Theorem 5 (ntr). It should be noted, however, that the failure of ltr-(c) applies only to infinitary trees. It is connected with the failure of the logical law

$$\forall x\ (C \vee B(x)) \rightarrow C \vee \forall x\ B(x);$$

but it may be shown by induction that

$$\forall m_{m \leq n}\ (C \vee B(m)) \rightarrow C \vee \forall m_{m \leq n}\ B(m).$$

Suppose that $\underline{T}$ is finitary, that $\underline{S}$ bars a, and that, for every b $\in$ $\underline{S}$, ltr(P, b), where P is atomic. Then, by the Fan Theorem, there is an upper bound k on the length of a path from a to a node in $\underline{S}$. Hence, by appeal to ltr-(b), the species $\underline{S}' = \{c \mid c \leq a$ and either c is terminal or c is k levels below a$\}$ is finite, bars a, and also has the property that ltr(P, c) for every c $\in$ $\underline{S}'$. We thus have

$$\forall c_{c \in \underline{S}'}\ (\langle \underline{T}, \underline{V}\rangle \text{ explodes or } \{d \mid \underline{V}(P, d)\} \text{ bars } c)$$

and so, since $\underline{S}'$ is finite,

$$\langle \underline{T}, \underline{V}\rangle \text{ explodes or } \forall c_{c \in \underline{S}'}\ \{d \mid \underline{V}(P, d)\} \text{ bars } c$$

whence, since $\underline{S}'$ bars a,

$$\langle \underline{T}, \underline{V}\rangle \text{ explodes or } \{d \mid \underline{V}(P, d)\} \text{ bars } a$$

and thus ltr(P, a). The five cases of the induction step present no difficulty, so that ltr-(c) holds for finitary trees. On the other hand, a dual tree generated by our particular procedure for constructing dual sequences will not always be finitary. If, for example, Δ_a contains a formula of the form $\exists x\ (B(x) \rightarrow C(x))$, then, for each n, it will contain $B(\bar{n}) \rightarrow C(\bar{n})$, and so, for each n, there will be a distinct node b immediately above a such that $B(\bar{n}) \in \Gamma_b$ and

$C(\bar{n}) \in \Delta_b$.

At this stage, we pause to compare the notions of truth defined here with those employed by the Nijmegen school. In 'Choice Sequences and Completeness of Intuitionistic Predicate Logic', Troelstra (not himself a member of that school) defines 'A is forced at p in a generalised Beth model' exactly as we have defined 'A is largely true at a', save that he does not impose condition (iii') on his generalized Beth models; it is easily seen that the relaxation of condition (iii') makes no effective difference if condition (i) holds ($\underline{V}$ is decidable) and $\underline{T}$ is finitary; Troelstra does not actually incorporate these conditions into his definition of a generalized Beth model, but 'restricts attention' to those that satisfy them. In 'Another Intuitionistic Completeness Proof', de Swart also considers generalized Beth trees that do not necessarily satisfy condition (iii'), and, for such a tree $\langle \underline{T}, \underline{V} \rangle$, gives the definition (in our notation):

$\underline{\underline{\mathrm{tr}}}(A, a, \underline{T}, \underline{V})$ iff either:

(0) $\langle \underline{T}, \underline{V} \rangle$ explodes;

(1) A is atomic and $\{b \mid \underline{V}(A, b)\}$ bars a;

(2) A is B & C and $\underline{\underline{\mathrm{tr}}}(B, a)$ and $\underline{\underline{\mathrm{tr}}}(C, a)$;

(3) A is $B \vee C$ and $\{b \mid \underline{\underline{\mathrm{tr}}}(B, b)$ or $\underline{\underline{\mathrm{tr}}}(C, b)\}$ bars a;

(4) A is $B \rightarrow C$ and, for every $b \leq a$, if $\underline{\underline{\mathrm{tr}}}(B, b)$, then $\underline{\underline{\mathrm{tr}}}(C, b)$;

(5) A is $\forall x\, B(x)$ and, for every n, $\underline{\underline{\mathrm{tr}}}(B(\bar{n}), a)$;

(6) A is $\exists x\, B(x)$ and $\{b \mid$ for some n, $\underline{\underline{\mathrm{tr}}}(B(\bar{n}), b)\}$ bars a.

Here clauses (2) - (6) are as usual. Given $\underline{V}$, let us set $\underline{V}'(A, a)$ iff $\underline{V}(A, a)$ or $\underline{V}(\bot, a)$. It is then easy to show that if $\underline{\underline{\mathrm{tr}}}(A, a, \underline{T}, \underline{V})$, then $\mathrm{ltr}(A, a, \underline{T}, \underline{V}')$. If $\underline{V}$ is decidable and $\underline{T}$ is finitary, then the converse holds also; but if T is not finitary, we could know that $\{b \mid \underline{V}'(P, b)\}$ barred a, for some atomic formula P, and hence that $\mathrm{ltr}(P, a)$, without knowing either that $\{b \mid \underline{V}(P, b)\}$ barred a or that $\langle \underline{T}, \underline{V} \rangle$ exploded, and hence that $\underline{\underline{tr}}(P, a)$. De Swart incorrectly states the property ltr-(c) (or rather $\underline{\underline{tr}}$-(c)) as part of a theorem, and later uses it in his proof of what corresponds to Theorem 4 (ltr).

In 'Intuitionistic Logic in an Intuitionistic Metalanguage', de Swart gives a quite different definition, namely one which takes as basic the notion of truth on a *path* of the tree. Where $\langle \underline{T}, \underline{V} \rangle$ is again a generalized Beth tree not necessarily satisfying condition (iii'), and α ranges over paths in $\underline{T}$, he defines:

$\mathrm{Tr}(A, \alpha, \underline{T}, \underline{V})$ iff either:

(0) $\exists n\ \underline{V}(\bot, \bar{\alpha}(n))$;

(1) A is atomic and $\exists n\ \underline{V}(A, \bar{\alpha}(n))$;

(2) A is B & C and $\mathrm{Tr}(B, \alpha)$ and $\mathrm{Tr}(C, \alpha)$;

(3) A is $B \vee C$ and either $\mathrm{Tr}(B, \alpha)$ or $\mathrm{Tr}(C, \alpha)$;

(4) A is $B \to C$ and $\exists k\ \ \forall \beta_{\ \beta \in \bar{\alpha}(k)}$ (if $\mathrm{Tr}(B, \beta)$ then $\mathrm{Tr}(C, \beta)$);

(5) A is $\forall x\ B(x)$ and $\exists k\ \forall m\ \forall \beta_{\ \beta \in \bar{\alpha}(k)}\ \mathrm{Tr}(B(\bar{m}), \beta)$;

(6) A is $\exists x\ B(x)$ and $\exists m\ \mathrm{Tr}(B(\bar{m}), \alpha)$.

This definition corresponds to the method of subsuming the Beth trees under the topological interpretation by treating them as topological spaces whose points are the paths in the trees. This notion of truth on a path corresponds to our notion, not of near truth or large truth, but of ordinary truth at a node: where $\underline{V}'$ is as before, we can show by induction:

$$\mathrm{Tr}(A, \alpha, \underline{T}, \underline{V}) \text{ iff } \exists n\ \mathrm{tr}(A, \bar{\alpha}(n), \underline{T}, \underline{V}') \ .$$

(This time the transition from $\underline{V}$ to $\underline{V}'$ causes no difficulty.)

At this stage it appears that we have to choose between having only a conditional completeness proof and having an unconditional one with respect to a notion of validity without intuitive significance. However, we have not yet exhausted our resources. We introduce yet another variant notion of truth at a node of a generalized Beth tree with the

Definition. Where $\langle \underline{T}, \underline{V} \rangle$ is a generalized Beth tree, a a node of $\underline{T}$, and A a formula, A is *very largely true* at a in $\langle \underline{T}, \underline{V} \rangle$ (in symbols vltr(A, a, $\underline{T}$, $\underline{V}$) or, for short, vltr(A, a)) iff either:

(1) A is atomic and $\{b \mid \underline{V}(A, b)$ or $\langle \underline{T}, \underline{V} \rangle$ explodes$\}$ bars a;

(2) A is B & C and vltr(B, a) and vltr(C, a);

(3) A is B $\vee$ C and $\{b \mid$ vltr(B, b) or vltr(C, b)$\}$ bars a;

(4) A is B $\rightarrow$ C and, for every b $\leq$ a, if vltr(B, b), then vltr(C, b):

(5) A is $\forall$x B(x) and, for every n, vltr(B($\bar{n}$), a);

(6) A is $\exists$x B(x) and $\{b \mid$ for some n, vltr(B($\bar{n}$), b)$\}$ bars a.

Clauses (2) - (6) are as usual. The notion of very large truth represents only a minor modification of that of large truth, but one adequate to rectify the failure of ltr-(c). Suppose that Q is an atomic formula, that $\underline{S}$ bars a and that, for all b $\in \underline{S}$, vltr(Q, b). Then, for each b $\in \underline{S}$, b is barred by $\{c \mid \underline{V}(Q, c)$ or $\langle \underline{T}, \underline{V} \rangle$ explodes$\}$. It is then obvious that a is also barred by $\{c \mid \underline{V}(Q, c)$ or $\langle \underline{T}, \underline{V} \rangle$ explodes$\}$, whence vltr(Q, a). The various cases of the induction step are straightforward, so vltr-(c) holds. vltr-(d) fails, however, for the same reason as ltr-(d).

We could also consider the notion of a formula's being *very nearly true* at a node (abbreviated vntr(A, a)), in the definition of which clause (1) was modified from that for *nearly true* in a similar way; that is, clause (1) would run:

A is atomic and $\{b \mid \underline{V}(A, b)$ or, for some c $\leq$ a, $\underline{V}(\perp, c)\}$ bars a,

while the inductive clauses (2) - (6) would be as usual.

vntr-(c) would then likewise hold; but the notion of very near truth would be of relatively little use to us, since vntr-(b) would still fail, for the same reason as ntr-(b).

In exactly the same way as before, we can prove:

Theorem 4 (vltr). If a is a node of a dual tree, then if $A \in \Gamma_a$, vltr(A, a), and if $A \in \Delta_a$ and vltr(A, a), then for some b, $\bot \in \Gamma_b$.

From this, as before, we can derive:

Theorem 5 (vltr). If A is very largely valid on generalized Beth trees, then $\vdash A$.

The point of introducing the notion of very large truth lies in the fact that, since vltr-(d) is the only one of our four general properties that fails for this notion, we are much closer to proving the analogue of Theorem 1 for it. It would, indeed, be very surprising if we were able to prove the analogue of Theorem 1 outright for the notion of very large truth, i.e. to prove:

If A is internally valid, then A is very largely valid on generalized Beth trees,

since then we should be able to derive, from Theorem 5 (vltr), an unconditional version of Theorem 7, namely the proposition:

If A is internally valid, then $\vdash A$;

and from this, by Theorem 8, we should be able to conclude to the validity of schema (9'). It is, however, entirely implausible that we should be able to deduce the validity of such a schema by means of such reasoning.

As already remarked, in proving the lemma to Theorem 1, we appealed only to the properties tr-(a), tr-(b), and tr-(c) of the notion of truth. Since vltr-(a), vltr-(b), and vltr-(c) all hold, we have no difficulty in proving the analogues, for the notion of very large truth, of the induction steps for the three binary connectives and the two quantifiers. We cannot, however, prove the induction step for negation: given that $B^*(\gamma)$ iff $\exists n$ vltr(B, $\bar{\gamma}(n)$), it does not in the least follow that if vltr($\neg B$, $\bar{\gamma}(n)$), i.e. vltr($B \to \bot$, $\bar{\gamma}(n)$), then $\neg B^*(\gamma)$; indeed, it is the whole point of generalized

Beth trees that it does *not* follow. $\perp$ may be an absurd statement under the interpretation on the Beth tree, i.e. the least likely to be true of all those we are interpreting; and it was emphasized that elementary logical considerations do not require that an absurd statement can never be proved. But, once we allow that $\perp$ can be true at a node, then, among statements *about* the tree, the statement that $\perp$ is somewhat true is not in the least absurd.

We have nevertheless gained a great deal; for, while we cannot prove the unrestricted analogue of the lemma to Theorem 1, we can prove the following restricted version:

<u>Lemma</u>. Let $\langle \underline{T}, \underline{V} \rangle$ be a generalized Beth tree, $\langle s, h \rangle$ a dressed spread which represents it, and γ a lawless element of s. For each atomic formula P, we set $P^*(\gamma)$ equivalent to $\exists n\, \mathrm{vltr}(P, \bar{\gamma}(n))$. Let A be any formula not containing $\perp$ (or $\neg$). Then $A^*(\gamma)$ iff $\exists n\, \mathrm{vltr}(A, \bar{\gamma}(n))$.

Provided that we accept the notion of a lawless sequence as intuitionistically meaningful, we can then derive from this lemma, as a restricted analogue of Theorem 1:

<u>Theorem 12</u>. If A is an internally valid formula not containing $\perp$ (or $\neg$), then A is very largely valid on generalized Beth trees.

And from this, in turn, we can derive a restricted, but unconditional, internal completeness theorem:

<u>Theorem 13</u>. If A is an internally valid formula not containing $\neg$ or $\perp$, then $\vdash A$; i.e. the negation-free fragment of IPC is internally complete for single formulas.

We cannot, however, derive from Theorem 13 the validity of any instances of schema (9') by the methods of Theorem 8: for inspection of the proof of Theorem 8 shows that it was essential that the formula B, constructed relative to a given primitive recursive predicate $A(\alpha, n)$, should contain negation (it was of the form $\neg(C \;\&\; \neg D)$).

Ironically, the Nijmegen school reject the concept of a lawless sequence, so that this reasoning is not available to them. Instead, they rely on the highly questionable claim that generalized Beth trees provide *the* correct semantics for intuitionistic logic. Even should this claim be accepted, it is certain that none of the deviant notions of truth at a node of such a tree that we have been considering correspond to the intuitive notion of the truth of an intuitionistic statement, while the unconditional completeness proof cannot be carried through with the ordinary notion of truth. To say that an atomic formula P is very largely true at a node a, for example, is in effect to say that, in the corresponding state of information, we can assert that we shall be in a position either to verify P* or to say that there has been (though may be no longer) a possibility of verifying ⊥, and this still involves a gross reinterpretation of the atomic formulas. The failure of vltr-(d) is itself highly counterintuitive: the status, as true or otherwise, of an intuitionistic statement cannot vary according to the moment when one came in, as it were, that is, according not to the stage that one is at but to the state of mathematical knowledge when the individual subject commenced mathematical activity. Nevertheless, the possibility of proving the lemma to Theorem 12 shows that, given this deviant interpretation of the atomic formulas, the notion of very large truth is faithful to the intended meanings of the logical constants other than negation; this by itself enables us to view the Nijmegen completeness proof, not as a conjuring trick without genuine significance for intuitionistic logic, but as a real unconditional completeness proof for a very large fragment of first-order logic.

De Swart has also shown, however, how to improve on this, by means of the following theorem.

Theorem 14. If A is valid on generalized Beth trees, then A is very largely valid on them.

Proof. Assume that A is valid on generalized Beth trees. Let $\langle \underline{T}, \underline{V} \rangle$ be any generalized Beth tree. Let $\langle \underline{T}', \underline{V}' \rangle$ be the corresponding tree with no terminal nodes (any terminal node b of $\underline{T}$ having appended to it an infinite chain $b_1, b_2, b_3, \ldots$ of nodes of $\underline{T}'$, where $\underline{V}'(P, b_i)$ iff $\underline{V}(P, b)$, and otherwise $\underline{V}'$ agrees with $\underline{V}$). Let $a_o, a_1, a_2, \ldots$ be an enumeration of all the nodes of $\underline{T}'$ such that if $a_j < a_i$, then $i < j$. We introduce $\underline{V}''$, for the nodes of $\underline{T}'$, by the definition: $\underline{V}''(P, a_j)$ iff $\underline{V}'(P, a_j)$ or, for some $i < j$, $\underline{V}'(\bot, a_i)$. We now claim that, for any A and i,

$$\text{tr}(A, a_i, \underline{T}', \underline{V}'') \text{ iff vltr}(A, a_i, \underline{T}', \underline{V}').$$

The proof is by induction on the complexity of A. Suppose, first, that A is atomic. Assume that tr(A, a_i, $\underline{T}'$, $\underline{V}''$). Then a_i is barred by a species $\underline{S}$ of nodes such that, for each $b \in \underline{S}$, $\underline{V}''(A, b)$. Hence, by the definition of $\underline{V}''$, for each $b \in \underline{S}$, either $\underline{V}'(A, b)$ or, for some c, $\underline{V}'(\bot, c)$. It follows that vltr(A, a_i, $\underline{T}'$, $\underline{V}'$).

Assume now (still for atomic A) that vltr(A, a_i, $\underline{T}'$, $\underline{V}'$). Then a_i is barred by a species $\underline{S}$ of nodes such that, for each $b \in \underline{S}$, either $\underline{V}'(A, b)$ or, for some c, $\underline{V}'(\bot, c)$. Let $\underline{J} = \{j \mid a_j \in \underline{S} \text{ and not } \underline{V}'(A, a_j)\}$. Then, for each $j \in \underline{J}$, there exists k such that $\underline{V}'(\bot, a_k)$, and, by the Axiom of Choice, there is a constructive function f such that, if $j \in \underline{J}$, then $\underline{V}'(\bot, a_{f(j)})$. For each $j \in \underline{J}$, let $\underline{M}_j = \{a_m \mid m > f(j) \text{ and } a_m \leq a_j\}$. Then, for $j \in \underline{J}$, $\underline{M}_j$ bars a_j; further, for $a_m \in \underline{M}_j$, $\underline{V}''(A, a_m)$. Put $\underline{S}' = \{b \mid b \in \underline{S} \text{ and } \underline{V}'(A, b)\}$, and $\underline{S}^* = \underline{S}' \cup \bigcup_{j \in \underline{J}} \underline{M}_j$. Then $\underline{S}^*$ bars a_i, and $\underline{V}''(A, b)$ for every $b \in \underline{S}^*$. Thus tr(A, a_i, $\underline{T}'$, $\underline{V}''$).

The other cases present no difficulty, since we are claiming an equivalence, and the inductive clauses in the definitions of tr and of vltr are equiform. The claim is therefore established. Since A was assumed valid on generalized Beth trees, we have, in particular, that tr(A, a_i, $\underline{T}'$, $\underline{V}''$), and hence that vltr(A, a_i, $\underline{T}'$, $\underline{V}'$), and so vltr(A, a_i, $\underline{T}$, $\underline{V}$). Since $\langle\underline{T}, \underline{V}\rangle$ was an arbitrary generalized Beth tree, and a_i an arbitrary node of it, it follows that A is very largely valid on generalized Beth trees.

Putting Theorem 14 together with Theorem 5 (vltr), we obtain:

Theorem 5 (gen). If A is valid on generalized Beth tree, then $\vdash A$.

There still appears no way to give a direct proof of this theorem, without a detour via a deviant notion of truth, which now appears as a mere technical device. The same obstacle as before remains to obtaining any strengthening of Theorem 12, and hence of Theorem 13: what significance one attaches to Theorem 14 depends entirely on how far one regards the generalized Beth trees, taken in relation to the standard notion of truth, as providing a reasonable semantics for intuitionistic logic. In any case, it is now apparent, as it was not previously, that the difficulty of arriving at an intuitionistic completeness proof turned on the peculiarities of intuitionistic negation; and we thus now have a good reason for the proposition, often advanced on unconvincing grounds, that negation is more problematic than the other intuitionistic logical operations.

5.8. COMPACTNESS

Relative to some given semantics, a branch or fragment $\mathcal{K}$ of intuitionistic logic is *sound* just in case, for any formula A and set Γ of formulas belonging to $\mathcal{K}$, if $\Gamma \vdash A$, then $\Gamma \models A$. $\mathcal{K}$ is *complete for single formulas* just in case, for any formula A belonging to $\mathcal{K}$, if $\models A$, then $\vdash A$. This is obviously equivalent to saying that $\mathcal{K}$ is *complete for finite sets*, i.e. that, for any formula A and finite set Γ of formulas belonging to $\mathcal{K}$, if $\Gamma \models A$, then $\Gamma \vdash A$. A stronger claim is made if $\mathcal{K}$ is asserted to be *complete for infinite sets*, the claim, namely, that, for any formula A and any set Γ of formulas belonging to $\mathcal{K}$, if $\Gamma \models A$, then $\Gamma \vdash A$. Likewise, a valuation system $\mathfrak{M}$ (or family $\mathbb{F}$ of valuation systems) is *finitely strictly characteristic* for $\mathcal{K}$ just in case, for any formula A and finite set Γ of formulas belonging to $\mathcal{K}$, $\Gamma \models_{\mathfrak{M}} A$ ($\Gamma \models_{\mathbb{F}} A$) iff $\Gamma \vdash A$; that is to say, just in case

$\mathcal{K}$ is sound and complete for infinite sets with respect to $\mathfrak{M}$ (to $\mathbb{F}$). In the preceding section, we considered only completeness and quasi-completeness for single formulas; and, in our previous discussion of valuation systems, we sidestepped the question whether the family of Beth trees (or of Kripke trees or of PO-spaces) was strictly characteristic for intuitionistic sentential logic, contenting ourselves with the demonstration that it was finitely strictly characteristic. The techniques described in the preceding two sections enable us to make an attack upon this question.

If, with respect to some given semantics, a certain fragment $\mathcal{K}$ of intuitionistic logic is both sound and complete for infinite sets, then it is also *compact* with respect to that semantics; that is to say, for any formula A and set Γ of formulas belonging to $\mathcal{K}$, $\Gamma \models A$ iff $\Gamma_0 \models A$ for some finite subset $\Gamma_0 \subseteq \Gamma$. For, if $\Gamma \models A$, then, by the completeness of $\mathcal{K}$, $\Gamma \vdash A$, whence, by the definition of $\vdash$, $\Gamma_0 \vdash A$ for some finite $\Gamma_0 \subseteq \Gamma$, and so, by the soundness of $\mathcal{K}$, $\Gamma_0 \models A$. Conversely, if we know that $\mathcal{K}$ is both compact and complete for finite sets, we may infer that it is complete for infinite sets. For, if $\Gamma \models A$, then, by compactness $\Gamma_0 \models A$ for some finite subset $\Gamma_0 \subseteq \Gamma$, whence, by completeness for finite sets, $\Gamma_0 \vdash A$, and so, by the definition of $\vdash$, $\Gamma \vdash A$.

We introduce the notion of an *infinite sequence* $\underline{T}_0, \underline{T}_1, \underline{T}_2, \ldots$ *of dual tree-trunks*, or *dual sequence*, *for* $\Gamma : A$, where A is a formula and Γ an infinite set of formulas of ICP: we specify such a dual sequence relative to some given enumeration $C_0, C_1, C_2, \ldots$ of the formulas in Γ. For each i, let d_i be the maximum degree of any formula in $\{A, C_0, \ldots, C_i\}$. We define

$$\chi(n) = \begin{cases} 2(2^{d_0+1} - 1) & \text{if } n = 0 \\ \\ \frac{n+2}{n}((n+1)^{d_n+1} - 1) & \text{if } n \geq 1. \end{cases}$$

We then put $\Theta(n) = \sum_{i=0}^{n} \chi(i)$. $\underline{T}_0$ is now the dual tree-trunk consisting of a single node with which is associated the sequent $C_0 : A$. For any i such that $i < \Theta(0) - 1$ or, for some m, $\Theta(m) \leq i < \Theta(m + 1) - 1$, $\underline{T}_{i+1}$ is formed from $\underline{T}_i$ in the usual way. For $i = \Theta(m) - 1$, we first form, in the usual way, a dual tree-trunk $\underline{T}^*_{i+1}$, supposing m to be the smallest number such that $i \leq \psi(m)$; we then let $\underline{T}_{i+1}$ have just the same nodes as $\underline{T}^*_{i+1}$, and, where a is any such node, we take $\Gamma_{a,i+1} : \Delta_{a,i+1}$ to be $\Gamma^*_{a,i+1}, C_{m+1} : \Delta_{a,i+1}$, where $\Gamma^*_{a,i+1} : \Delta_{a,i+1}$ is the sequent associated with a in $\underline{T}^*_{i+1}$.

By a *proof-tree for* $\Gamma : A$ is meant whatever is, for some finite $\Gamma_0 \subseteq \Gamma$, a proof-tree for $\Gamma_0 : A$; similarly, by a *proof tree-trunk for* $\Gamma : A$ is meant whatever is, for some finite $\Gamma_0 \subseteq \Gamma$, a proof tree-trunk for $\Gamma_0 : A$. A *dual tree for* $\Gamma : A$ is one formed as before from a dual sequence for $\Gamma : A$; and a *refutation tree-trunk* or *refutation tree* is, as before, a dual tree-trunk or dual tree such that for no node a and formula B do we have $B \in \Gamma_a \cap \Delta_a$.

It is now a routine matter to prove, by exactly the same methods as before, propositions corresponding to Theorems 2 and 3 of section 5.6, applied to the pair $\Gamma : A$, with Γ infinite, rather than to a sequent. Furthermore, by appeal to the function Θ in place of ψ, we can prove, for a dual tree for $\Gamma : A$, the proposition corresponding to the lemma for Theorem 4 of section 5.6, and hence also the proposition corresponding to Theorem 4 itself. Accordingly, by the same methods as those by which we obtained Theorems 5, 6, and 7 of section 5.6, we now obtain:

<u>Theorem 1</u>. If schema (9) holds, ICP is complete for infinite sets, and, if schema (11) holds, it is quasi-complete for infinite sets, both internally and with respect to Beth trees.

Let us say that a branch or fragment $\mathcal{K}$ of intuitionistic logic is *quasi-compact*, relative to a given notion of an interpretation, just in case, for every formula A and set Γ of formulas belonging to $\mathcal{K}$, if $\Gamma \models A$, then it is not the case that, for every finite subset $\Gamma_0 \subseteq \Gamma$, not $\Gamma_0 \models A$. Note that quasi-compactness is related to quasi-completeness exactly as compactness is related to completeness. $\mathcal{K}$ is quasi-complete for infinite sets just in case, for every Γ and A belonging to $\mathcal{K}$, if $\Gamma \models A$, then not $\Gamma \nvdash A$. Assume that $\mathcal{K}$ is sound and quasi-complete for infinite sets, and that $\Gamma \models A$. Now suppose that, for every finite $\Gamma_0 \subseteq \Gamma$, $\Gamma_0 \not\models A$. Then, since $\mathcal{K}$ is sound, for every finite $\Gamma_0 \subseteq \Gamma$, $\Gamma_0 \nvdash A$, and so $\Gamma \nvdash A$, which contradicts the quasi-completeness of $\mathcal{K}$. Hence, if $\mathcal{K}$ is sound and quasi-complete for infinite sets, $\mathcal{K}$ is quasi-compact. Now assume, conversely, that $\mathcal{K}$ is quasi-compact and quasi-complete for finite sets, and that $\Gamma \models A$. Suppose that $\Gamma \nvdash A$. Then $\Gamma_0 \nvdash A$ for any finite $\Gamma_0 \subseteq \Gamma$, whence, since $\mathcal{K}$ is quasi-complete for finite sets, $\Gamma_0 \not\models A$ for any finite $\Gamma_0 \subseteq \Gamma$. This contradicts the quasi-compactness of $\mathcal{K}$, and thus the quasi-completeness of $\mathcal{K}$ for infinite sets follows from its being quasi-compact and quasi-complete for finite sets.

Consequently, assuming the soundness of ICP, we have, as a direct corollary of Theorem 1:

Theorem 2. If schema (9) holds, ICP is compact, and, if schema (11) holds, it is quasi-compact, with respect both to internal interpretations and to Beth trees.

In view of Theorem 8 of section 5.6, it is evident that Theorem 1 cannot be improved. It is not immediately obvious, on the other hand, that Theorem 2 cannot be improved: but it happens that the proof of Theorem 8 of section 5.6 can be adapted to show that we can get no proof of compactness, even for IC, under any hypothesis weaker than the validity of schema (9').

Theorem 3. If, with respect to internal interpretations, IC is compact, then schema (9') holds, and, if it is quasi-compact, schema (11') holds.

Proof. Let $A(\alpha, n)$ be any primitive recursive predicate, let $\Phi(\alpha, n)$ be its characteristic functional, and let $\underline{E}$ be a set of equations defining $\Phi(\alpha, n)$ primitive recursively, and containing function-symbols $f_o, f_1, \ldots, f_k$, all as in the proof of Theorem 8 of section 5.6. We begin by considering a first-order language containing a numeral $\bar{n}$, as an individual constant, for each natural number n, the two-place predicate-symbol = , the one-place predicate-letter Q, and, for each i, $0 \leq i \leq k$, an $(r + 1)$-place predicate-letter F_i, where r is the number of arguments of the function-symbol f_i in $\underline{E}$. We first construct an infinite set Γ' of quantifier-free sentences of this language. For each i, $1 \leq i \leq k$, $\underline{E}$ must contain an equation or pair of equations of one of the forms (i) to (v) listed in the proof of Theorem 8 of section 5.6. Γ' then consists of the following closed formulas:

$\bar{n} = \bar{n}$	for every n
$\neg \bar{n} = \bar{m}$	for every n and m, $n \neq m$
$F_i\bar{n}_1 \ldots \bar{n}_r\bar{n} \rightarrow \neg F_i\bar{n}_1 \ldots \bar{n}_r\bar{m}$	
	for every i, $0 \leq i \leq k$, and for every $n_1, \ldots, n_r$, n and m such that $n \neq m$
$Q\bar{n} \rightarrow F_o\bar{n}0$	for every n
$\neg Q\bar{n} \rightarrow F_o\bar{n}1$	for every n
$F_i\bar{n}0$	for every n and every i such that $\underline{E}$ contains an equation of type (i)
$F_i\bar{n}\overline{(n+1)}$	for every n and every i such that $\underline{E}$ contains an equation of type (ii)
$F_i\bar{n}_1 \ldots \bar{n}_r\bar{n}_j$	for every $n_1, \ldots, n_r$ and every i such that $\underline{E}$ contains an equation of type (iii)

$$F_{s_1}\overline{n_1}\ldots\overline{n_r}\,\overline{m_1} \;\&\; \ldots \;\&\; F_{s_q}\overline{n_1}\ldots\overline{n_r}\,\overline{m_q} \;\&\; F_{s_0}\overline{m_1}\ldots\overline{m_q}\,\overline{k} \rightarrow$$

$$F_i\overline{n_i}\ldots\overline{n_r}\,\overline{k}$$

for every $n_1, \ldots, n_r, m_1, \ldots m_q$ and k and every i such that E contains an equation of type (iv)

$F_r\bar{n}\bar{k} \rightarrow F_i 0\bar{n}\bar{k}$ and

$F_i\bar{m}\bar{n}\bar{k} \;\&\; F_s\bar{m}\bar{n}\bar{k}\bar{h} \rightarrow F_i\overline{(m+1)}\bar{n}\bar{h}$

for every n, k, m and h, and every i such that E contains a pair of equations of type (v)

$Q\bar{n} \vee \neg Q\bar{n}$ for every n

$\neg F_k\bar{n}0$ for every n.

Given a denumerable set $\{P_0, P_1, P_2, \ldots\}$ of sentence-letters, we set up an effective one-one map of the atomic sentences of the first-order language on to these sentence-letters; for each such atomic sentence A, let P[A] be the sentence-letter which is its image under this mapping. The mapping induces, in an obvious way, a mapping of all quantifier-free sentences of the language on to sentential formulas. We now let Γ be the set consisting of all the images of members of Γ' under this mapping.

Lemma 1. If $\forall\alpha_{\alpha \in b}\neg\neg\exists n\, A(\alpha, n)$, then there is no internal interpretation which brings out true every formula in Γ.

Proof. An internal interpretation of formulas of IC is a mapping * which associates with each sentence-letter P_i a determinate proposition $P_i{}^*$ of intuitionistic mathematics, and hence associates with every sentential formula A the proposition A* compounded out of the $P_i{}^*$ as A is compounded out of the P_i. Let us suppose that * is such an interpretation, and that, for each formula $A \in \Gamma$, A^* is a true proposition. We now define predicates I^*, Q^*, F_o^*, $F_1^*, \ldots,$ F_k^* of natural numbers by setting:

$$I^*(n, m) \text{ iff } (P[\bar{n} = \bar{m}])^* \text{ is true}$$

$$Q^*(n) \text{ iff } (P[Q\bar{n}])^* \text{ is true}$$

$$F_i^*(n_1, \ldots, n_r) \text{ iff } (P[F_i\overline{n_1}\ldots\overline{n_r}])^* \text{ is true.}$$

We also set

$$\alpha^*(n) = \begin{cases} 0 & \text{if } Q^*(n) \\ 1 & \text{if } \neg Q^*(n). \end{cases}$$

Since $P[Q\bar{n}] \vee \neg P[Q\bar{n}] \in \Gamma$ for every n, $(P[Q\bar{n}])^* \vee \neg(P[Q\bar{n}])^*$ is true for every n, and so $\forall n(Q^*(n) \vee \neg Q^*(n))$; α^* is therefore well-defined.

Now assume that $\forall \alpha_{\alpha \in b} \neg\neg \exists n\, A(\alpha, n)$. Then, in particular, $\neg\neg\exists n\, A(\alpha^*, n)$.

For each i, $0 \leq i \leq k$, let h_i be the function such that, for all $n_1, \ldots, n_r$ and m, $h_i(n_1,\ldots,n_r) = m$ iff the equation $f_i\overline{n_1}\ldots\overline{n_r} = \bar{m}$ is derivable from $\underline{E}$ together with sufficiently many equations of the form $f_o\bar{n} = \bar{j}$, where $j = \alpha^*(n)$. We can now establish by induction that, for each i, $0 \leq i \leq k$, and for each $n_1,\ldots,n_r$ and m, $h_i(n_1,\ldots,n_r) = m$ iff $F_i^*(n_1,\ldots,n_r,m)$. Hence in particular, for every n, $h_k(n) = 0$ iff $F_k^*(n, 0)$. Since, for every n, $\neg(P[F_k\bar{n}0])^*$ belongs to Γ, we have $\forall n \neg F_k^*(n, 0)$, and so $h_k(n) \neq 0$ for every n. Now h_k is $\lambda n.\ \Phi(\alpha^*, n)$, so that $h_k(n) = 0$ iff $A(\alpha^*, n)$. It follows that $\neg\exists n\, A(\alpha^*, n)$, contrary to the hypothesis. We conclude that Γ is not satisfiable by any internal interpretation.

Lemma 2. For each $\alpha \in b$, and each finite subset $\Gamma_o \subseteq \Gamma$, if Γ_o is not satisfiable, then $\exists n\, A(\alpha, n)$.

Proof. Suppose Γ_o is a finite subset of Γ, and that Γ_o is not satisfiable.

For given $\alpha \in b$, put $h_o(n) = \alpha(n)$ for each n, and, as before, let $h_1,\ldots,h_k$ be the functions such that, for each $n_1,\ldots,n_r$ and m, and each i such that $1 \leq i \leq k$, $h_i(n_1,\ldots n_r) = m$ iff $f_i\bar{n}_1\ldots\bar{n}_r = \bar{m}$ is derivable from $\underline{E}$ together with sufficiently many equations of the form $f_o\bar{n} = \bar{j}$, where $j = \alpha(n)$ $(1 \leq i \leq k)$.

We define an interpretation * as follows. For each n and m, we take $(P[\bar{n} = \bar{m}])^*$ as the statement that $n = m$. For each n, we take $(P[Q\bar{n}])^*$ as the statement that $\alpha(n) = 0$. For each $n_1,\ldots,n_r$ and m, and each i such that $0 \leq i \leq k$, we take $(P[F_i\bar{n}_1\ldots\bar{n}_r\bar{m}])^*$ as the statement that $h_i(n_1,\ldots,n_r) = m$. It is then easily seen that, for every $A \in \Gamma$ not of the form $\neg P[F_k\bar{n}0]$, A^* is true. Let $\neg P[F_k\bar{n}_1 0],\ldots, \neg P[F_k\bar{n}_q 0]$ be all the members of Γ_o of the form $\neg P[F_k\bar{n}0]$. Then, since Γ_o is not satisfiable, $\neg(P[F_k\bar{n}_1 0])^*,\ldots, \neg(P[F_k\bar{n}_q 0])^*$ cannot all be true, and thus $\neg F_k{}^*(n_1,0),\ldots, \neg F_k{}^*(n_q,0)$ cannot all be true. But, as before, for every n, $F_k{}^*(n,0)$ iff $A(\alpha, n)$; moreover, for each n, $A(\alpha, n) \vee \neg A(\alpha, n)$. It follows that, for some i, $1 \leq i \leq q$, $A(\alpha, n_i)$, and therefore $\exists n\, A(\alpha, n)$.

To prove the theorem, assume, for any given primitive recursive $A(\alpha, n)$, that $\forall\alpha_{\alpha \in b}\neg\neg\exists n\, A(\alpha, n)$. Then, by lemma 1, Γ is not satisfiable, where Γ is the set of sentential formulas corresponding to the predicate $A(\alpha, n)$. Hence, for an arbitrary formula B, $\Gamma \models B \,\&\, \neg B$.

Suppose, first, that IC is compact. Then, for some finite subset $\Gamma_o \subseteq \Gamma$, $\Gamma_o \models B \,\&\, \neg B$, and Γ_o is therefore not satisfiable. Hence, by Lemma 2, for each α, we have

$\exists n\, A(\alpha, n)$. This shows that compactness of IC implies, for each primitive recursive $A(\alpha, n)$, that
$\forall\alpha_{\alpha \in b} \neg\neg\exists n\, A(\alpha, n) \rightarrow \forall\alpha_{\alpha \in b}\exists n\, A(\alpha, n)$.

Suppose, secondly, that IC is quasi-compact. It follows that it is not the case that, for every finite subset $\Gamma_o \subseteq \Gamma$, not $\Gamma_o \models B \,\&\neg B$, i.e. that not every finite subset of Γ is not unsatisfiable. Lemma 2 tells us that, for each $\alpha \in b$ and each finite subset Γ_o of Γ, if Γ_o is unsatisfiable, then $\exists n\, A(\alpha, n)$. It follows that if there is a finite $\Gamma_o \subseteq \Gamma$ which is unsatisfiable, then $\forall\alpha_{\alpha \in b}\exists n\, A(\alpha, n)$, whence, if not every finite subset of Γ is not unsatisfiable, then $\neg\neg\forall\alpha_{\alpha \in b}\exists n\, A(\alpha, n)$. We have thus shown that quasi-compactness of IC implies, for each primitive recursive predicate $A(\alpha, n)$, that

$\forall\alpha_{\alpha \in b} \neg\neg\exists n\, A(\alpha, n) \rightarrow \neg\neg\forall\alpha_{\alpha \in b}\exists n\, A(\alpha, n)$.

6
SOME FURTHER TOPICS

6.1. INTUITIONISTIC FORMAL SYSTEMS

So far we have discussed various principles which may be taken as axiomatic in intuitionistic mathematics, without attempting to delineate any actual formalizations of intuitionistic theories. This is appropriate to the subject. For the Hilbert school, and for formalists properly so called, formalization is integral to an exact treatment of mathematics; but the original impulse to formalization did not come from them, but from the logicists, for whom the formalization of a theory was a necessary means of identifying its basic principles, so that they could then show these to be derivable from pure logic. The intuitionists, on the other hand, were from the start hostile to formalization: for them, it is highly unlikely that the mental constructions intuitively recognizable as proving a statement of a given theory should be isomorphic to the formal proofs of any calculus, recognizable as such by a mechanical procedure making no appeal to meaning. There is therefore some irony in the intensive study that has been made by logicians of intuitionistic formal systems; but it can reasonably be retorted that, just as Gödel's incompleteness results did not destroy the interest in investigating proof-theoretical questions relating to classical theories, so the fact that we never expect to have a complete formalization of any intuitionistic theory should not deter us from studying similar questions in this area.

We begin with the system HA of intuitionistic first-order arithmetic. This is usually considered as having as primitives, not only =, 0, ', +, and . , but also a symbol for each primitive recursive function. As axioms we take the third and fourth Peano axioms (the first two being redundant since the variables are construed as ranging only over $\underline{N}$), each instance of the axiom schema of induction, and the recursion equations for +, . , and all other primitive recursive functions (see section 2.1). We note, mostly without proof, a number of properties of HA.

(i) HA has the *explicit definability property* that if $\vdash_{HA} \exists x\, A(x)$, where x is the only free variable in A(x), then $\vdash_{HA} A(\bar{n})$ for some n.

(ii) HA has the *disjunction property* that if $\vdash_{HA} A \vee B$, where A and B are closed, then $\vdash_{HA} A$ or $\vdash_{HA} B$. This follows from (i) by the equivalence of $A \vee B$ with $\exists x[(x = 0 \to A) \;\&\; (x \neq 0 \to B)]$.

(iii) HA satisfies the *independence of premisses rule* that if $\vdash_{HA} \neg A \to \exists x\, B(x)$, where x does not occur free in A, then $\vdash_{HA} \exists x(\neg A \to B(x))$.

(iv) However, the corresponding schema, even in its most restricted form, does not hold in HA; there exists a formula B(x) in which x is the only free variable, and a primitive recursive formula A(y) in which y is the only free variable, such that $\nvdash_{HA} (\forall y\, A(y) \to \exists x\, B(x)) \to \exists x(\forall y\, A(y) \to B(x))$.

(v) From (i) and (iii) it follows that if $\vdash_{HA} \neg A \to \exists x\, B(x)$, then $\vdash_{HA} \neg A \to B(\bar{n})$ for some n.

(vi) Not even the weakest form of Markov's principle holds in HA. By Gödel's Incompleteness Theorem, there exists a closed formula $\forall x\; \neg C(x)$, with C(x) primitive recursive, which is formally undecidable in HA. By (v), if $\vdash_{HA} \neg\, \forall x\; \neg C(x) \to \exists x\; C(x)$, then $\vdash_{HA} \neg \forall x\; \neg C(x) \to C(\bar{n})$ for some n. Since C(x) is decidable, $\vdash_{HA} C(\bar{n})$ or $\vdash_{HA} \neg C(\bar{n})$; if $\vdash_{HA} C(\bar{n})$, then $\vdash_{HA} \neg \forall x\; \neg C(x)$, and if $\vdash_{HA} \neg C(\bar{n})$, then by contraposition $\vdash_{HA} \neg\neg \forall x\; \neg C(x)$, and so, equivalently, $\vdash_{HA} \forall x\; \neg C(x)$, contradicting, in either case, the undecidability of $\forall x\; \neg C(x)$ in HA. Thus $\nvdash_{HA} \neg \forall x\; \neg C(x) \to \exists x\; C(x)$.

(vii) However, the rule corresponding to Markov's principle holds in HA; i.e. if $\vdash_{HA} \forall x\; (A(x) \vee \neg A(x)) \;\&\; \neg \forall x\; \neg A(x)$, then $\vdash_{HA} \exists x\; A(x)$.

(viii) From (vii) it follows that in HA and PA (classical arithmetic) the same numbers can be proved to be Gödel numbers of general recursive functions. For if $\vdash_{PA} \exists y\, T_1(\bar{n}, x, y)$, then, by the Godel translation of PA into HA, $\vdash_{HA} \neg \forall y \neg T_1(\bar{n}, x, y)$, whence, since $T_1(u, x, y)$ is primitive recursive, $\vdash_{HA} \exists y\, T_1(\bar{n}, x, y)$ by (vii).

(ix) A theorem of de Jongh states that if $A(p_1,\ldots,p_n)$ is an unprovable formula of intuitionistic sentential logic IC with sentence-letters $p_1,\ldots,p_n$, then there exist sentences $B_1,\ldots,B_n$ of HA such that $\nvdash_{HA} A(B_1,\ldots,B_n)$. The choice of the B_i can, moreover, be made independently of the particular sentential formula A.

In formalizations of intuitionistic analysis, we need variables for constructive functions and/or choice sequences as well as for natural numbers; in the following, x, y, z, ... are to be numerical variables, f, g, h, ... variables for constructive unary functions of natural numbers, and α, β, γ, ... variables for choice sequences. Economy is attained by coding finite sequences of natural numbers as natural numbers. The intuitively simplest such code is that used by Kleene, under which the finite sequence $\vec{u} = \langle u_0,\ldots,u_{\ell-1}\rangle$ is represented by the numbers $\prod_{i<\ell} p_i^{u_i+1}$, where p_i is the i-th prime (2 being p_0); the null sequence is represented by 1. Where, for any number n, $(n)_i$ is the exponent of p_i in the prime factorization of n, if n represents the finite sequence $\vec{u}$, u_i will be $(n)_i - 1$. This coding has the advantage of making it easy to define the operation which, when applied to numbers n and m representing finite sequences $\vec{u}$ and $\vec{v}$, yields the number representing $\vec{u} * \vec{v}$, and, equally, the operation which applied to a number n representing $\vec{u}$, yields the length ℓ of $\vec{u}$. It has the disadvantage that the coding is not on to $\underline{N}$; not every natural number represents a finite sequence (for instance, 10 does not, since it is divisible by 5 but not by 3). For this reason, Kreisel and Troelstra employ a different coding, under which the null sequence is represented by 0 and every natural number represents a finite sequence; the idea underlying it is intuitively straightforward, but it necessitates a very complicated definition of the arithmetical operation corresponding to concatentation. It is of minor importance which coding is adopted, provided that it is effective, that every finite sequence is coded uniquely, and that there are constructive functions corresponding to determining the length of a finite sequence, to extracting its i-th term, and to concatenating two finite sequences. In the following, we shall continue to use our variables $\vec{u}$, $\vec{v}$, $\vec{w}$, ... for finite sequences, but these are now to be understood as number-variables, restricted, if necessary, to those numbers which represent finite sequences; similarly, symbols such as $*$ for operations on finite sequences are now to be understood as denoting the corresponding arithmetical operations. Further where $t_o,\ldots,t_{k-1}$ are numerical terms, the notation $\langle t_o,\ldots,t_{k-1}\rangle$ will be taken as representing a complex term, containing $t_o,\ldots,t_{k-1}$, for the number which codes the k-tuple of their values; similarly, notations like $\bar{\alpha}(x)$ will be taken as constituting an expression for the number representing an initial segment of a choice sequence, rather than for that initial segment itself.

In most systems of intuitionistic analysis, except the theory of lawless sequences, it is convenient to be able to form complex terms (functors) for constructive functions or choice sequences; we therefore employ the notation of λ-abstraction, taking $\lambda x.\, t$ as a functor whenever t is a numerical term. In systems in which there are distinct sorts of variable for constructive functions and for choice sequences, we shall need in our primitive notation two distinct λ-symbols, one to form functors for constructive functions and the other to form functors for choice sequences. In stating the formation rules, we shall in any case need a simultaneous inductive definition of (numerical) *term* and of *functor*, one clause of which will provide that, whenever ϕ is a functor and t a term, $\phi(t)$ will be a term. For each λ-operator, our axioms will then include the axiom schema of λ-conversion, namely

$$(\lambda x.\ s)\ (t) = s[t]\ ,$$

where s and t are terms such that t is free for x in s, and s[t] is the result of replacing every free occurrence of x in s by t. The equality symbol = is permitted to stand between functors, either as a part of primitive notation or as a definitional abbreviation, in either case expressing extensional equivalence, so that we can derive

$$\phi = \phi \leftrightarrow \forall x\ \phi(x) = \Phi(x)\ .$$

In some systems there will also be a symbol ≡ for intensional identity.

We begin by considering the system IDB_1 of Kreisel and Troelstra, which formalizes the theory of constructive functions and of what have come to be known as 'Brouwer-operations'. (The term 'neighbourhood function' is used for those constructive functions which represent continuous functionals and belong to the species $\underline{K}_o = \{e \mid \forall\alpha\ \exists n\ \exists k\ (\forall m_{m<n}\ e(\bar{\alpha}(m)) = 0\ \&\ \forall m_{m \geq n}\ e(\bar{\alpha}(m)) = k + 1)\}$, while 'Brouwer-operation' is used for those belonging to the inductively defined species $\underline{K}$; the principle of K-Induction thus says that all neighbourhood functions are Brouwer-operations -- see section 3.5. Since IDB_1 contains no variables for choice sequences, $\underline{K}_o$ cannot be defined in it; the axioms embody the inductive definition of $\underline{K}$.) Besides numerical variables and variables for constructive functions, there is also in this system a special sort of variables e, e', ... for Brouwer-operations; correspondingly, besides the ordinary abstraction operator λ for forming functors for constructive functions (F-functors), there is also the operation λ' for forming functors for Brouwer-operations (K-functors). The reason for thus distinguishing between F-functors and K-functors is purely technical: the notations e(f) and e|f for applying a Brouwer-

operation to a constructive function to yield, respectively, a natural number and a constructive function are primitive in IDB_1, but could not be taken as always defined if e were permitted to range over all constructive functions; the formation rules therefore provide that, where η is a K-functor and ϕ an F-functor, $\eta(\phi)$ is a term and $\eta|\phi$ an F-functor. For any term t, $\lambda x.\, t$ is an F-functor; the operator λ', on the other hand, can be applied only to terms of very restricted forms, which will not be enumerated here, the restrictions being devised to guarantee that, for all values of the variables, each K-functor denotes a Brouwer-operation.

Besides the equality symbol, the primitive symbols of HA, the two λ-operators, and the ordinary operation of functional application, IDB_1 has as primitives the unary predicate K, taking constructive functions as argument, and the two above-mentioned operations e(f) and e|f. The axioms governing the latter are:

$$e(f) = x \;\&\; e(\bar{f}(y)) = z + 1 \rightarrow z = x$$

$$(e|f)\;(x) = (\lambda' y.\; e(\langle x\rangle * y))\;(f).$$

As explained earlier, Kleene takes, as representing continuous functionals, only those functions e such that, for each infinite sequence α, there is only one n such that $e(\bar{\alpha}(n))$ is positive; Kreisel and Troelstra make the slightly more convenient requirement -- embodied in the definition of $\underline{K}_0$ -- that, if $m > n$ and $e(\bar{\alpha}(n))$ is positive, $e(\bar{\alpha}(m)) = e(\bar{\alpha}(n))$. The axioms of IDB_1 governing K are accordingly:

(1) $f = \lambda y.\; x + 1 \rightarrow K(f)$.

(2) $f(0) = 0 \;\&\; \forall x\, K(\lambda y.\; f(\langle x\rangle * y)) \rightarrow K(f)$

(3) $\forall x\; (f = \lambda y.\; x + 1 \rightarrow A(f)) \;\&$
$\forall f\; (f(0) = 0 \;\&\; \forall x\, A(\lambda y.\; f(\langle x\rangle * y)) \rightarrow A(f)) \rightarrow$
$\forall f\; (K(f) \rightarrow A(f))$,
where A(f) is any formula (induction axiom schema).

(4) $\forall f\; (K(f) \rightarrow \exists e\; f = e) \;\&\; \forall e\; K(\lambda y.\; e(y))$.

Axiom (4) provides that every Brouwer-operation (element of the domain of the variable e) is extensionally equivalent to a constructive function satisfying K and conversely.

In IDB_1 all contexts are extensional, and it is possible to prove the principle of extensionality for functions as a theorem schema:

$$f = g \ \& \ A(f) \rightarrow A(g) \ .$$

The Axiom of Choice in IDB_1 is that of the form $\forall x \ \exists f$, namely the schema:

$$\forall x \ \exists f \ A(x, f) \rightarrow \exists g \ \forall x \ A(x, (g)_x) \ ,$$

where $(g)_x = \lambda y. \ g(\langle x, y \rangle)$, i.e., on Kleene's coding, $\lambda y. \ g(2^{x+1}.3^{y+1})$. From this we can of course prove the $\forall x \ \exists y$ form.

Going to the opposite extreme, we next consider Kreisel's system FC, which has, besides numerical variables, choice-sequence variables intended to range only over lawless sequences (originally called absolutely free choice sequences), namely choice sequences upon the choice of whose terms no restriction is at any time placed. There is no λ-operator, and the formula $\alpha = \beta$ is defined to mean $\forall x \ \alpha(x) = \beta(x)$; there is also a symbol $\equiv$ for intensional identity, which stands between choice-sequence variables, and is subject to the axiom schema of substitutivity:

$$\alpha \equiv \beta \ \& \ A(\alpha) \rightarrow A(\beta).$$

Here $A(\alpha)$ is *any* formula containing the variable α; in all other schemata, however, we shall follow the practice of assuming that the only free variables are those explicitly shown in the schema (since the strength and plausibility of principles governing choice sequences are sensitive to the presence or absence of parameters for choice sequences).

The axioms for FC are as follows.

$$\alpha \equiv \beta \lor \neg \alpha \equiv \beta \ .$$

Strict, i.e. intensional, identity is always decidable in intuitionistic mathematics, since it depends solely upon how the objects are given, and we must be able to tell whether the way in which an object is presented to us on one occasion is or is not the same as that in which an object is presented to us on another.

$$\forall \vec{u}\ \exists \alpha\ \alpha \in \vec{u},$$

where, of course, $\alpha \in \vec{u}$ is defined to mean $\exists x\ \bar{\alpha}(x) = \vec{u}$. If the range of the choice-sequence variables were to consist of those lawless sequences generated by some empirically identifiable sequences of unrestricted choices, we could assert only that, for each finite sequence, it was not excluded that it should be the initial segment of some choice sequence, i.e. that $\forall \vec{u}\ \neg\neg \exists \alpha\ \alpha \in \vec{u}$; we should have no guarantee that any particular choice sequence would in fact start in a prescribed way. This axiom therefore suggests that we are quantifying over all possible lawless sequences. Since the notion of all possible lawless sequences is a hard one, we may prefer to say that the variables range, not over absolutely unrestricted choice sequences, but over those that may be restricted by requiring them to have any specified initial segment, but thereafter are generated by wholly unrestricted choices of their terms.

$$A(\alpha, \beta_1, \ldots, \beta_n) \ \&\ \neg\alpha \equiv \beta_1 \ \&\ \ldots \ \&\ \neg\alpha \equiv \beta_n \rightarrow$$
$$\exists x\ \forall \gamma(\gamma \in \bar{\alpha}(x) \ \&\ \neg\gamma \equiv \beta_1 \ \&\ \ldots \ \&\ \neg\gamma \equiv \beta_n \rightarrow$$
$$A(\gamma, \beta_1, \ldots, \beta_n)).$$

This axiom schema is called the *principle of open data*; it says that the truth of any statement made about a lawless sequence α intensionally distinct from all other lawless sequences mentioned in the statement can depend only upon some initial segment of α; that is, it must hold good for any other lawless sequence with that initial segment and distinct from the other lawless sequences mentioned. If the condition requiring that α be intensionally distinct from the other lawless sequences mentioned were not included, we should be able to infer

$$\alpha = \beta \rightarrow \exists x\ \forall\gamma(\gamma \in \bar{\alpha}(x) \rightarrow \gamma = \beta),$$

and hence, by putting α for β,

$$\exists x\ \forall\gamma(\gamma \in \bar{\alpha}(x) \rightarrow \gamma = \alpha),$$

which contradicts the second axiom. As a special case of the principle of open data, we have:

$$A(\alpha) \rightarrow \exists x\ \forall\gamma(\gamma \in \bar{\alpha}(x) \rightarrow A(\gamma)),$$

where, in accordance with our convention, α is the only free variable for choice sequences occurring in $A(\alpha)$. We can also prove that lawless sequences are extensionally equivalent only when they are intensionally identical. For suppose that $\alpha = \beta\ \&\ \neg\alpha \equiv \beta$. By the principle of open data,

$$\forall\gamma(\gamma \in \bar{\alpha}(x)\ \&\ \neg\gamma \equiv \beta \rightarrow \gamma = \beta)$$

for some x. By the second axiom, there exists $\gamma \in \bar{\alpha}(x)^\frown(\alpha(x) + 1)$. Then $\gamma \neq \alpha$, whence $\gamma \neq \beta$, whence $\neg\gamma \equiv \beta$; therefore $\gamma = \beta$, a contradiction. We have thus shown that $\alpha = \beta \rightarrow \neg\neg\alpha \equiv \beta$, whence, by the decidability of $\equiv$ (first axiom), we have:

$$\alpha = \beta \rightarrow \alpha \equiv \beta\ .$$

The theory FC has now been replaced, as a formalization of the theory of lawless sequences, by the theory LS, in which there is no symbol for intensional identity, and the foregoing axioms are assumed with = replacing $\equiv$. In addition, the theory of Brouwer-operations is incorporated, so as to permit the adoption of an axiom schema of continuity, which reduces, for the case without parameters, to:

$$\forall\alpha\ \exists x\ A(\alpha, x) \rightarrow \exists e\ \forall\alpha\ A(\alpha, e(\alpha)).$$

For systems with variables ranging over choice sequences of a more general kind, a greal deal is known about the deductive relations between various forms of the principles, such as continuity, governing them; we shall not, however, attempt to summarize this information here, but merely look at some salient ideas for formalizing the theory of choice sequences. In the system FIM of Kleene and Vesley, there are only two sorts of variables, numerical variables and variables for choice sequences; constructive functions are thus treated as a particular kind of choice sequence. The intention is that a formula of the form $\exists\alpha\ A(\alpha)$ (where again α is the only choice-sequence variable occurring free in $A(\alpha)$) shall be so understood as to be true only if there is a constructive function satisfying $A(\alpha)$. Further, we can express in FIM the statement that α is general recursive, namely by the formula $\exists z\ \forall x\ \exists y\ (T_1(z, x, y)\ \&\ U(y) = \alpha(x))$. If we accept Church's Thesis, and identify constructive functions with general recursive ones, we can then obtain the effect of restricting the choice-sequence variables to constructive functions by means of this predicate. But, if we do not accept Church's Thesis, there is no means in the system for expressing that γ is a constructive function; and hence, while we can express the statement that there is a constructive function satisfying a given (absolute) condition, we cannot express the statement that something holds good for all constructive functions. As noted by Myhill, this has the consequence that the notion of a spread in FIM differs from the usual intuitionistic notion. We can in the usual way express the condition that γ satisfies the requirements for being a spread-law, by means of the formula $\forall\vec{u}(\gamma(\vec{u}) = 0 \leftrightarrow \exists x\ \gamma(\vec{u}^\frown x) = 0)$ (in FIM a spread may be empty); we can then, as usual, express membership of the spread determined by a given law by defining $\alpha \in \gamma$ to hold just in case $\forall x\ \gamma(\bar{\alpha}(x)) = 0$. Under these definitions, however, any choice sequence satisfying the above condition determines a spread, regardless of whether or not it is constructive, whereas on Brouwer's conception it is an essential part of the notion of a spread that the spread-law be given by a constructive function.

For any principle of intuitionistic analysis (Axiom of Choice, Continuity Principle, Bar Induction, etc.), let us say that it holds in a restricted form when the relevant schema holds without choice-sequence parameters, and that it holds in a generalized form when the schema holds with such parameters. Since any finite number $\beta_1, \ldots, \beta_n$ of choice sequences can always be coalesced into one by setting $\beta = \lambda x. \langle \beta_1(x), \ldots, \beta_n(x)\rangle$, and again recovered therefrom, it is always sufficient to consider only a single parameter. FIM has = as a primitive only between terms; between functors, it is defined to express extensional equivalence, and there is no symbol for intensional identity. Besides the axioms for equality and λ-conversion, and those for HA, FIM has the axiom schema for the generalized $\forall x\, \exists\alpha$ form of Axiom of Choice, that for generalized Bar Induction for a decidable bar-predicate R, and that for the generalized form of $\forall\alpha\, \exists\beta$-Continuity Principle. Using our convention for schemata, these therefore run as follows:

Axiom of Choice.

$$\forall x\, \exists\alpha\, A(x, \alpha, \beta) \rightarrow \exists\gamma\, \forall x\, A(x, (\gamma)_x, \beta),$$

where again $(\gamma)_x = \lambda y.\, \gamma(\langle x, y\rangle)$.

Bar Induction.

$$\begin{aligned}
&\forall\vec{u}\, (R(\vec{u}, \beta) \vee \neg R(\vec{u}, \beta))\ \& \\
&\quad \forall\alpha\exists x\, R(\bar{\alpha}(x), \beta)\ \& \\
&\qquad \forall\vec{u}\, (R(\vec{u}, \beta) \rightarrow A(\vec{u}, \beta))\ \& \\
&\qquad \forall\vec{u}\, (\forall x\, A(\vec{u}^\frown x, \beta) \rightarrow A(\vec{u}, \beta)) \rightarrow \\
&\qquad A(\langle\ \rangle, \beta).
\end{aligned}$$

In FIM, the null sequence is represented by 1 rather than by 0, and therefore is here denoted by $\langle\ \rangle$ to avoid confusion. R and A are both syntactic variables.

Continuity.

In formulating this axiom-schema, I use KL(η) as an abbreviation for the formula expressing that η is a neighbourhood function, on Kleene's way of construing these, viz. the formula $\forall\alpha\ \exists!x\ \eta(\bar{\alpha}(x)) > 0$.

$$\forall\alpha\ \exists\beta\ B(\alpha, \beta, \delta) \rightarrow \exists\eta\ (KL(\eta)\ \&$$

$$\forall\alpha\ \forall\gamma\ (\forall x\ \exists y\ \eta(\langle x\rangle * \bar{\alpha}(y)) = \gamma(x) + 1 \rightarrow B(\alpha, \gamma, \delta))).$$

The condition that B(α, β, δ) be extensional is not here given as one of the hypotheses, because it represents a formula of FIM, the language of which is purely extensional. The last clause says that B(α, η|α, δ) for all α, but the notation η|α is not used in FIM.

From these axioms, it is of course possible to derive the generalized ∀x ∃y form of Axiom of Choice, the generalized ∀α ∃x-Continuity Principle and generalized Bar Induction for monotonic R; the necessity for formulating these principles as axiom schemata is due, naturally, to the absence of any variables for species. To formalize intuitionistic analysis without either a special sort of variables for constructive functions or a predicate of choice sequences picking out those that are lawlike is in any case a *tour de force*, since the notion of a constructive function appears integral to intuitionistic mathematics. We therefore now turn to the system CS of Kreisel and Troelstra.

CS results from adding to IDB_1 a theory of choice sequences. It therefore contains four sorts of variables: number-variables, variables for constructive functions, variables for Brouwer-operations and variables for choice sequences. Correspondingly, it contains, as primitive, three λ-operators, and three types of functor: F-functors for constructive functions, K-functors for Brouwer-operations and C-functors for choice sequences; λ is used to form F-functors, λ' to form K-functors, and λ" to form C-functors. λ" may be applied to any numerical term: in particular, where f is a variable for a constructive function, we may form the C-functor λ"x. f(x), so that it is trivially provable that there is always a choice sequence extensionally equivalent to any constructive function. On the other hand, we do not want to obtain the converse result, since otherwise the distinction between constructive functions and choice sequences becomes otiose; the formation of such F-functors as $\lambda x.\ \alpha(x)$ has therefore to be barred. Essentially, we wish to allow λ to be applied only to terms not containing any variables for choice sequences. More exactly, we demand that a variable for a choice sequence occurring in an F-functor can occur only as part of a numerical term no number-variable in which is bound by λ; i.e. that $\lambda x.\ t$ is an F-functor only if t is the result of replacing, in a term $t'(x_1, \ldots, x_n)$ which contains no choice-sequence variables, each number-variable x_i by a term s_i. (The reason for this is as follows. If we have a formula $\forall y\ A(y)$, we wish to be able to infer $A(t)$ for any term t, in particular $A(\alpha(z))$, provided that t is free for y in $A(y)$. Suppose, however, that $A(y)$ contains the F-functor $\lambda x.\ (y + x)$; then the formulation of $A(\alpha(z))$ will transform this into $\lambda x.\ (\alpha(z) + x)$, which is intuitively unobjectionable, but would be ill-formed if we ruled out all F-functors containing choice-sequence variables.)

CS has = as primitive only between terms, and as defined, to express extensional equivalence, between functors. There is no symbol for intensional identity, all the vocabulary is extensional, and the principle of extensionality is derivable as a theorem schema. It contains the axioms of HA, the axiom schemata of λ-conversion for the three λ-operators and the axioms of IDB_1 governing K. Its most distinctive axiom is the axiom schema (AD) which expresses the *principle of analytic data*:

$$\text{(AD)} \qquad A(\alpha) \rightarrow \exists e\ (\exists \beta\ \alpha = e|\beta\ \&\ \forall \gamma\ A(e|\gamma)).$$

(AD) states, in effect, that a statement about a choice sequence α can be known to be true only on the basis of the knowledge that α was obtained by applying a certain continuous operation to some other choice sequence; just as the principle of open data asserted that a statement about a lawless sequence could be known to be true only on the basis of the knowledge that it had a certain initial segment. Note that (AD) is implied by, though it does not imply, the principle (SD):

$$\text{(SD)}\quad A(\alpha) \to \exists f\ (spr(f)\ \&\ \alpha \in f\ \&\ \forall\beta\ (\beta \in f \to A(\beta))),$$

which says that a statement about a choice sequence can be known to be true only on the basis of the knowledge that it belongs to a particular spread ($spr(f)$ here expresses that f is a spread(-law)): for, given any spread f, we can easily construct an e such that, for every α, $e|\alpha \in f$, while, for $\beta \in f$, $e|\beta = \beta$. (AD) is equivalent to the schema (AD'):

$$\text{(AD')}\quad \forall e\ (\forall\gamma\ A(e|\gamma) \to \forall\gamma\ B(e|\gamma)) \to \forall\alpha(A(\alpha) \to B(\alpha)).$$

For assume $A(\alpha)$, and assume also the antecedent of (AD'). By (AD), there exist e and β such that $\alpha = e|\beta$ and $\forall\gamma\ A(e|\gamma)$. Hence $\forall\gamma\ B(e|\gamma)$, and therefore $B(e|\beta)$, i.e. $B(\alpha)$. Conversely, assume $A(\alpha)$, and take $B(\alpha)$ as $\exists e\ (\exists\beta\ \alpha = e|\beta\ \&\ \forall\gamma\ A(e|\gamma))$; we wish, by appeal to (AD'), to prove $B(\alpha)$. Now, for arbitrary e', assume $\forall\gamma\ A(e'|\gamma)$. It follows that $\forall\gamma\ \exists e\ (\exists\beta\ e'|\gamma = e|\beta\ \&\ \forall\gamma\ A(e|\gamma))$, i.e. $\forall\gamma\ B(e'|\gamma)$. Hence $B(\alpha)$ by (AD').

(AD) yields as a consequence (SP):

$$\text{(SP)}\quad \exists\alpha\ A(\alpha) \to \exists f\ A(\lambda''x.\ f(x))\ ,$$

which says that, whenever we can assert the existence of a choice sequence satisfying some (absolute) condition, then we can say that there is such a choice sequence extensionally equivalent to a constructive function (this expresses the meaning attached by Kleene to the quantifier $\exists\alpha$). The implication is trivial: if $\exists\alpha\ A(\alpha)$, then by (AD) there exists e such that $\forall\alpha\ A(e|\alpha)$, whence $A(e|\lambda''x.\ x)$, for example. By our convention for schemata, α is the only free variable for choice sequences in $A(\alpha)$, and we see from the principle (SP) that we could not generalize (AD) by replacing $A(\alpha)$ by $C(\alpha,\beta)$ without trivializing the theory; for then, by taking $C(\alpha,\beta)$ as $\alpha = \beta$, we should obtain, by (SP), $\forall\beta\ \exists f\ f = \beta$. (SP) does, however, yield $\forall\beta\neg\neg\exists f\ f = \beta$; for if $\neg\exists f\ f = \beta$, then $\exists\alpha\neg\exists f\ f = \alpha$, whence by (SP) $\exists g\ \neg\exists f\ f = g$, which is absurd.

Before stating the other axioms of CS, it is worth reviewing a selection from the wide variety of principles of intuitionistic analysis. There are, first, five forms of the Axiom of Choice (the asterisk indicates a generalized version of a restricted form):

$(\forall x\ \exists y)$ $\quad \forall x\ \exists y\ A(x, y) \to \exists f\ \forall x\ A(x, f(x))$

$(\forall x\ \exists y)^*$ $\quad \forall x\ \exists y\ A(x, y, \alpha) \to \exists \beta\ \forall x\ A(x, \beta(x), \alpha)$

$(\forall x\ \exists f)$ $\quad \forall x\ \exists f\ A(x, f) \to \exists g\ \forall x\ A(x, (g)_x)$

$(\forall x\ \exists f)^*$ $\quad \forall x\ \exists f\ A(x, f, \alpha) \to \exists \beta\ \exists g\ \forall x\ (A(x, (g)_{\beta(x)}, \alpha)$

$(\forall x\ \exists \alpha)$ $\quad \forall x\ \exists \alpha\ A(x, \alpha, \beta) \to \exists \gamma\ \forall x\ A(x, (\gamma)_x, \beta).$

Then there is Bar Induction, stated as for a monotonic bar-predicate R, in its restricted and generalized forms:

$$(BI_M)\quad \forall \vec{u}\ \forall x\ (R(\vec{u}) \to R(\vec{u}^\frown x))\ \&\ \forall \alpha\ \exists x\ R(\bar{\alpha}(x))\ \&$$
$$\forall \vec{u}\ (R(\vec{u}) \to A(\vec{u}))\ \&\ \forall \vec{u}(\forall x\ A(\vec{u}^\frown x) \to A(\vec{u}))$$
$$\to A(0) \qquad [\text{here } 0 \text{ codes } \langle\ \rangle]$$

$$(BI_M)^*\quad \forall \vec{u}\ \forall x\ (R(\vec{u}, \beta) \to R(\vec{u}^\frown x, \beta))\ \&$$
$$\forall \alpha\ \exists x\ R(\bar{\alpha}(x), \beta)\ \&$$
$$\forall \vec{u}\ (R(\vec{u}, \beta) \to A(\vec{u}, \beta))\ \&$$
$$\forall \vec{u}\ (\forall x\ A(\vec{u}^\frown x, \beta) \to A(\vec{u}, \beta))$$
$$\to A(0, \beta).$$

Next, there are three restricted forms of Continuity Principle:

$(\forall \alpha\ \exists x)$ $\quad \forall \alpha\ \exists x\ A(\alpha, x) \to \exists e\ \forall \alpha\ A(\alpha, e(\alpha))$

$(\forall \alpha\ \exists f)$ $\quad \forall \alpha\ \exists f\ A(\alpha, f) \to \exists e\ \exists g\ \forall \alpha\ A(\alpha, (g)_{e(\alpha)})$

$(\forall \alpha\ \exists \beta)$ $\quad \forall \alpha\ \exists \beta\ A(\alpha, \beta) \to \exists e\ \forall \alpha\ A(\alpha, e|\alpha).$

As in FIM, these require no explicit extensionality condition in their antecedents; their generalized forms cannot be expressed by means of a quantifier of the form $\exists e$, since the presence of a choice-sequence parameter deprives us of a guarantee that there exists a *constructive* neighbourhood function. We therefore define

$K_0(\eta) \leftrightarrow \forall\alpha\ \exists x\ \exists z\ \forall y\ ((y < x \rightarrow \eta(\bar{\alpha}(y)) = 0)\ \&$
$(y \geq x \rightarrow \eta(\bar{\alpha}(y)) = z + 1))$, and formulate the generalized principles as follows:

$$(\forall\alpha\ \exists x)^* \quad \forall\alpha\ \exists x\ A(\alpha, x, \beta) \rightarrow \exists\eta\ \forall\alpha\ (K_0(\eta)\ \&\ \forall x\ \forall y\ (\eta(\bar{\alpha}(x)) = y + 1 \rightarrow A(\alpha, y, \beta)))$$

$$(\forall\alpha\ \exists f)^* \quad \forall\alpha\ \exists f\ A(\alpha, f, \beta) \rightarrow \exists\eta\ \exists g\ \forall\alpha\ (K_0(\eta)\ \&\ \forall x\ \forall y\ (\eta(\bar{\alpha}(x)) = y + 1 \rightarrow A(\alpha, (g)_y, \beta)))$$

$$(\forall\alpha\ \exists\beta)^* \quad \forall\alpha\ \exists\beta\ A(\alpha, \beta, \delta) \rightarrow \exists\eta\ \forall\alpha\ \forall\gamma\ (K_0(\eta)\ \&\ (\forall x\ \exists y\ \eta(\langle x\rangle * \bar{\alpha}(y)) = \gamma(x) + 1 \rightarrow A(\alpha,\gamma,\delta))).$$

For each version of the Axiom of Choice and of the Continuity Principle, we may also consider the weakened forms in which, in the antecedent, $\exists$ is replaced by $\exists!$.

In CS all these principles are derivable. Those actually assumed as axiom schemata, apart from (AD), are $\forall\alpha\ \exists\beta$-Continuity, $\forall\alpha\ \exists!f$-Continuity, and the $(\forall x\ \exists!y)$ and $(\forall x\ \exists!y)^*$ forms of the Axiom of Choice (if one may properly apply that title to principles relating to unique existence); but the remainder can be derived from these. In particular, it was shown in Section 3.5 that Bar Induction for decidable R (BI_D) is equivalent to K-Induction, without appeal to any continuity principle. The situation is slightly different in CS, since K is taken as primitive, so that, by themselves, the axioms governing it merely characterize it: where K_0 is defined as above, but as a predicate of constructive functions, what was called in section 3.5 K-Induction must be expressed as $\forall f\ (K(f) \leftrightarrow K_0(f))$. If (BI_M) is assumed as an axiom, this statement is derivable from it; conversely, from the axioms governing K together with $\forall\alpha\ \exists x$-Continuity, even restricted to antecedents of the form $\forall\alpha\ \exists x\ A(\bar{\alpha}(x))$, (BI_M) can be derived. It is worth observing also that, from $\forall\alpha\ \exists f$-Continuity, we can prove that $\neg\forall\alpha\ \exists f\ \alpha = f$. For suppose that $\forall\alpha\ \exists f\ \alpha = f$. Then, by the Continuity Principle, there exist e and g such that $\forall\alpha\ \alpha = (g)_{e(\alpha)}$. Hence, if $e(\vec{u}) = y + 1$ and $h = (g)_y$,

$\forall\alpha(\alpha \in \vec{u} \to \alpha = h)$, which is absurd. The details of the derivation of the various principles listed above from those assumed axiomatically will not be pursued here.

The most interesting proof-theoretic result concerning CS is the 'elimination of choice sequences': a mapping τ is defined which carries every formula of CS containing no free variables for choice sequences into an equivalent formula of IDB_1, that is, into a formula containing no choice-sequence variables at all. This mapping is effected as follows. First, we transform every subformula of the form $\neg A$ into $A \to 0 = 1$, and every subformula of the form $A \vee B$ into $\exists x((x = 0 \to A)\ \&\ (x \neq 0 \to B))$; we also replace every subformula of the form $K(\phi)$, where ϕ is an F-functor, by the formula $\exists e\ \forall x\ (e(x) = \phi(x))$. We then repeatedly apply to subformulas not containing free variables for choice sequences the transformation $\mapsto$ given by the clauses listed below; it can be shown that the process terminates, and that the result is unique up to change of bound variables.

(i) $\forall\alpha\ t[\alpha] = s[\alpha] \mapsto \forall f\ t'[f] = s'[f]$, where $t[\alpha]$ and $s[\alpha]$ are terms, and $t'[f]$ and $s'[f]$ result from them by replacing free occurrences of α by f and λ'' by λ;

(ii) $\forall\alpha(A(\alpha)\ \&\ B(\alpha)) \mapsto \forall\alpha A(\alpha)\ \&\ \forall\alpha B(\alpha)$;

(iii) $\forall\alpha\ \forall x\ A(\alpha, x) \mapsto \forall x\ \forall\alpha\ A(\alpha, x)$;

(iv) $\forall\alpha\ \forall f\ A(\alpha, f) \mapsto \forall f\ \forall\alpha\ A(\alpha, f)$;

(v) $\forall\alpha\ \forall e\ A(\alpha, e) \mapsto \forall e\ \forall\alpha A(\alpha, e)$;

(vi) $\forall\alpha\ \forall\beta\ A(\alpha, \beta) \mapsto \forall e\ \forall e'\ \forall\alpha\ A(e|\alpha, e'|\alpha)$;

(vii) $\forall\alpha\ (A(\alpha) \to B(\alpha)) \mapsto \forall e\ (\forall\alpha\ A(e|\alpha) \to \forall\alpha\ B(e|\alpha))$;

(viii) $\forall\alpha\ \exists x\ A(\alpha, x) \mapsto \exists e\ \forall\alpha\ A(\alpha, e(\alpha))$;

(ix) $\forall\alpha\ \exists f\ A(\alpha, f) \mapsto \exists e\ \exists f\ \forall\alpha\ A(\alpha, \lambda\vec{u}.\ f(\langle e(\alpha)\rangle * \vec{u}))$;

(x) $\forall\alpha\ \exists e\ A(\alpha, e) \mapsto \exists e\ \exists e'\ \forall\alpha\ A(\alpha, \lambda'\vec{u}.\ e'(\langle e(\alpha)\rangle * \vec{u}))$;

(xi) $\forall\alpha\ \exists\beta\ A(\alpha, \beta) \mapsto \exists e\ \forall\alpha\ A(\alpha, e|\alpha)$;

(xii) $\exists\alpha\ A(\alpha) \mapsto \exists f\ A(\lambda''x.\ f(x))$.

The clauses given here for (viii), (ix), and (x) are actually those given by Kreisel and Troelstra for their 'alternative translation'. The main forms given by them depend upon defining, relative to (a number representing) a finite sequence $\vec{u}$, a K-functor $\{\vec{u}\}$ such that, for each α, $(\{\vec{u}\}|\alpha)(x)$ is the x-th term of $\vec{u}$ if $x < \ell h(\vec{u})$ and is $\alpha(x)$ if $x \geq \ell h(\vec{u})$. The clauses then take the form:

(viii)' $\forall\alpha\ \exists x\ A(\alpha, x) \mapsto$
$\exists e\ \forall\vec{u}\ (e(\vec{u}) \neq 0 \rightarrow \forall\alpha\ A(\{\vec{u}\}|\alpha, e(\vec{u}) \dot{-} 1))$;

(ix)' $\forall\alpha\ \exists f\ A(\alpha, f) \mapsto$
$\exists e\ \exists f\ \forall\vec{u}\ (e(\vec{u}) \neq 0 \rightarrow \forall\alpha\ A(\{\vec{u}\}|\alpha, \lambda\vec{v}.\ f(\langle e(\vec{u}) \dot{-} 1\rangle * \vec{v})))$;

(x)' $\forall\alpha\ \exists e\ A(\alpha, e) \mapsto$
$\exists e'\ \exists e\ \forall\vec{u}\ (e'(\vec{u}) \neq 0 \rightarrow \forall\alpha\ A(\{\vec{u}\}|\alpha, \lambda'\vec{v}.\ e(\langle e'(\vec{u}) \dot{-} 1\rangle * \vec{v})))$.

Where $\tau(A)$ is the end-product of repeated applications of the transformation $\mapsto$ to a formula A to which the preliminary modifications have been applied, we have:

(a) if A contains no free variables for choice sequences, then $\vdash_{CS} A \leftrightarrow \tau(A)$; and

(b) it can be finitistically proved that $\vdash_{CS} A$ iff $\vdash_{IDB_1} \tau(A)$.

It would be a mistake to react to the elimination of choice sequences by concluding that what appeared to be one of the new ideas contributed to mathematics by intuitionism has depressingly proved not to be a fundamental idea at all, but one resoluble into the notions of finite sequence and of constructive function, for two reasons. First, the elimination is not definitive, resting essentially, as it does, upon the dubious $\forall\alpha\ \exists\beta$-Continuity Principle; although it figures as an axiom schema in both FIM and CS, it is far from plain that it is intuitionistically correct. Secondly, even if the elimination can be carried out, that does not mean that the idea of choice sequences has been dissolved. On the contrary, without the intuitive notion of a choice sequence, no one would think of viewing the formulas of IDB_1 into which those of CS translate as having the quantificational structure of the original CS formulas; nor would anyone seek to construct, e.g., a theory of real numbers in terms of real number generators taken as choice sequences if these appeared only in the thorough disguise provided by the translation into IDB_1. In so far as the translation is correct, the creative ingredient of the notion of a choice sequence survives intact, while the translation serves to guarantee the coherence of the notion. However, as already remarked, there is room for serious doubt whether it is correct.

6.2. REALIZABILITY

The notion of realizability was originally devised by Kleene to provide an interpretation of statements of intuitionistic mathematics in terms of recursive functions. Since the intuitive meanings of the intuitionistic logical constants are given in terms of the notion of a construction and of the relation which holds between a construction and a statement when the former is a proof of the latter, the idea was to represent constructions by natural numbers, and to define a relation expressed by 'n realizes A' (abbreviated as 'n r A'), considered as holding when the construction represented by the number n is a proof of the statement A. Since the notion of a construction requires that a construction constituting a proof of a universally quantified or of a conditional statement be applicable to a natural number or to a proof to yield a proof, we need a conception of applying a natural number to a natural number to yield a natural number, that is, an association of natural numbers with numerical functions; since the process of application must be effective, the obvious correlation is that of a natural number to the partial recursive function of which it is the Gödel number. We shall use the standard notation whereby the partial recursive function with Gödel number n is symbolized by $\{n\}$, so that $\{n\}(m)$ is defined iff $\exists r\, T_1(n, m, r)$, and $\{n\}(m) = k$ iff

$\exists r\ (T_1(n,m,r)\ \&\ U(r) = k)$. This then yields the following inductive definition of the relation n **r** A, where A is a *closed* formula of HA:

(i) if A is atomic, n **r** A iff A is true;

(ii) n **r** A & B iff $n = 2^a.3^b$ for some a and b such that a **r** A and b **r** B;

(iii) n **r** A ∨ B iff either $n = 3^a$ for some a such that a **r** A or $n = 2.3^b$ for some b such that b **r** B;

(iv) n **r** A → B iff, for every m such that m **r** A, {n}(m) is defined and {n}(m) **r** B;

(v) n **r** ¬A iff n **r** A → 0 = 1;

(vi) n **r** ∀x A(x) iff, for every m, {n}(m) is defined and {n}(m) **r** $A(\bar{m})$);

(vii) n **r** ∃x A(x) iff $n = 2^m.3^a$ for some m and a such that a **r** $A(\bar{m})$.

Then, for any formula A of HA, we say that n **r** A iff n realizes the universal closure of A. A formula A is called *realizable* iff, for some n, n **r** A.

The intention of this definition was that a formula should be realizable just in case it is intuitionistically true; if this held classically, we should have a classical explanation of intuitionistic truth for arithmetical statements. There is, however, a twofold defect in such an explanation. First, if we are going to allow ourselves classical reasoning (and, in so far as it was hoped that realizability would provide an interpretation of intuitionistic mathematics in purely classical terms, there is no reason why we should not), we are entitled to assert that any formula either is or is not realizable. It follows that for every closed formula A, A ∨ ¬A is realizable. For, if A is realizable, then 3^n **r** A ∨ ¬A, where n **r** A; if, on the other hand, A is not realizable, then 0 **r** ¬A, since, trivially, for every m such that m **r** A, {0}(m) **r** 0 = 1, and hence 2 **r** A ∨ ¬A. This obviously prevents us from equating realizability with intuitionistic truth; we might, however, still hope that we shall be able to cite a specific number which realizes a given formula when and only when that formula is intuitionistically true. However, the more serious defect is that the interpretation makes no allowance for the requirement that it be effectively decidable whether or not a given construction is a proof of a given statement; it is for this reason that clause (iii) could not be stated in the simple form:

$$n \mathbf{r} A \vee B \text{ iff } n \mathbf{r} A \text{ or } n \mathbf{r} B,$$

corresponding to the intuitive explanation:

> a construction is a proof of A ∨ B iff it is a proof of A or a proof of B.

Instead, clause (iii) had to be given in such a form that we can tell, by inspecting a number n which realizes A ∨ B, whether it yields a realization of A or of B. This would correspond to an intuitive explanation of the form:

> a construction is a proof of A ∨ B iff either it is an ordered pair ⟨0,a⟩, where a is a proof of A, or an ordered pair ⟨1,b⟩, where b is a proof of B;

and this is quite unnecessary if we can effectively recognize, for any construction, whether or not it is a proof of A or of B. We cannot, of course, effectively decide, for a given number n, whether or not it is the Gödel number of a general recursive function, and so certainly cannot in general decide whether or not $n \mathbf{r} \forall x\, A(x)$. For this reason, the notion of realizability diverges very considerably from the intended meanings of the intuitionistic logical constants; and this quickly became apparent when it was proved by G. F. Rose that there exists an intuitionistically invalid formula of sentential logic every (closed or open) substitution instance of which in the language of HA can be shown, using classical reasoning, to be realizable.

The idea that realizability supplies a formulation, in classical terms, of the intended meanings of statements of intuitionistic arithmetic was, therefore, rapidly abandoned. Realizability and its cognate notions have nevertheless proved an extremely useful tool for proving underivability and relative consistency results for intuitionistic formal systems, since it is usually straightforward to prove (using only intuitionistic reasoning) that the axioms of a system are realizable and that the rules of inference preserve realizability; the problem is only to formulate the right notion of realizability for showing a given formula to be unrealizable, and therefore underivable, or to be realizable and therefore capable of being consistently added to the axioms.

Suppose that we wish to show that HA has the explicit definability property, namely that if $\vdash_{HA} \exists x\, A(x)$, where x is the only variable occurring free in A(x), then $\vdash_{HA} A(\bar{m})$ for some m. It is obvious from the definition of realization that if $\exists x\, A(x)$ is realizable, then, for some m, $A(\bar{m})$ is realizable; but this does not give us what we need, since, while we can easily show that, if $\vdash_{HA} C$, then C is realizable, we cannot show the converse. To overcome this difficulty, Kleene introduced a modification of the original notion of realization, by invoking provability in HA in certain clauses of its definition; we symbolize the new relation by 'n q A'. The definition, which is again given in the first place for *closed* formulas, runs as follows:

(i) If A is atomic, n q A iff A is true;

(ii) n q A & B iff $n = 2^a.3^b$ for some a and b such that a q A and b q B;

(iii) n q A ∨ B iff either $\vdash_{HA} A$ and $n = 3^a$ for some a such that a q A, or $\vdash_{HA} B$ and $n = 2.3^b$ for some b such that b q B;

(iv) n q A → B iff for every m such that $\vdash_{HA} A$ and m q A, {n}(m) is defined and {n}(m) q B;

(v) n q ¬A iff n q A → 0 = 1;

(vi) n q ∀x A(x) iff, for every m, {n}(m) is defined and {n}(m) q $A(\bar{m})$;

(vii) n q ∃x A(x) iff $n = 2^m.3^a$ for some m and a such that $\vdash_{HA} A(\bar{m})$ and a q $A(\bar{m})$.

If n q A, we also say that n $\vdash$-*realizes* A. As before, for an open formula A, we define 'n q A' to mean that n $\vdash$-realizes the universal closure of A; and, for any A, we say that A is $\vdash$-*realizable* iff, for some n, n q A. The notion of $\vdash$-realization may also be relativized to a set Γ of closed formulas to obtain the notion of Γ$\vdash$-realization, which we symbolize by 'n q_Γ A': in the above definition, we replace '$\vdash_{HA}$C' throughout by 'Γ $\vdash_{HA}$C', and, of course, 'n q C' by 'n q_Γ C'.

It is now immediate from the definition that if ∃x A(x) is $\vdash$-realizable, where x is the only variable occurring free in A(x), then $\vdash_{HA}$A($\bar{m}$) for some m; hence by proving that every axiom of HA is $\vdash$-realizable and that the rules of inference preserve $\vdash$-realizability, we establish that HA has the explicit definability property (and hence also the disjunction property). It is obvious that we cannot extend this to Γ$\vdash$-realizability for every Γ (for instance, not if Γ consists of a single closed undecidable formula of the form ∃x B(x)): but by proving the extended result for certain sets Γ, we can obtain strengthened forms of the explicit definability and disjunction properties.

It is subsequently struck Kleene, however, that the essential features of the notion of Γ$\vdash$-realizability which are made use of in the proof of such results are the inductive structure of the definition and the appeal to derivability in HA: the specific conception of realization by natural numbers, and, in particular, the use in clauses (iv), (v) and (vi) of the notion of a (partial) recursive function, are a hangover from the original notion of realizability, and play no essential role. He therefore defined a new relation, generally called the slash and symbolized by | , by 'simply omitting the realizability from Γ$\vdash$-realizability' (compare the omission of recursion from the definition of 'primitive recursive' to obtain that of 'general recursive'). Better still, what was omitted was the notion of realiz*ation*, since the resulting notion corresponds to Γ$\vdash$-realiz*ability*, but is not defined by existentially quantifying any notion corresponding to Γ$\vdash$-realiz*ation*: what is directly defined is simply a relation between a set Γ of closed formulas and a closed formula A. The definition runs as follows:

(i) if A is atomic, $\Gamma|A$ iff $\Gamma \vdash_{HA} A$;

(ii) $\Gamma|A \mathbin{\&} B$ iff $\Gamma|A$ and $\Gamma|B$;

(iii) $\Gamma|A \vee B$ iff either $\Gamma|A$ and $\Gamma \;_{HA} A$ or $\Gamma|B$ and $\Gamma \vdash_{HA} B$;

(iv) $\Gamma|A \rightarrow B$ iff, if $\Gamma|A$ and $\Gamma \vdash_{HA} A$, then $\Gamma|B$;

(v) $\Gamma|\ A$ iff $\Gamma|A \rightarrow 0 = 1$;

(vi) $\Gamma|\forall x\, A(x)$ iff, for every $m, \Gamma|A(\bar{m})$;

(vii) $\Gamma|\exists x\, A(x)$ iff, for some $m, \Gamma|A(\bar{m})$ and $\Gamma \vdash_{HA} A(\bar{m})$.

As usual, if $x_1,\ldots,x_n$ are all the variables free in $A(x_1,\ldots,x_n)$, $\Gamma|A(x_1,\ldots,x_n)$ iff $\Gamma|\forall x_1 \ldots \forall x_n\, A(x_1,\ldots,x_n)$.

Obviously, all thought of giving an interpretation of formulas of HA has now been relinquished, and we are left with a notion useful as an auxiliary in obtaining proof-theoretic results. Equally obviously, the significance of the definition depends upon whether we construe the informal logical constants classically or intuitionistically: since we are no longer concerned to give a classical interpretation of intuitionistic statements we shall interpret them intuitionistically, i.e. only intuitionistic reasoning will be needed for the proofs of the theorems to be cited in this section.

It is now completely straightforward to prove, by induction on the length of the derivation in HA of A from Γ:

Theorem 1. If Γ is a set of closed formulas such that $\Gamma|C$ for every $C \in \Gamma$, and A is any formula such that $\Gamma \;_{HA} A$, then $\Gamma|A$.

In the proof, we shall need to appeal to the readily verified facts that, if t is any closed term, B(t) a closed formula, and k the denotation of t under the intended interpretation of HA, then $\Gamma \vdash_{HA} B(t)$ iff $\Gamma \vdash_{HA} B(\bar{k})$ and $\Gamma|B(t)$ iff $\Gamma|B(\bar{k})$.

By taking $\Gamma = \emptyset$, we can at once deduce that HA has the explicit definability property; but by choosing sets Γ which satisfy the hypothesis of Theorem 1, we can strengthen this result. In particular let us write 'C|B' for '{C}|B', and give the

Definition. A formula C is *detachable* iff it is closed and, for every formula A(x) in which x is the only free variable, if $\vdash_{HA} C \to \exists x\, A(x)$, then, for some m, $\vdash_{HA} C \to A(\bar{m})$.

We can then assert

Lemma 1. If C is closed and C|C, C is detachable.

Proof. Suppose C is closed, C|C, x is the only variable free in A(x), and $\vdash_{HA} C \to \exists x\, A(x)$. By modus ponens, $C \vdash_{HA} \exists x\, A(x)$. Hence, by Theorem 1, $C|\exists x\, A(x)$. By the definition of |, $C \vdash_{HA} A(\bar{m})$ for some m, and so, by the Deduction Theorem, $\vdash_{HA} C \to A(\bar{m})$.

To apply Lemma 1, we need to determine a class of closed formulas C for which we can establish that C|C. A partial solution is given by:

Lemma 2. If C is closed, $\neg C|\neg C$.

Proof. By the definition of |, $\neg C|\neg C$ iff $\neg C|C \to 0 = 1$. Suppose that $\neg C|C$ and $\neg C \vdash_{HA} C$. Then, by sentential logic, $\neg C \vdash_{HA} 0 = 1$, and hence, by clause (i) of the definition of |, $\neg C|0 = 1$. Therefore, by clause (iv) of the definition of |, $\neg C|C \to 0 = 1$.

In order to improve on Lemma 2, we first prove a strengthened converse of Lemma 1:

Lemma 3. If C is a detachable formula, then for every closed formula D such that $C \vdash_{HA} D$, C|D.

Proof. By induction on the complexity of D. Assume that C is detachable, D is closed and $C \vdash_{HA} D$.

(i) If D is atomic, C|D by the definition of |.

(ii) If D is E & F, then $C \vdash_{HA} E$ and $C \vdash_{HA} F$, and so C|E and C|F by the induction hypothesis, whence C|D by the definition of |.

(iii) If D is E ∨ F, then $C \vdash_{HA} \exists x\, [(x = 0 \to E)\ \&$

$(x \neq 0 \rightarrow F)]$. By the detachability of C, $C \vdash_{HA} E$ or $C \vdash_{HA} F$, and, by the induction hypothesis, $C|E$ or $C|F$ respectively. Hence $C|D$ by the definition of $|$.

(iv) If D is $E \rightarrow F$, suppose that $C \vdash_{HA} E$. Then $C \vdash_{HA} F$, and so, by the induction hypothesis, $C|F$, whence $C|D$ by the definition of $|$.

Case (v) [D is $\neg E$] is a subcase of case (iv); case (vi) [D is $\forall x\, E(x)$] is like case (ii); and case (vii) [D is $\exists x\, E(x)$] is like case (iii).

Lemma 1 and 3 together yield:

Lemma 4. If C is closed, then C is detachable iff $C|C$. This allows us to show:

Lemma 5. If C is closed, $C|C$ and $\vdash_{HA} C \leftrightarrow D$, then $D|D$.

Proof. If $C|C$, then by Lemma 4 C is detachable. Hence, if $\vdash_{HA} C \leftrightarrow D$, D is detachable, and so, by Lemma 4, $D|D$.

We recall that a *negative* formula is one built up from negations of atomic formulas (in HA, from atomic formulas) without the use of $\vee$ or $\exists$. A wider class of formulas is that of *Harrop* formulas, defined inductively by:

Definition.

(i) An atomic formula is a Harrop formula;

(ii) if A and B are Harrop formulas, so is A & B;

(iii) if A is a Harrop formula, so is $\forall x A$;

(iv) if B is a Harrop formula and A is any formula, $A \rightarrow B$ is a Harrop formula;

(v) if A is any formula, $\neg A$ is a Harrop formula.

We have:

Lemma 6. Every Harrop formula is stable in HA.

Proof. Let A be a Harrop formula. We have to show that $\vdash_{HA} A \leftrightarrow \neg\neg A$. We argue by induction on the complexity of A.

(i) Every atomic formula is stable in HA.

(ii) If A is B & C, then by sentential logic $\vdash_{HA} \neg\neg A \leftrightarrow (\neg\neg B \,\&\, \neg\neg C)$. By the induction hypothesis $\vdash_{HA} B \leftrightarrow \neg\neg B$ and $\vdash_{HA} C \leftrightarrow \neg\neg C$, whence $\vdash_{HA} \neg\neg A \leftrightarrow A$.

(iii) If A is $\forall x\ B(x)$, then by predicate logic $\vdash_{HA} \neg\neg A \to \forall x\ \neg\neg B(x)$. By the induction hypothesis, $\vdash_{HA} B(x) \leftrightarrow \neg\neg B(x)$, so that $\vdash_{HA} \neg\neg A \to A$.

(iv) If A is $B \to C$, then by sentential logic $\vdash_{HA} \neg\neg A \leftrightarrow (B \to \neg\neg C)$. By the induction hypothesis, $\vdash_{HA} C \leftrightarrow \neg\neg C$, so that $\vdash_{HA} \neg\neg A \leftrightarrow A$.

(v) If A is $\neg B$, A is stable by sentential logic alone. We may now finally conclude:

Theorem 2. If C is a closed Harrop formula, C is detachable.

Proof. Let C be a closed Harrop formula. By Lemma 6, $\vdash_{HA} C \leftrightarrow \neg\neg C$. By Lemma 2, $\neg\neg C | \neg\neg C$. By Lemma 5, $C|C$. By Lemma 1, C is detachable. Note that it can be shown that there are Harrop formulas which are not provably equivalent in HA to any negative formula. There are obviously detachable formulas which are not Harrop formulas, since any formally decidable closed formula is detachable; but it can also be shown that there are detachable formulas which are not provably equivalent in HA to any Harrop formula. If C is detachable, then evidently the 'independence of premisses' rule holds in relation to it in the form:

If $\vdash_{HA} C \to \exists x\ A(x)$, then

$$\vdash_{HA} \exists x\ (C \to A(x)),$$

where x is the only free variable in A(x). However, if C is taken as a Harrop formula, this is no stronger than the rule:

If $\vdash_{HA} \neg B \to \exists x\ A(x)$, then

$$\vdash_{HA} \exists x\ (\neg B \to A(x)),$$

where B and $\exists x\ A(x)$ are closed, since Harrop formulas are stable in HA.

De Jongh introduced an internalized version of the slash: for any formula E (not necessarily closed), we define a mapping which takes every formula A of HA into a formula $E \,I\, A$ of HA as follows:

<u>Definition</u>.

(i) If A is atomic, $E \mathbin{I} A$ is $E \to A$:

(ii) $E \mathbin{I} (A \mathbin{\&} B)$ is $(E \mathbin{I} A) \mathbin{\&} (E \mathbin{I} B)$;

(iii) $E \mathbin{I} (A \vee B)$ is $[(E \mathbin{I} A) \mathbin{\&} (E \to A)] \vee [(E \mathbin{I} B) \mathbin{\&} (E \to B)]$;

(iv) $E \mathbin{I} (A \to B)$ is $[(E \mathbin{I} A) \mathbin{\&} (E \to A)] \to (E \mathbin{I} B)$;

(v) $E \mathbin{I} \neg A$ is $[(E \mathbin{I} A) \mathbin{\&} (E \to A)] \to \neg E$;

(vi) $E \mathbin{I} \forall x\, A(x)$ is $\forall x\, (E \mathbin{I} A(x))$;

(vii) $E \mathbin{I} \exists x\, A(x)$ is $\exists x\, [(E \mathbin{I} A(x)) \mathbin{\&} (E \to A(x))]$.

By mimicking the proof of Theorem 1, we may then show:

<u>Theorem 3</u>. If $E \vdash_{HA} A$, then $(E \mathbin{I} E) \vdash_{HA} (E \mathbin{I} A)$, for any formulas E and A.

From this we can derive, for any formulas A and B(x):

<u>Theorem 4</u>. If $\vdash_{HA} A \to \exists x\, B(x)$, then $(A \mathbin{I} A) \vdash_{HA} \exists x(A \to B(x))$; in particular, if $\vdash_{HA} \neg A \to \exists x\, B(x)$, then $\vdash_{HA} \exists x(\neg A \to B(x))$.

Proof. If $\vdash_{HA} A \to \exists x\, B(x)$, then $A \vdash_{HA} \exists x B(x)$, and hence by Theorem 3, $(A \mathbin{I} A) \vdash_{HA} (A \mathbin{I} \exists x B(x)$. By the definition of $\mathbin{I}$,

$(A \mathbin{I} A) \vdash_{HA} \exists x[(A \mathbin{I} B(x)) \mathbin{\&} (A \to B(x))]$, and so a fortiori

$(A \mathbin{I} A) \vdash_{HA} \exists x(A \to B(x))$.

For the second half, note that by the definition of $\mathbin{I}$, $(\neg A \mathbin{I} \neg A)$ is

$$[(\neg A \mathbin{I} A) \mathbin{\&} (\neg A \to A)] \to \neg\neg A,$$

so that $\vdash_{HA} (\neg A \mathbin{I} \neg A)$ for every A.

What we have gained by this device is the generalization of the 'independence of premisses' rule (with a negation as premiss) to open formulas. Note that we have as a corollary that if $\vdash_{HA} A \to B \vee C$, then

$$(A \mathbin{I} A) \vdash_{HA} (A \to B) \vee (A \to C),$$

and if $\vdash_{HA} \neg A \to B \vee C$, then

$$\vdash_{HA} (\neg A \to B) \vee (\neg A \to C).$$

In a similar way, we can give an internalized version of the original notion of realizability: as in the case of the slash, we must, in order to handle the quantifiers, specify the mapping directly for all formulas, and not only for closed ones. Where A is any formula of HA, and x is any variable not occurring free in A, we specify a mapping which takes A into a formula $[x \mathbf{r} A]$ of HA which has as free variables all those of A together with x, as follows:

(i) If A is atomic, $[x \mathbf{r} A]$ is $x = x \,\&\, A$;

(ii) $[x \mathbf{r} A \,\&\, B]$ is $[(x)_0 \mathbf{r} A] \,\&\, [(x)_1 \mathbf{r} B] \,\&\, \forall y (p_y | x \to y \leq 1)$;

(iii) $[x \mathbf{r} A \vee B]$ is $((x)_0 = 0 \to [(x)_1 \mathbf{r} A]) \,\&\, ((x)_0 > 0 \to [(x)_1 \mathbf{r} B]) \,\&\, \neg 4 | x \,\&\, \forall y\ (p_y | x \to y \leq 1)$;

(iv) $[x \mathbf{r} A \to B]$ is $\forall u\ ([u \mathbf{r} A] \to \exists y (T_1(x,u,y) \,\&\, [U(y) \mathbf{r} B]))$;

(v) $[x \mathbf{r} \neg A]$ is $[x \mathbf{r} A \to 0 = 1]$;

(vi) $[x \mathbf{r} \forall y\ A(y)]$ is $\forall y \exists z (T_1(x,y,z) \,\&\, [U(z) \mathbf{r} A(y)])$;

(vii) $[x \mathbf{r} \exists y\ A(y)]$ is $[(x)_1 \mathbf{r} A((x)_0)] \,\&\, \forall z (p_z | x \to z \leq 1)$.

Here p_y is the formal term of HA which denotes the y-th prime, and $(x)_y$ that which denotes the exponent of the y-th prime in the factorization of x.

We can now show:

<u>Theorem 5</u>. If A is a closed formula such that $\vdash_{HA} A$, then for some n $\vdash_{HA} [\bar{n} \mathbf{r} A]$. Further, if Γ is a set of closed formulas such that, for each $C \in \Gamma$, $\vdash_{HA} \exists x\ [x \mathbf{r} C]$, and A is a closed formula such that $\Gamma \vdash_{HA} A$, then $\Gamma \vdash_{HA} \exists x\ [x \mathbf{r} A]$.

The proof of the main part of the theorem proceeds by a

detailed verification that each of the axioms of HA is provably realizable, and that the rules of inference preserve provable realizability. The second part follows from the fact that if $\Gamma \vdash_{HA} A$, then $\vdash_{HA} B \rightarrow A$, where B is a conjunction of members of Γ. By the first part of the theorem, for some n $\vdash_{HA} [\bar{n} \ \mathbf{r} \ B \rightarrow A]$, i.e. $\vdash_{HA} \forall u\, [u \ \mathbf{r} \ B] \rightarrow \exists y(T_1(\bar{n}, u, y)$ & $[U(y) \ \mathbf{r} \ A]))$, and the conclusion follows from the fact that $\vdash_{HA} \exists x\, [x \ \mathbf{r} \ B]$.

As a simple example of an application of realizability, we show that Church's Thesis is provably realizable. In HA Church's Thesis may be expressed by the schema:

$$\forall x\, \exists y\, A(x, y) \rightarrow \exists z\, \forall x\, \exists w(T_1(z, x, w) \ \& \ A(x, U(w))),$$

where A(x, y) may contain other free variables than those shown. We assert:

Theorem 6. If C is the universal closure of an instance of the schema expressing Church's Thesis, then, for some n, $\vdash_{HA} [\bar{n} \ \mathbf{r} \ C]$.

Proof. For simplicity, we consider an instance of the schema without additional parameters. In HA, we carry out the reasoning of which the following is the informal representation. Suppose

$$[u \ \mathbf{r} \ \forall x\, \exists y\, A(x, y)].$$

Then $\forall x\, [\{u\}(x) \ \mathbf{r} \ \exists y\, A(x, y)]$,

i.e. $x\, [(\{u\}(x))_1 \ \mathbf{r} \ A(x, (\{u\}(x))_0)]$.

In what follows, we use the Λ notation: if $\phi(x_1, \ldots, x_r, y_1, \ldots, y_s)$ is a partial recursive function with Gödel number g, $\Lambda x_1 \ldots x_r.\ \phi(x_1, \ldots, x_r, y_1, \ldots, y_s)$ is, for each $y, \ldots y_s$, the Gödel number h of a partial recursive function such that $\{h\}(x_1, \ldots, x_r) \simeq \phi(x_1, \ldots, x_r, y_1, \ldots, y_s)$ (h is given as the value of a primitive recursive function for the arguments $y_1, \ldots, y_s$ and g). We now abbreviate

$\Lambda x.\ (\{u\}(x))_0$ as $c(u)$

$\min v\ [T_1(c(u), x, v)]$ as $d(u, x)$

and $2^{d(u, x)}\ 3^{3^{(\{u\}(x))_1}}$ as $b(u, x)$.

Since $\{u\}$ is general recursive, so is $\{c(u)\}$, and hence

$$T_1(c(u), x, d(u, x))$$

for every x. Treating this as an atomic formula, we therefore have

$$[0 \mathbf{r} T_1(c(u), x, d(u, x))],$$

i.e. $[0 \mathbf{r} T_1(c(u), x, (b(u, x))_0)]$.

Also

$$[(\{u\}(x))_1 \mathbf{r} A(x, (\{u\}(x))_0)],$$

i.e. $[(\{u\}(x))_1 \mathbf{r} A(x, U((b(u, x))_0))]$.

Hence

$$[3^{(\{u\}(x))_1} \mathbf{r} T_1(c(u), x, (b(u, x))_0) \ \& \ A(x, U((b(u, x))_0))],$$

i.e. $[(b(u, x))_1 \mathbf{r} T_1(c(u), x, (b(u, x))_0) \ \& \ A(x, U((b(u, x))_0))]$.

Therefore

$$[b(u, x) \mathbf{r} \exists w(T_1(c(u), x, w) \ \& \ A(x, U(w)))].$$

Since this holds for every x, if we write $\Lambda x.\ b(u, x)$ as $a(u)$ we have:

$$[a(u) \mathbf{r} \forall x \exists w(T_1(c(u), x, w) \ \& \ A(x, U(w)))].$$

Writing $2^{c(u)} \cdot 3^{a(u)}$ as $m(u)$, we have:

$$[m(u) \mathbf{r} \exists z \forall x \exists w(T_1(z, x, w) \ \& \ A(x, U(w)))].$$

Hence, finally, by putting $n = \Lambda u.\ m(u)$,

$$[n \mathbf{r} \forall x \exists y A(x, y) \rightarrow \exists z \forall x \exists w(T_1(z, x, w) \ \& \ A(x, U(w)))].$$

We can conclude from Theorem 6 that the schema for Church's Thesis may consistently be added to the axioms of HA. In fact, a stronger result along these lines is obtainable. A formula is *almost negative* if it does not contain $\vee$ and contains $\exists$ only in contexts of the form $\exists x\ t(x) = s(x)$. Then we can show that if A is almost negative, $\vdash_{HA} A \leftrightarrow \exists x\ [x\ \mathbf{r}\ A]$. Further, if C is the universal closure of an instance of the schema for 'Extended Church's Thesis':

$$\forall x\ (A(x) \rightarrow \exists y\ B(x, y)) \rightarrow \exists z\ \forall x(A(x) \rightarrow \exists w(T_1(z, x, w)\ \&\ B(x, U(w)))),$$

where A(x) is almost negative, then, for some n, $\vdash_{HA}[n\ \mathbf{r}\ C]$. Moreover, where HA^{ECT} is the system obtained by adding the above schema to the axioms of HA, $\vdash_{HA^{ECT}} A \leftrightarrow \exists x[x\ \mathbf{r}\ A]$ for every formula A. From this fact, the provability in HA of the realizability of the closure of each instance of the schema, and the second part of Theorem 5, it follows that, for any A, $\vdash_{HA} \exists x\ [x\ \mathbf{r}\ A]$ iff $\vdash_{HA^{ECT}} A$. This result gives an exact condition for a formula of arithmetic to be realizable, if its realizability is to be demonstrated intuitionistically. The Extended Church's Thesis is not intuitionistically plausible, and may be shown, by appeal to a variant notion of realizability, to be unprovable in HA^{CT} (the system obtained by adding Church's Thesis to the axioms of HA); thus even intuitionistically demonstrable realizability cannot be identified with intuitionistic truth.

Under the original version of realizability, formulas are realized by numbers, but by numbers considered primarily under the aspect of Godel numbers of partial recursive functions. An important variant of the notion is obtained by defining realization as a relation between arbitrary unary number-theoretic functions and formulas; to play the role of the application of one function to another, we take the operation $\eta|\phi$, understood in the sense in which η represents a continuous functional with functions as values (relative to some given coding of finite sequences as natural numbers), and taken as undefined when η does not in fact constitute a neighbourhood function. In this way we obtain the notion of ^{1}realization. In the following definition, for closed formulas A, of 'η ^{1}realizes A' (abbreviated '$\eta\ {}^{1}\mathbf{r}$ A'), we write κ_n for the function $\lambda m.\ n$ with constant value n; as usual, $(n)_i$ is the exponent of the i-th prime in the prime factorization of n, and, in the present context, $(\eta)_i$ is $\lambda m.(\eta(m))_i$. The definition then runs:

(i) if A is atomic, $\eta\ {}^{1}\mathbf{r}\ A$ iff A is true;

(ii) $\eta\ {}^{1}\mathbf{r}\ A \,\&\, B$ iff $(\eta)_0\ {}^{1}\mathbf{r}\ A$ and $(\eta)_1\ {}^{1}\mathbf{r}\ B$;

(iii) $\eta\ {}^{1}\mathbf{r}\ A \vee B$ iff either $(\eta(0))_0 = 0$ and $(\eta)_1\ {}^{1}\mathbf{r}\ A$ or $(\eta(0))_0 > 0$ and $(\eta)_1\ {}^{1}\mathbf{r}\ B$;

(iv) $\eta\ {}^{1}\mathbf{r}\ A \to B$ iff, for every ϕ, if $\phi\ {}^{1}\mathbf{r}\ A$, then $\eta|\phi$ is defined and $\eta|\phi\ {}^{1}\mathbf{r}\ B$;

(v) $\eta\,{}^{1}\mathbf{r}\ \neg A$ iff $\eta\,{}^{1}\mathbf{r}\ A \to 0 = 1$;

(vi) $\eta\ {}^{1}\mathbf{r}\ \forall x\, A(x)$ iff, for every n, $\eta|\kappa_n$ is defined and $\eta|\kappa_n\ {}^{1}\mathbf{r}\ A(\bar{n})$;

(vii) $\eta\ {}^{1}\mathbf{r}\ \exists x\, A(x)$ iff $(\eta)_1\ {}^{1}\mathbf{r}\ A(\overline{(\eta(0))_0})$.

A closed formula A is then said to be ^{1}realizable iff $\eta\ {}^{1}\mathbf{r}\ A$ for some general recursive η.

The notion of ^{1}realizability was also introduced by Kleene, and, as he has emphasized, it is clause (iv) of the definition that makes the effective difference between it and the original kind of realizability. Since 0 = 1 is unrealizable, it follows, for every n and each closed A, that $n\ \mathbf{r}\ \neg A$ iff A is unrealizable, so that $\neg A$ is realizable iff A is unrealizable; hence $\neg\neg A$ is realizable iff A is not unrealizable. We can therefore never show $\neg\neg A \to A$ to be unrealizable when A is closed. In the same way, when A is closed, for every function ϕ, ϕ ^{1}realizes $\neg A$ iff no function ^{1}realizes A, and hence, for every ϕ,ϕ ^{1}realizes $\neg\neg A$ iff not every function fails to ^{1}realize A. But to say that A is not ^{1}realizable means that no *general recursive* function ^{1}realizes A. It is therefore possible that neither A nor $\neg A$ is ^{1}realizable; and, in such a case, since no function will ^{1}realize $\neg A$, $\neg\neg A$ will be ^{1}realizable although A is not.

Kleene's example for this is the formula $\forall x\ (\exists y\ T_1(x, x, y) \vee \neg\exists y\ T_1(x, x, y))$. This formula (call it A) is neither realizable nor ^{1}realizable. For suppose that n **r** A, and let θ be $(\{n\})_o$. If $\theta(m) = 0$, $(\{n\}(m))_1$ **r** $\exists y\ T\ (\bar{m}, \bar{m}, y)$, and so $\mathcal{T}_1\ (m, m, ((\{n\}(m))_1)_o)$ (where $\mathcal{T}_1$ is the informal predicate which T_1 formally expresses). If $\theta(m) = 1$, $(\{n\}(m))_1$ **r** $\neg\exists y\ T\ (\bar{m}, \bar{m}, y)$, and so not $\mathcal{T}_1\ (m, m, k)$ for any k. Thus θ is the characteristic function of the species $\underline{I} = \{m \mid \text{for some } k, \mathcal{T}_1\ (m, m, k)\}$; but this is impossible, since θ is general recursive. Thus $\neg A$ is realized by every number, while A and $\neg\neg A$ are both unrealizable. By exactly parallel reasoning, A is not ^{1}realizable, i.e. for no general recursive η does η ^{1}realize A. However, classically, in terms of the characteristic function of $\underline{I}$ we can easily define a function which ^{1}realizes A, though it is not general recursive; hence $\neg A$ is not ^{1}realizable, and so $\neg\neg A$ is ^{1}realized by every function.

The notion of ^{1}realizability was used by Kleene in order to obtain a definition of realizability for formulas of analysis, specifically for his system FIM. In this case, it is no longer possible to define the relation of ^{1}realizing only for closed formulas in the first instance, since we do not have an adequate supply of closed functors to replace the free variables for choice sequences: hence what is defined is the ^{1}realization by a function of an open formula relative to an assignment of numbers and functions to its free variables, expressed by '$\eta\ {}^1\mathbf{r}\text{-}\Psi\, A$', where Ψ is an assignment to the free variables of A. It is plain how clauses (i) - (vii) of the definition of $^1\mathbf{r}$ may be adapted for this purpose, and how to formulate additional clauses for the quantifiers $\forall\alpha$ and $\exists\alpha$. Further, where $\underline{P}$ is any class of functions containing any function general recursive in members of $\underline{P}$, we may generalize the definition to obtain a definition of 'η $\underline{P}/{}^1$realizes A relative to the assignment Ψ', by restricting the variables η and ϕ to $\underline{P}$, and requiring the functions assigned by Ψ to be members of $\underline{P}$. It may then be shown that, if $\Gamma \vdash_{FIM} A$, and C is ^{1}realizable for every $C \in \Gamma$, then A is ^{1}realizable; further that, where FIM^- is FIM without the axiom schema of Bar Induction, and $\underline{P}$ is a class of functions closed under general recursiveness, if $\Gamma \vdash_{FIM^-} A$ and C is $\underline{P}/{}^1$realizable for every $C \in \Gamma$, then A is $\underline{P}/{}^1$realizable. On the other hand, the example given in section 3.2 of a predicate $R(\vec{u})$ such that, for every general recursive binary function α, $R(\bar{\alpha}(n))$ for some n, but, for every m, there exists a general recursive binary function α, $R(\bar{\alpha}(n))$ for some n, but, for every m, there exists a general recursive binary α such that not $R(\bar{\alpha}(n))$ for any $n \leq m$, shows that the Fan Theorem fails when the choice-sequence variables are taken to range only over general recursive functions; it may be adapted to show that, where $\underline{P}$ is the class of general recursive functions, an instance of the axiom schema of Bar Induction is not $\underline{P}/{}^1$realizable, and hence that it is independent of the other axioms.

The notion of ^{1}realizability will not serve to yield the independence of all formulas for which we should like to demonstrate it, in particular not that of Markov's principle. Suppose that we tried to show that $\forall x\, M(x)$ was not ^{1}realizable, where $M(x)$ is

$$\neg\forall y \neg T_1(x, x, y) \rightarrow \exists y\, T_1(x, x, y).$$

We might argue as follows. If $\eta\ {}^1\mathbf{r}\ \forall x\, M(x)$, then, for each n, $\eta|\kappa_n$ is defined and ^{1}realizes $M(x)$ relative to the assignment of n to x. If for some k $\mathcal{T}_1(n, n, k)$, then every ϕ ^{1}realizes $\neg\forall y \neg T_1(x, x, y)$ relative to that assignment; so, where ϕ is any fixed general recursive function, for each n such that $\mathcal{T}_1(n, n, k)$, $(\eta|\kappa_n)|\phi$ is defined and ^{1}realizes $\exists y\, T_1(x, x, y)$ when n is assigned to x, so that

$\mathcal{T}_1(n, n, (((\eta|\kappa_n)|\phi)(0))_0)$. Now if we could be sure that $(\eta|\kappa_n)|\phi$ was defined for every n, this would yield a general recursive function χ such that $\mathcal{T}_1(n, n, \chi(n))$ iff $n \in \underline{I}$, which is impossible: we should thus have shown that no η can ^{1}realize $\forall x\ M(x)$. However, the definition of '^{1}realizes' only requires $(\eta|\kappa_n)|\phi$ to be defined when ϕ ^{1}realizes $\neg\forall y \neg T_1(x, x, y)$ relative to the assignment of n to x, and so we can assert no more than the obvious fact that there is a *partial* recursive function χ such that $n \in I$ iff $\chi(n)$ is defined and $\mathcal{T}_1(n, n, \chi(n))$.

The remedy is to appeal to a variant kind of realizability, under which functions are assigned to types corresponding to the logical structure of formulas, and to allow a function to realize a formula only if it is of the type corresponding to the structure of that formula. Where $\phi[\psi]$ represents that notion of the application of one function to another that is being used in the definition of the given kind of realizability (in the way that was used in defining ^{1}realizability), and where σ and τ are the types corresponding to the formulas A and B, it will then be required, for a function ϕ to realize $A \rightarrow B$, that for every function ψ of type σ, $\phi[\psi]$ is defined and is of type τ, irrespective of whether ψ in fact realizes A. This is the fundamental idea underlying the various notions of 'modified realizability' and 'special realizability', introduced by Kreisel and Kleene respectively, which enable us to show the independence of Markov's principle from FIM and other systems of intuitionistic analysis; but we shall not go any further into the details of these variant notions.

6.3. THE CREATIVE SUBJECT

Brouwer, in various of his later writings, made use of a device to construct counter-examples to certain intuitionistically unacceptable statements that has given rise to much controversy. This device involves supposing time to be divided into discrete stages, and then defining a choice sequence in terms of whether or not some given statement has been proved at each stage.

To do this is, of course, to introduce temporal reference into mathematics, and, to many, this appears shocking; the question is whether it is illegitimate. For a platonist, a mathematical statement is rendered true or false by a reality which lies outside time and is therefore not subject to change; hence, for him, no mathematical statement can involve any temporal reference, explicitly or implicitly. Even he must, indeed, allow that a mathematical statement may enter into a more complex statement which does involve temporal reference, such as 'Lindemann proved in 1882 that π is transcendental'; but, just for that reason, such a statement must be regarded as an empirical one and not as itself belonging to the domain of mathematical discourse.

This explanation of the tenselessness of mathematical statements is not available to an intuitionist, since for him a mathematical statement is rendered true or false by a proof or disproof, that is, by a construction, and constructions are effected in time. This fact supplies a prima facie ground for thinking that it may be possible intuitionistically to introduce temporal reference into mathematical statements. On the other hand, it is a feature of mathematical statements that they are not construed as being significantly tensed, even when they are understood intuitionistically rather than classically; and, if we are to introduce temporal reference into mathematics, this feature of mathematical statements must be preserved if we are, by this means, to generate further mathematical statements, and not merely empirical ones, such as may be formed even on a platonistic view. If we regard a mathematical statement as becoming true only when it is proved, then the predicate ' ... is true' is significantly tensed; the statement 'π is transcendental', for example, has been true since 1882 and was not true before that year. But, for all that, we shall not want to say that π has been transcendental since 1882, but was not transcendental before that; and, for that reason, when ' ... is true' is understood in this way, a mathematical statement A will not be equivalent to 'It is true that A', and an attribution of truth-value to a mathematical statement will not itself be a mathematical statement.

Our reluctance to say that π was not transcendental before 1882, or, more generally, to construe mathematical statements as significantly tensed, is not merely a lingering effect of platonistic misconceptions; it is, rather, because to speak in this way would be to admit into mathematical statements a non-intuitionistic form of negation, as will be apparent if one attempts to assign a truth-value to 'π is not algebraic', considered as a statement made in 1881. This is not because the 'not' which occurs in ' ... is not true' or ' ... was not true' is non-constructive: we may reasonably view it as decidable whether or not a given statement has been proved at a given time. But, though constructive, this is an empirical type of negation, not the negation that occurs in statements of intuitionistic mathematics. The latter relates to the impossibility of ever carrying out a construction of some fixed type, the former to the outcome, at variable times, of some observation or inquiry.

Platonistically considered, mathematical statements have fixed, invariant truth-values, *true* or *false*, and it is this which makes them inaccessible to temporal qualification. Intuitionistically, however, the tenselessness of mathematical statements is due to their having the characteristic that, though they often lack a truth-value, once they have acquired one they retain it. From this, two conclusions follow. First, that any temporal reference we introduce into mathematical statements must be non-indexical, i.e. it cannot be effected by means of explicit or implicit temporal adverbs like 'now' whose reference varies with the occasion of utterance: 'it is true that ... ', construed as suggested above, is significantly present-tensed, and thus contains 'now' implicitly, and therefore cannot be used to form *mathematical* statements. Secondly, negation (and perhaps other logical constants) cannot be taken as commuting with temporal adverbs: we shall want to allow the truth of 'Not (in 1881, π was transcendental)', but not that of 'In 1881, π was not transcendental'.

In accordance with these ideas, we may regard time, from some fixed point on, as being divided into denumerably many discrete stages, and introduce the symbol '$\vdash_n$' as a sentential operator, where '$\vdash_n$ A' is to have the meaning 'At the n-th stage we have a proof that A'. Here for the first time we encounter a non-extensional context; it would not be reasonable to require

(i) $(A \leftrightarrow B) \,\&\, (\vdash_n A) \rightarrow (\vdash_n B)$.

One axiom schema that we evidently may adopt as governing $\vdash_n$ is:

(1) $\forall n\, ((\vdash_n A) \vee \neg(\vdash_n A))$;

it is reasonable to suppose that we can recognize, at any given stage, whether or not a particular statement A has been proved at that stage. (Where A contains parameters, we mean, as usual, to assume the universal closure of the axiom; e.g., if A is B(m), we assert:

$$\forall m\, \forall n\, ((\vdash_n B(m)) \vee \neg(\vdash_n B(m))).)$$

Since we do not intend to envisage the possibility that a proof, once obtained, will be subsequently lost, another straightforward axiom schema is:

(2) $\forall n\, \forall m\, (n \leq m \,\&\, (\vdash_n A) \rightarrow (\vdash_m A))$.

With this machinery, we may define the choice sequence used by Brouwer to give counter-examples. For instance, if F is Fermat's Last Theorem, we may set:

$$\alpha(n) = \begin{cases} 0 & \text{if } \neg(\vdash_n F) \\ 1 & \text{if } \vdash_n F. \end{cases}$$

Or, again, we might put:

$$\beta(n) = \begin{cases} 0 \text{ if } \neg(\vdash_n (F \vee \neg F)) \\ 1 \text{ if } \vdash_n (F \vee \neg F), \end{cases}$$

i.e. $\beta(n)$ is 0 so long as F has not been decided, and 1 as soon as it has been.

What further axioms are we entitled to assume? The following can hardly be controverted:

$$(3) \qquad \forall n\ ((\vdash_n A) \rightarrow A);$$

a statement proved at any stage is true; or, more exactly, from a proof that A has been proved at some stage we can obtain a proof of A. (This requires us to assume that a demonstration that a statement has been proved at some stage involves a citation of, or at least an effective means of determining, the proof then given.) From (3) we can of course derive:

$$(ii) \qquad \neg A \rightarrow \neg \exists n\ (\vdash_n A)\ ;$$

we shall never be able to prove a false statement (one that has been disproved). Recalling that, intuitively, $\neg A$ is equivalent to 'It can never be proved that A', it may seem plausible to assert the converse:

$$(iii) \qquad \neg \exists n\ (\vdash_n A) \rightarrow \neg A,$$

which is equivalent to:

$$(4) \qquad A \rightarrow \neg\neg \exists n\ (\vdash_n A);$$

we can never show, of a true statement, that it can never be proved; more exactly, given a proof of A, we can derive a contradiction from the supposition that it will not be proved at any stage. (3) and (4) together yield:

$$(iv) \qquad \neg A \leftrightarrow \neg \exists n\ (\vdash_n A)\ ;$$

A is false just in case it can never be proved. Note that, for the β defined above, we have:

$$\forall n\ \beta(n) = 0 \rightarrow \neg \exists n\ (\vdash_n (F \vee \neg F)),$$

which, by (iv), implies:

$$\forall n\ \beta(n) = 0 \rightarrow \neg (F \vee \neg F).$$

and hence:

$$\neg \forall n \; \beta(n) = 0.$$

The axiom schemata (1) - (4) are in fact enough to enable us to construct the counter-examples more informally given by Brouwer. However, before considering these applications, it is worth while, since some have looked askance at the whole theory of the creative subject, to consider a little further what other axioms we might adopt, in order to test whether we really have hold of a coherent notion. Some such further possible axioms may be derived by reflection on the intended meanings of the logical constants. Thus a proof of A & B is, itself, a proof of A and of B, and this suggests the axiom schema:

$$\text{(v)} \qquad \forall n \; ((\vdash_n (A \,\&\, B)) \rightarrow (\vdash_n A) \,\&\, (\vdash_n B)).$$

Likewise, a proof of A v B is, itself, a proof either of A or of B, so that we might adopt:

$$\text{(vi)} \qquad \forall n \; ((\vdash_n (A \vee B)) \leftrightarrow (\vdash_n A) \vee (\vdash_n B)),$$

and similarly for ∃:

$$\text{(vii)} \qquad \forall n \; ((\vdash_n \exists m \, A(m)) \leftrightarrow \exists m \; (\vdash_n A(m))).$$

The converse of (v);

$$\text{(viii)} \qquad \forall n \; ((\vdash_n A) \,\&\, (\vdash_n B) \rightarrow (\vdash_n (A \,\&\, B)))$$

is less evident. The existence of proofs of A and of B does not of itself guarantee that we have a proof of A & B, unless any two constructions automatically coalesce to form a joint construction; intuitively expressed, the fact that we have proved both A and B does not guarantee that we have explicitly noticed that they are both provable. But even (vi) and (vii), and (v) itself, are subject to similar doubts. It may be that one and the same construction can serve as a proof of $A(\bar{k})$ & B, of $A(\bar{k})$, of $A(\bar{k}) \vee C$, and of $\exists m \, A(m)$; but are we to count $\vdash_n D$ as true simply on the ground that at stage n we have effected a construction which would constitute a proof of D, or is it necessary that we should have consciously registered the fact that it is a proof of that statement? The stronger requirement seems the more reasonable, and is certainly a possible one; but, if we make it, then none of the axiom schemata (v) - (vii) is strictly cogent.

As already observed in our preliminary discussion, the operator $\vdash_n$ cannot be held to commute with negation. The principle in one direction, viz.

$$\text{(ix)} \qquad \forall n\ ((\vdash_n \neg A) \rightarrow \neg(\vdash_n A))$$

is, indeed, already derivable from (3); but its converse

$$\text{(x)} \qquad \forall n\ (\neg(\vdash_n A) \rightarrow (\vdash_n \neg A))$$

is not only flagrantly false intuitively, but, taken together with (1) and (3), would imply

$$\forall n\ \forall m\ (n \leq m\ \&\ (\vdash_m A) \rightarrow (\vdash_n A)),$$

and hence, with (2):

$$\forall n\ \forall m\ ((\vdash_m A) \leftrightarrow (\vdash_n A)).$$

The corresponding law for $\rightarrow$:

$$\text{(xi)} \qquad \forall n\ (((\vdash_n A) \rightarrow (\vdash_n B)) \rightarrow \vdash_n (A \rightarrow B))$$

is equally implausible on any interpretation of '$\vdash_n$'. That for $\forall$:

$$\text{(xii)} \qquad \forall n\ (\forall m\ (\vdash_n A(m)) \rightarrow (\vdash_n \forall m\ A(m)))$$

is less obviously wrong, but should surely be rejected on the ground that the antecedent would be true if, at some stage later than n, we came to recognize that we had, by stage n, already proved $A(\bar{k})$ for each k, whereas it does not seem that the consequent need hold in such circumstances. Some have advocated an interpretation of '$\vdash_n$' under which the converse of (xii):

(xiii) $\forall n\ ((\vdash_n \forall m\ A(m)) \rightarrow \forall m\ (\vdash_n A(m)))$

would come out true, that is, one under which a proof of $\forall m\ A(m)$ is itself to count as being, for each k, a proof of $A(\bar{k})$. The standard explanation of the intuitionistic meanings of the logical constants does not provide the same warrant for (xiii) as for (v), in that, while a proof of A & B is said itself to be a proof both of A and of B, a proof of $\forall m\ A(m)$ is not said in itself to be a proof of $A(\bar{k})$ for each k, but only an effective means of obtaining such proofs. Acceptance of (xiii) would therefore rest upon a weaker requirement on our explicit awareness of what we are to be taken as having proved than acceptance of (v), (vi) and (vii). It is difficult to see how, on such a conception, we could resist the converse of (xi):

(xiv) $\forall n\ ((\vdash_n (A \rightarrow B)) \rightarrow ((\vdash_n A) \rightarrow (\vdash_n B)))$;

if we have already proved A, then a subsequent proof of $A \rightarrow B$ ought to count as being, simultaneously, a proof of B, while, if we have already proved $A \rightarrow B$, a subsequent proof of A ought to count as being simultaneously, a proof of B. However, on a stricter conception of what constitutes proof, neither (xiii) nor (xiv) is plausible. If we have a proof of $\forall m\ A(m)$, then we have an effective means for obtaining, for given k, a proof of $A(\bar{k})$, but we need not yet have obtained it; likewise, if we have proofs of $A \rightarrow B$ and of A, we have the means to obtain a proof of B, but we need not yet have obtained one.

These reflections have not yielded any indisputably valid new axiom schema, and the acceptability of those suggested evidently depends upon exactly how we construe the notion of having a proof of a statement. At one extreme, it may be demanded that we have explicitly recognized the statement as having been proved, that is, that we have not only effected a construction constituting a proof of it, but are consciously aware that we have done so. Under this, the strictest, interpretation, none of the proposed axiom schemata is valid (save for (ix), which is already provable). At the other extreme, we may be regarded as having proved a statement provided that we have explicitly proved one or more statements from which it follows very directly. Under this, the most lenient, interpretation, axiom schemata (v), (vi), (vii), (viii), (xii), (xiii), and (xiv) may all be taken as valid. An intermediate position would be that we have proved a statement just in case we have effected a construction which would, by itself, be a proof of that statement, whether or not we have noticed that it is so. This would validate axiom schemata (v), (vi), and (vii), but not any of the others.

Probably the most sensible attitude is that it is indeterminate which of these interpretations of '$\vdash_n$ A' should be adopted, and that we are free to choose between them. There are, however, difficulties both about the lenient and about the intermediate interpretations. The lenient interpretation is afflicted with the problem that attaches to all attempts to distinguish between immediate and remote consequences, namely that a remote consequence is reached by means of a chain of immediate consequences. Axiom schema (xiv) admittedly states merely that, when we have proved any statement, we have thereby proved all its *perceived* consequences (not those which, in some objective sense, are consequences, whether we realize it or not). The remaining axiom schemata acceptable on this interpretation embody some, but cannot embody all, of the basic rules of inference of a natural deduction calculus (in the sense that $\vdash_n$ is closed under these rules); they cannot embody rules which discharge hypotheses, since we have provided no sense for '$\Gamma \vdash_n$ A'. However, if we are held to have proved implicitly every instance of a universally quantified statement which we have proved explicitly, by the same token we may be held to have proved implicitly every instance of a schema whose validity we have explicitly established; hence if at any stage we may be credited with having established all the axiom schemata of some axiomatic formalization of first-order logic, with modus ponens as the only rule of inference, then, from that stage on, with every statement that we prove we shall thereby have proved all its first-order consequences. In any case, there would be no obstacle to giving a sense to expressions of the form '$B_1, \dots, B_r \vdash_n A$', understood as meaning 'At stage n we have effected a construction which constitutes a proof of A from $B_1, \dots, B_r$ as hypotheses'; and, if this were interpreted in the same lenient fashion, then there would be no reason not to accept axiom schemata embodying all the natural deduction rules, with the same effect. All this is not to say that we could not adopt a lenient interpretation of '$\vdash_n$' without going to these lengths; merely that any such interpretation which allows, as having been proved, some immediate consequences of what has been proved explicitly, but not others, is not intuitively stable, while, if every immediate consequence is allowed, then also every remote consequence is thereby allowed, at least if we consider only first-order consequences.

The intermediate interpretation is also unstable, since it depends upon rather arbitrary decisions about the form which a construction must take to be a proof of a given statement. From an intuitive point of view, it makes no difference whatever whether we say that a proof of A & B must actually be a proof both of A and of B, or that it should be compounded out of them in some manner unique to conjunctions; but it makes all the difference to the acceptability, under the intermediate interpretation, of axiom schema (v). Exactly the same holds for proofs of disjunctive and of existential statements, *vis-à-vis* axiom schemata (vi) and (vii). It would be perfectly possible so to frame the requirement on a construction, for it to be a proof, that no construction could be a proof of more than one statement; and then the intermediate interpretation would collapse into the strict one.

On the strict interpretation, since each construction proves no more than one statement, no two distinct statements can be proved simultaneously; so we may as well suppose that the division of time into stages is carried out in such a way that there is exactly one new statement proved at each stage, that is, for each n, there is just one statement A satisfying

$$(\vdash_n A) \ \& \ \forall m_{m < n} \neg (\vdash_m A).$$

(On this interpretation, we might, therefore, endorse axiom schema (xii) on the ground that its antecedent was invariably false, and it accordingly true vacuously, since, at any stage, we can have explicitly proved only finitely many statements.) On this basis, we could accept the axiom schema:

$$\text{(xv)} \qquad (\vdash_n A) \ \& \ (\vdash_n B) \ \& \ \forall m_{m < n} \neg ((\vdash_m A) \vee (\vdash_m B)) \rightarrow (A \leftrightarrow B).$$

In fact, we could, if we wished, introduce a new operator, by labelling as $P^{(n)}$ the statement newly proved at the n-th stage, subject to the axiom schema:

$$\text{(xvi)} \qquad (\vdash_n P^{(n)}) \ \& \ \forall m_{m < n} \neg (\vdash_m P^{(n)}) \ \& \ ((\vdash_n B) \ \& \ \forall m_{m < n} \neg (\vdash_m B) \rightarrow (B \leftrightarrow P^{(n)})).$$

From this Troelstra has constructed a paradox purporting to show the strict interpretation of '$\vdash_n$' to be incoherent. Some of the statements $P^{(n)}$ will assert, of some infinite sequence, that it is constructive (lawlike); and, since it is reasonable to suppose that we can effectively recognize, of any given statement, whether it makes an assertion of this form or of some other, we can enumerate, as $P^{(b(0))}$, $P^{(b(1))}$, $P^{(b(2))}$,..., those statements $P^{(n)}$ which do so. Hence we can define a constructive binary function h such that, for each n, if $P^{(b(n))}$ is the statement 'α is a lawlike sequence', then $h(n, m) = \alpha(m)$ for each m. From this a contradiction follows by diagonalization: if c is taken as $\lambda n.\ h(n, n) + 1$, then c is a lawlike sequence (constructive unary function); moreover, the statement 'c is a lawlike sequence' is provable, and must therefore be $P^{(b(k))}$ for some k; and from this we obtain $h(k, k) = c(k) = h(k, k) + 1$.

However, as Troelstra himself points out, this paradox depends upon a great many assumptions besides the strict interpretation of '$\vdash_n$', and can be resolved without rejecting that interpretation. Its resolution turns on our imposing a hierarchy upon constructions and, simultaneously, upon statements. A statement which does not involve the notion expressed by '$\vdash_n$' is of level 0, and we assume that we have in the first place a conception of a range of constructions, of level 0, such that any statement of level 0 which can be proved at all can be proved by means of a construction of level 0. This enables us to give a sense to '$\vdash_n$' as applied to statements of level 0, thereby obtaining statements of level 1: time is understood to be divided into stages punctuated by our effecting constructions, of level 0, which prove statements of level 0. We may now consider a new range of constructions, namely those effected by appeal to the notion expressed by '$\vdash_n$' as applied to statements of level 0: these are the constructions of level 1, and we again assume that, if a statement of level 1 can be proved at all, it can be proved by means of a construction of level 1. Now we may suppose time to be divided into stages punctuated by our effecting constructions, of level 1, which prove statements of level 1: and so we may give a sense to '$\vdash_n$' when applied to statements of level 1, thus obtaining statements of level 2. In general, a construction which may be effected by appeal to the notion expressed by '$\vdash_n$' as applied only to statements of level $\leq p$ is of level $p + 1$, and a statement which involves only constructions of level $\leq p$ and subordinate statements of the form $\vdash_n A$, for A of level $\leq p$, is, again, of level $p + 1$, and, if provable at all, provable by a construction of level $p + 1$. The diagonalization is now blocked. If $P^{(b(0))}$, $P^{(b(1))}$, $P^{(b(2))}$, ... is taken to be an enumeration of those statements $P^{(n)}$ of level $p + 1$ which assert that some infinite sequence is governed by a law of level p, then the unary function c will be of level $p+1$, and the statement 'c is a lawlike sequence' will be of level $p+2$, and hence will not occur in the enumeration.

Not only does this hierarchy provide an adequate

resolution of the paradox, but, even if it were not needed for this purpose, there would be a compelling reason to impose it. The operator '$\vdash_n$' may be intelligibly introduced as applying to statements which we already understand; but to suppose that we can ever arrive at a range of statements closed under the application of this operator is to invoke an intolerably impredicative notion. The introduction of choice sequences defined effectively in terms of the notion expressed by '$\vdash_n$', as required by Brouwer's method of producing counter-examples, represents an *extension* of the notion of a constructive or lawlike sequence: although it is a legitimate extension, we have no right to pretend that we have a grasp of any domain of sequences or functions closed under the application of this device.

It was not the adoption of the strict interpretation of '$\vdash_n$' that was responsible for the appearance of paradox, but this impredicativity. If we view mathematical constructions as being effected in time, then this must apply not only to proofs but to definitions; and, in the case of an inductive definition, at any given temporal stage the definition may have been effected only for a part of the domain. Suppose, now, that we are defining inductively a spread-law c which determines a subspread of the full binary spread b. At any given temporal stage m, c may have been defined only over certain finite sequences, say those whose length falls below some bound $\ell + 1$; so we may consider the relativized spread-law c_m^* which admits a finite sequence $\vec{u}$ just in case (i) $\vec{u}$ is admissible under b and (ii) every initial segment of $\vec{u}$ for which c is defined at stage m is admissible under c. If we construe '$\vdash_n$' impredicatively, we must regard as well-defined that spread-law c such that, where $\vec{u}$ is admissible under c, then so is $\vec{u}^\frown 1$, and, further, $\vec{u}^\frown 0$ is admissible iff, where $m = \ell h(\vec{u})$, $\vdash_m \neg \exists \alpha_{\alpha \in c_m^*} \exists n\, \alpha(n) = 0$. Now suppose that, for some $\vec{u}$ of length m, $\vec{u}^\frown 0$ were admissible under c. Then $\vdash_m \neg \exists \alpha_{\alpha \in c_m^*} \exists n\, \alpha(n) = 0$, and hence, a fortiori, $\neg \exists \alpha_{\alpha \in c} \exists n\, \alpha(n) = 0$, which contradicts the assumption that $\vec{u}^\frown 0$ is admissible. We have thus shown that a finite sequence is admissible under c iff it consists entirely of 1's. If we are presently at stage k, c has been defined for all finite sequences at stage k, and moreover we have proved that $\neg \exists \alpha_{\alpha \in c} \exists n\ \alpha(n) = 0$, and so $\vdash_k \neg \exists \alpha_{\alpha \in c} \exists n\ \alpha(n) = 0$. By the definition of c, it follows that the finite sequence consisting of k 1's followed by 0 is admissible, and we have arrived at a contradiction. This contradiction springs solely from the impredicative character we are attributing to the notion expressed by '$\vdash_n$', for which we have not here assumed the strict interpretation; of course, if we construe '$\vdash_n$' predicatively, then c is not properly defined at all.

The motivation for adopting a lenient interpretation of '$\vdash_n$' under which such a schema as (xiii) would come out valid, and hence for trying to find a ground for rejecting the strict interpretation, derives from what appear to be counter-intuitive consequences of strengthening axiom schema (4) to:

(4*) $$A \rightarrow \exists n\ (\vdash_n A),$$

which can be read as saying that any true statement will at some time be proved, and yields, in conjunction with (3):

(xvii) $$A \leftrightarrow \exists n\ (\vdash_n A),$$

which says that a statement is true iff it is at some time proved, and converts '$\exists n\ (\vdash_n \ldots)$' into a sort of redundant truth-operator. To some, (4*) has appeared tolerable only on a lenient interpretation of '$\vdash_n$', since it has the consequence

(xviii) $$\forall m\ B(m) \rightarrow \forall m\ \exists n\ (\vdash_n B(m)).$$

If axiom schema (xiii) holds, this is quite unsurprising; but if we are so interpreting '$\vdash_n$' that only one statement is proved at any stage, and something counts as a proof of a statement only if it is explicitly a proof of that statement and of no other, schema (xviii) appears to have the consequence that, once we have proved a universal statement, we have thereby committed ourselves to explicitly deriving each instance of it at some future time.

(4*) itself, read as saying that each true statement is at some time proved, may appear unreasonable if we surreptitiously advert to some quasi-platonistic notion of truth; but, intuitionistically, what notion do we have of a statement's being true other than that of its at some time being proved? Paying closer attention to the intuitionistic meaning of $\rightarrow$, we shall read (4*) as saying that we have a method of transforming any proof of A into a proof that A has been proved at some particular time: and this we surely have, provided that we make the harmless assumption that any mathematical construction we effect can be recognized as being effected at some specific temporal stage. This argument in defence of (4*) does not depend upon adopting the lenient interpretation of '$\vdash_n$'; so it is unlikely that we really need this interpretation to avoid deriving implausible consequences from the axiom. By taking A in (4*) as B(m) and then applying universal generalization, we obtain

(xix) $\forall m\ (B(m) \rightarrow \exists n\ (\vdash_n B(m))),$

which entails no principle stronger than (4*) itself. To pass from (xix) to (xviii) requires only the logical principle

(xx) $\forall x\ (C(x) \rightarrow D(x)) \rightarrow (\forall x\ C(x) \rightarrow \forall x\ D(x)).$

This principle is unquestionably valid, since it says that if we have a means, for each individual, of transforming a proof that it satisfies C(x) into a proof that it satisfies D(x), then we also have a means of converting a method of finding, for each individual, a proof that it satisfies C(x) into a method of finding, for each individual, a proof that it satisfies D(x). In our case, since we have a means, for each natural number, of transforming a proof that it satisfies B(x) into a proof that we can at some time prove that it satisfies B(x), we also have a means of converting a method of finding, for each natural number, a proof that it satisfies B(x) into a method of finding, for each natural number, a proof that we can at some time prove that it satisfies B(x). Hence, if the passage from (xix) to (xviii) is to be valid, we must understand (xviii) in accordance with those readings of the logical constants required for the validity of (xx). That is to say, we must read (xviii) as follows: Suppose that we have a proof of $\forall m\, B(m)$, i.e. a means of finding, for each m, a proof of B(m); then we shall be able, for each m, to find a proof of $\exists n\, (\vdash_n B(m))$, viz. by first effecting a proof of B(m), and then noting at which stage this proof was effected.

Under this interpretation, (xviii) does not say that, if we have proved $\forall m\, B(m)$, we shall as a matter of fact construct, for each k, an explicit proof of $B(\bar{k})$, but merely that we shall have the means at our disposal to construct such a proof whenever we wish. So interpreted, it is no longer in the least counter-intuitive, even on the strict interpretation of '$\vdash_n$', and so the objection to (4*) evaporates; but, equally, it is the only legitimate way of reading (xviii), since it is the only one which validates the reasoning by which we derived it from (4*). Here we see a subtler application of the principle which we encountered at the outset with negation: if we are to understand the operator '$\vdash_n$', not merely as constructively significant, but as yielding, when applied to mathematical statements, statements that still belong to *mathematical* discourse, we must understand the logical constants, when applied to statements involving that operator, not as having a constructive empirical meaning, but in the same way as they are understood in other statements of intuitionistic mathematics. In the present case, (xviii) appeared implausible only so long as we insisted on interpreting the existential quantifier '$\exists n$', when binding a variable appearing in '$\vdash_n$', in an empirical manner, as meaning that we can identify a particular temporal stage n at which such-and-such a proof is, in historical reality, carried out. But, if we are, by means of such quantification, to obtain mathematical rather than historical statements, we must interpret the existential quantifier as meaning that we have an effective means of bringing it about that there is such a stage n; and, as in other mathematical cases, we may permanently possess an effective means of carrying out a construction independently of whether, as a matter of empirical fact, we ever apply it.

We have so far been concerned only with the foundations of the theory; because it has appeared so dubious to many, it was worth while to satisfy ourselves that it is, after all, coherent before turning to its applications, as we now do. We may reasonably accept the axiom schemata (1), (2), (3), and (4*); for practical purposes, it is sufficient to assume the weaker (4) in place of (4*). For any given statement A (not containing '$\vdash_n$'), we may set β so that

$$\beta(n) = \begin{cases} 0 & \text{if } \neg(\vdash_n A) \\ 1 & \text{if } \vdash_n A. \end{cases}$$

By this means we obtain what is known as 'Kripke's schema'. If we assume only axiom schemata (1), (2), (3), and (4), this takes the weak form:

$$\text{(KS)} \quad \exists\beta[\forall n\, \forall m\, (n \le m \to \beta(n) \le \beta(m) \le 1)\ \&$$
$$\forall n\, (\beta(n) = 1 \to A)\ \&\ (\neg A \leftrightarrow \forall n\, \beta(n) = 0)].$$

If, in place of (4), we assume (4*), then the schema takes the stronger form:

$$\text{(KS*)} \quad \exists\beta[\forall n\, \forall m\, (n \le m \to \beta(n) \le \beta(m) \le 1)\ \&$$
$$(A \leftrightarrow \exists n\, \beta(n) = 1)].$$

Kripke's schema has the advantage of being expressed without explicit use of '$\vdash_n$', and from it all the applications of the theory of the creative subject can be derived. As an illustration, consider Brouwer's counter-example to the proposition.

$$x \neq 0 \to x \# 0.$$

Take any closed statement B (not containing '$\vdash_n$') such that we have not proved $B \vee \neg B$; take A as $B \vee \neg B$, and let β be a choice sequence satisfying (KS) for this A. By means of a suitable correlation law, we define a r. n. g. $\langle r_n \rangle$ in terms of this β, such that

$$r_n = \begin{cases} 0 \text{ if } \beta(n) = 0 \\ 2^{-m} \text{ if } m \leq n,\ \beta(m) = 1 \text{ and } \beta(k) = 0 \text{ for all } k < m. \end{cases}$$

If x is the real number given by $\langle r_n \rangle$, then $x = 0$ iff $\forall n\ \beta(n) = 0$, i.e. just in case $\neg A$ holds; and since in fact $\neg\neg A$ is logically true, we have $x \neq 0$. On the other hand, $x \# 0$ iff $\exists n\ \beta(n) = 1$, and so we cannot prove that $x \# 0$ until we have established A, i.e. until we have proved or refuted B.

As it stands, this is a counter-example only of the special, weak, intuitionistic type, namely it is an example, not of a case in which the proposition is false, but of one in which we can recognize that we cannot at present prove it. Let us, however, specify B to be the statement $\exists n\ \gamma(n) = 0$, where γ is a parameter for a choice sequence. As we have seen, for the x given in terms of this B, we have $x \neq 0$, and, further,

$$\begin{aligned} x \# 0 &\leftrightarrow B \vee \neg B \\ &\leftrightarrow \exists n\ \gamma(n) = 0 \vee \neg \exists n\ \gamma(n) = 0. \end{aligned}$$

It follows that

$$\forall x(x \neq 0 \rightarrow x \# 0) \rightarrow \forall \gamma(\exists n\ \gamma(n) = 0 \vee \neg \exists n\ \gamma(n) = 0).$$

We know, however, that the $\forall\alpha\ \exists!n$-continuity principle implies

$$\neg \forall \gamma(\exists n\ \gamma(n) = 0 \vee \neg \exists n\ \gamma(n) = 0),$$

and hence, in the presence of (KS), it implies

$$\neg \forall x\ (x \neq 0 \rightarrow x \# 0)\ .$$

Brouwer gave a number of other weak counter-examples to classical laws by means of the theory of the creative subject, all of which may be treated in a closely similar manner.

If we suppose that, for each statement A, we can label a choice sequence β satisfying (KS), then we can easily define the operator '$\vdash_n$' so as to render axiom schemata (1) - (4) valid. Let us introduce the notation 'β_A' as forming, from any formula A (not containing either '$\vdash_n$' or the operator 'β_B' itself), a functor subject to the axiom schema:

$$\text{(xxi)} \qquad \forall n\, \forall m\, (n \leq m \rightarrow \beta_A(n) \leq \beta_A(m) \leq 1)\ \&$$
$$\forall n\, (\beta_A(n) = 1 \rightarrow A)\ \&\ (\neg A \leftrightarrow \forall n\, \beta_A(n) = 0).$$

By so doing, we obtain a conservative extension of any theory in which (KS) was assumed (or was provable); and, within this extension, we may interpret '$\vdash_n$' by means of the definition:

$$\text{(xxii)} \qquad \vdash_n A \leftrightarrow \beta_A(n) = 1.$$

By the use of (xxi) and (xxii), each of the axiom schemata (1) - (4) becomes derivable. (If we strengthen (xxi) to:

$$\text{(xxi*)} \qquad \forall n\, \forall m\, (n \leq m \rightarrow \beta_A(n) \leq \beta_A(m) \leq 1)\ \&$$
$$(A \leftrightarrow \exists n\, \beta_A(n) = 1)\ ,$$

then (4*) becomes derivable also.)

The result just stated was obtained by van Dalen, who also proposed a means whereby the operator '$\vdash_n$' may be handled on Beth trees: we simply take $\vdash_n$ A to be true at a node a iff a is barred by a species of nodes, of length exactly n, at each of which A is true (this of course includes the case in which a is of length $\geq$ n, and A is true at the initial segment of a of length n). It is plain that on this interpretation each instance of axiom schemata (2), (3), and (4*) holds at the vertex of any Beth tree. However, although, from a classical standpoint, each instance of axiom schema (1) will also hold, we cannot demonstrate this intuitionistically, save for those Beth trees for which the property of being true at a node is decidable.

Because of the way in which the truth at a node of a Beth tree is defined, this accords with a lenient, not with the strict, interpretation of '$\vdash_n$'. Principles (v), (viii), (xii), (xiii), and (xiv) all hold on Beth trees, with '$\vdash_n$' so interpreted with respect to them. Owing to the particular interpretation of $\vee$ and of $\exists$ adopted for Beth trees, principles (vi) and (vii) will not hold in full, but only in one direction, from right to left:

(vi') $\quad \forall n\ ((\vdash_n A) \vee (\vdash_n B) \rightarrow (\vdash_n (A \vee B)))$

and

(vii') $\quad \forall n\ (\exists m\ (\vdash_n A(m)) \rightarrow (\vdash_n \exists m\ A(m)))$.

Naturally, these results merely reflect the constitution of Beth trees, and, in particular, the fact that they are set up in such a way that a logical consequence of what is true at any node will also be true at that node.

(KS) is, of course, to be understood as comprising those cases in which there are parameters in A, and therefore yields, as a special case:

(xxiii) $$\forall\alpha\ \exists\beta[\forall n\ \forall m\ (n \le m \rightarrow \beta(n) \le \beta(m) \le 1)\ \&$$
$$\forall n\ (\beta(n) = 1 \rightarrow \forall m\ \alpha(m) = 0)\ \&$$
$$(\neg\forall m\ \alpha(m) = 0 \leftrightarrow \forall n\ \beta(n) = 0)].$$

It was first observed by Myhill that this constitutes a counter-example to $\forall\alpha\ \exists\beta$-continuity. Let us write (xxiii) as $\forall\alpha\ \exists\beta\ C(\alpha, \beta)$; then, although $C(\alpha, \beta)$ is extensional, β cannot depend continuously upon α. For suppose that Φ is a continuous functional, and that we have

(xxiv) $\quad \forall\alpha\ C(\alpha, \Phi(\alpha))$.

Now assume that, for some arbitrary given α and n, $(\Phi(\alpha))(n) = 1$. Since Φ is continuous, there exists k such that

$$\forall\gamma_{\gamma\in\bar{\alpha}(k)}\ (\Phi(\gamma))(n) = 1$$

and hence

$$\forall\gamma_{\gamma\in\bar{\alpha}(k)}\ \forall m\ \gamma(m) = 0.$$

But this is absurd, since we can choose $\gamma \in \bar{\alpha}(k)$ such that $\gamma(k) = 1$. It follows that $(\Phi(\alpha))(n) = 0$ for all α and n, and from this we can infer that $\forall\alpha \neg \forall m\ \alpha(m) = 0$, which is again absurd, since α may be taken as $\lambda n.\ 0$. There is therefore no continuous functional Φ satisfying (xxiv).

It is due to this observation of Myhill's that the $\forall\alpha\ \exists\beta$-continuity principle, once widely accepted, is now generally rejected. It is also due to this observation that there is a greater awareness of the need to impose the hypothesis of extensionality as a condition for the validity of any continuity principle (although, indeed, once the point has been raised, it is apparent that no continuity principle has any intuitive plausibility if asserted with respect to a *non*-extensional relation). For the $\forall\alpha\ \exists\beta$ Axiom of Choice ought in any case to be accepted; hence, if (KS) is valid, there will be *some* functional Φ, although not a continuous one, for which (xxiv) holds. But, as Myhill pointed out, for this Φ the trivially true statement

$$\forall\alpha\ \exists!\beta\ \beta = \Phi(\alpha)$$

will supply a counter-example to the principle of $\forall\alpha\ \exists!\beta$ - continuity considered as asserted without the hypothesis of extensionality. Kleene showed, however, that $\forall\alpha\ \exists!\beta$ - continuity is derivable from $\forall\alpha\ \exists n$-continuity: it is therefore essential to impose the extensionality condition even on the $\forall\alpha\ \exists n$-continuity principle.

Suppose that, where f is a constructive function, we take A in (KS) as $\forall m\ f(m) = 0$: shall we be entitled to assert that the β given by (KS) for this choice of A is itself a constructive function? That is, shall we be justified in affirming

$$\exists g[\forall n\ \forall m\ (n \le m \to g(n) \le g(m) \le 1)\ \&$$
$$\forall n\ (g(n) = 1 \to \forall m\ f(m) = 0)\ \&$$
$$(\neg\forall m\ f(m) = 0 \leftrightarrow \forall n\ g(n) = 0)] \quad ?$$

More generally, shall we be able to replace the '$\exists\beta$' of (KS) by '$\exists g$' whenever A contains no parameter for a choice sequence, but only lawlike parameters? It is true that the process of determining the values of a function β satisfying (KS) is perfectly effective, in view of axiom schema (1): no free choices are involved, as they are in determining the terms of a choice sequence. For all that, the foregoing discussion of the illegitimacy of an impredicative interpretation of '$\vdash_n$' should make it evident that we cannot suppose that the functions defined by reference to the notion expressed by '$\vdash_n$' are already contained within the domain of variables for constructive functions as these occur in the statements to which '$\vdash_n$' is in the first place applied. We require a distinction between constructive functions in the original sense and functions determined effectively but by reference to the sequence of proofs effected by the creative subject. It has become usual to refer to the former as 'mathematical' and to the latter as 'empirical', taking the term 'lawlike' to embrace both. This terminology is somewhat unfortunate, since, as emphasized in the earlier part of this section, the whole point of the theory of the creative subject is to interpret the operator '$\vdash_n$' in such a way that, by applying it, we still obtain statements that belong to mathematical discourse; no one has ever doubted that *empirical* statements may be made about mathematical proofs and their discovery, but the dubious question is how far such considerations may be imported into mathematics itself. But such a distinction is certainly required; indeed, if '$\vdash_n$' is to be reiterated, or applied to statements involving 'empirical' functions, we shall, as previously indicated, need to acknowledge a whole hierarchy of functions. Even if '$\vdash_n$' is restricted to apply only to statements in no way involving the notion it expresses, we shall need to distinguish between the so-called 'mathematical' and 'empirical' functions, either by using different styles of variables, or, as Myhill does, by means of unary predicates.

There is, however, no justification for using this distinction for any purpose distinct from that which motivated its introduction. For instance, Kreisel considered the relation A(m, n) which holds when, for some particular intuitionistically sound formal system, either m is the Godel number of a proof in that system of a closed statement of the form $\exists x\, B(x)$ (x a numerical variable) and n is the value of x yielded by that proof, or m is not the Godel number of such a proof and n = 0. Evidently, we have $\forall m\, \exists n\, A(m, n)$, and hence, by the Axiom of Choice, $\forall m\, A(m, f(m))$ for some f; but there is no reason to expect f to be recursive. (The determination of n from m depends upon an intuitive understanding of the formal proofs, one that is effective if such understanding is present, but need not be reducible to any mechanical procedure.) Myhill comments that this example reflects less upon Church's Thesis than upon the assumption that the choice function yielded by the $\forall m\, \exists n$ Axiom of Choice is always 'mathematical' whenever the relation in question is 'mathematical', and uses this as a ground for requiring only that, in this case, the choice function should be lawlike; for, he says, the method of finding n from m, although *lawlike*, is not *mechanical*. Indeed it is not mechanical; but that consideration belongs to a wholly different circle of ideas, namely that of classical recursion theory; there never was any assumption that the domain of functions considered as constructive or lawlike from an intuitionistic standpoint should be computable by means of a merely mechanical procedure, that is, one for which understanding is unnecessary. The procedure by which we find, for given m, an n satisfying A(m, n) involves reflection upon a proof; but it does not require any consideration of the temporal stage at which that proof was effected, and so it is not, in the sense involved, 'empirical'.

Do we need a parallel distinction between ordinary choice sequences and those generated by reference to the notion expressed by '$\vdash_n$'? At first sight, we do not, because the notion of a choice sequence is so general: one might resolve to determine the terms of a choice sequence by reference to some meteorological or astronomical phenomenon, for example; and, if so, why not also by reference to our own mathematical activities? On reflection, however, it is apparent that we may refrain from distinguishing choice sequences according to their mode of generation only so long as no reference to their mode of generation occurs within the mathematical statements that we make about them; as soon as it does, we may need to make such distinctions if we are to avoid vicious circles, for instance one constructed in exact analogy with Troelstra's paradox. In that paradox, we considered ' ... is a constructive function' as a mathematical predicate; we may equally well consider 'α is a choice of sequence' as a form of mathematical statement, perhaps construing it as equivalent to '$\forall m\ \exists! n\ \alpha(m) = n$'. Hence, if we adopt the strict interpretation of '$\vdash_n$', and therefore have an enumeration $P^{(0)}, P^{(1)}, P^{(2)}, \ldots$ of the (unique) statements proved at each temporal stage, we can also effectively enumerate the statements $P^{(d(0))}, P^{(d(1))}, P^{(d(2))}, \ldots$ which we prove and which are of the form $\forall m\ \exists! n\ \alpha(m) = n$. Hence, again, if we do not distinguish between levels of choice sequences, there will be a choice sequence χ such that, where j is a pairing function, for each n, if $P^{(d(n))}$ is the statement '$\forall m\ \exists! n\ \alpha(m) = n$', then for every m $\chi(j(n, m)) = \alpha(m)$, and we shall obtain a contradiction by diagonalization as before. Admittedly, this argument applies only under the strict interpretation of '$\vdash_n$', and some may choose to see it as further evidence of the illicit character of that interpretation. But it has been argued in this section that, if there is a coherent interpretation of '$\vdash_n$' at all, then a strict interpretation is as legitimate as, and rather more intuitively clear than, a lenient one; and, if this is so, then the correct inference is that, as soon as we permit the operator '$\vdash_n$' to appear explicitly in our mathematical statements, we must distinguish between those choice sequences which are generated by reference to it and those generated independently of it.

The theory of the creative subject attempts to exploit notions which one might expect to remain decently hidden within the foundations of intuitionistic mathematics. It is for this reason that it is so interesting philosophically; it is for the same reason that it appears constantly to tremble on the edge of absurdity or paradox. We surely may expect it to survive, in some form or another, in whatever formulation of intuitionistic mathematics appears finally satisfactory; but the form in which it survives may be very different from the present formulation.

7
CONCLUDING PHILOSOPHICAL REMARKS

7.1. THE PHILOSOPHICAL FOUNDATION OF CONSTRUCTIVE MATHEMATICS

As Kreisel has emphasized, the intuitionistic philosophy of mathematics comprises two theses: a positive one and a negative one. The positive one is to the effect that the intuitionistic way of construing mathematical notions and logical operations is a coherent and legitimate one, that intuitionistic mathematics forms an intelligible body of theory. The negative thesis is to the effect that the classical way of construing mathematical notions and logical operations is incoherent and illegitimate, that classical mathematics, while containing, in distorted form, much of value, is, nevertheless, as it stands unintelligible. The negative thesis of course lends support to the positive one: if there is a flaw at the heart of classical mathematics, then, even if the intuitionistic reconstruction of mathematics is not correct in every detail, something along those general lines must be right, unless, as is surely unthinkable, all but the most elementary part of mathematics are totally delusory. If, on the other hand, there is nothing wrong with classical mathematics as such, then there is no particular reason to suppose that there is any fully coherent constructive reinterpretation of mathematics: if it should prove that it is impossible to give a philosophical account of the basic notions of intuitionistic mathematics that is stable and free of vicious circularity, this may be not because there has been some mistake in detail in the execution of the intuitionistic programme, but because the very project of rebuilding mathematics on a constructivistic basis was misbegotten. Some, such as Kreisel himself, nevertheless prefer to adopt an eclectic position: to accept the positive intuitionistic thesis, but reject the negative one. In this way, one can admit both classical and intuitionistic mathematics in a peaceful coexistence.

I know of no argument against such eclecticism, other than the arguments which intuitionists use against classical mathematics in general; but it remains that, if classical mathematics is intelligible, then, while intuitionistic mathematics may be intelligible also, it loses much of its point. By admitting as legitimate the classical explanations of the sentential operators and the quantifiers, one does not rule out of order the intuitionistic explanations of them; but it is not very clear why, save as an exercise, one should then be interested in using sentences whose logical constants are to be understood according to their intuitionistic senses. It was emphasized in Chapter 1 that the notion of constructive proof is, in itself, one that can be accommodated within classical mathematics. Classical mathematicians are, quite rightly, interested in discovering constructive proofs; and there is no reason why, if he wished, a classical mathematician might not use symbols for disjunction and existential quantification, understood constructively, alongside the ordinary classical logical constants. But the appropriate notion of a constructive proof would not coincide with that of an intuitionistic proof. This is most easily seen in the case of Markov's principle. If '$P(x)$' is a decidable predicate of natural numbers, and we believe that we have a proof of '$\neg \forall n \neg P(n)$', classically understood, then we have no reason not to claim to have an effective method of finding a number that satisfies '$P(x)$'; but, unless we know more about the way in which '$\neg \forall n \neg P(n)$' was proved, we have no guarantee that '$\exists n\, P(n)$' is provable intuitionistically. The intuitionist is restricted in what he will acknowledge as an effective method of finding a number of a given kind precisely because he cannot grasp the senses which the classical mathematician wants to attach to the logical constants; but, if someone thinks that he can grasp those senses, there is no reason why he should regard as having any special interest those methods which can be recognized as effective even by someone who cannot.

Intuitionism thus raises two philosophical questions. (1) Do intuitionists succeed in conferring a coherent meaning on the expressions used in intuitionistic mathematics, and, in particular, on the logical constants? (2) Is there a ground for thinking that classical mathematicians *fail* to confer an intelligible meaning on the logical constants, and on mathematical expressions in general, as they use them? Of these two questions, the second has, in a sense, the priority, and is certainly the deeper.

The negative intuitionistic thesis must, evidently, be based upon a critique of the manner in which meaning is supposed to be conferred on the expressions of classical mathematics. Obviously, for any such critique to be possible, it cannot be the case that any established usage is justified by the mere fact that it is commonly observed; in particular, it must be erroneous to suppose this of any given set of generally accepted rules of inference. It is, of course, apparent that the meanings of a set of logical constants and the rules of inference which govern them are interdependent: it does not follow that any arbitrary (consistent) set of rules of inference admits a range, let along a unique range, of meanings for the logical constants involved under which those and only those rules of inference that are derivable from that set are valid. That would necessarily be so if a grasp of the meaning of a logical constant consisted solely in a readiness to acknowledge as correct those inferences involving it which exemplified one of the rules in some suitable basic set of such rules. Such an idea is one that may tempt a logician, since he is prone to think of logical constants as inhabiting only logical calculi; but, of course, they do not, since their whole point is to be able to be used in actual sentences, and such sentences are not primarily employed in setting out deductive arguments; deductive argument is and must be subservient to the primary purpose for which sentences are uttered. No one could think that the grasp of the meaning of an arbitrary sentence consisted solely in a knowledge of the ways in which it might figure in an inference, as premiss or conclusion; in so far as such an idea would be more plausible for mathematical sentences than for any others, this would be so only to the extent that inferential power is what, in their case, a more general conception of what understanding a sentence consists in reduces to. If we take it as the primary function of a sentence to convey information, then it is natural to view a grasp of the meaning of a sentence as consisting in an awareness of its *content*; and this amounts to knowing the conditions under which an assertion made by uttering it is correct. Since Frege, it has been generally acknowledged that we must view an understanding of a word or sign as a knowledge of that which determines its contribution to the meaning of any sentence in which it may occur. This must apply to the logical constants as much as to other words; the meanings of the logical constants must therefore consist in their contribution to determining the conditions for the correctness of an assertion made by means of a sentence involving them, and not directly in the validity or invalidity of possible forms of inference. Rather, the naive picture of the matter is also the correct one: rules of inference are justified or not according as they do or do not always carry us from sentences that could be correctly asserted to sentences having the same property, according to the meanings of the logical constants in question.

It does not follow from this that the supposition that some given set of rules of inference determines a range of meanings for the logical constants is false. For this supposition to hold, two requirements must be satisfied. First, the condition for the correctness of an assertion made by means of a sentence containing a logical constant must always coincide with the existence of a deduction, by means of those rules of inference, to that sentence from correct premisses none of which contains any of the logical constants in question. Secondly, there must not be any deduction from premisses of the same kind, via sentences involving the logical constants, to a conclusion also containing no logical constant whose assertion would not itself be correct. This second requirement is, in effect, the requirement that the addition of the logical constants to that fragment of the language which lacks them is a conservative extension of that fragment (with respect to the property of being correctly assertible). The first requirement is necessary if the rules of inference are to suffice, given the meanings of sentences that do not contain logical constants, to determine the condition for the correct assertibility of sentences that do; the second requirement is demanded if we are to be able to take the meanings of sentences not containing them as given. Whether these requirements are satisfied or not depends, of course, both upon the meanings of sentences not containing logical constants and upon the mode of employment of those sentences which do contain them, and therefore cannot be judged by inspection of the rules of inference alone. What is certainly ruled out is the assumption that the enunciation of *any* consistent set of rules of inference, considered against the background of *any* language, serves of itself to determine the meanings of the logical constants.

Our question is how it is possible to criticize generally accepted rules of inference. If any arbitrary set of rules of inference determined a set of meanings for the logical constants, or a range of admissible sets of meanings for them, such criticism would be impossible; but the converse does not follow. Suppose that we have a language for which the first of our two requirements fails, relative to the set of rules of inference generally accepted by its speakers. In that case, it has to be acknowledged that to know the meanings of the logical constants one must know more than just the rules of inference: one must also know of conditions for the correctness of assertions made by means of sentences containing logical constants that cannot be stated in terms of those rules. It does not follow at all that the rules of inference themselves are subject to criticism. What, then, if the second requirement also fails? If we take the foregoing statement of the requirement quite literally, it involves a severe defect in the rules of inference, since by means of them we can construct a deductive chain leading from correct initial premisses to an incorrect conclusion. But, unless the introduction of those rules of inference actually renders the language inconsistent, we do not need to view the matter in this way: we may interpret it, instead, as entailing that, with respect to the full language (including the logical constants), the conception of the condition for the correct assertability of a sentence free of logical constants with which we were working was an inadequate one, demanding supplementation by reference to the possibility of a deduction of this kind. So interpreted, the failure of the second re-requirement means only that we cannot envisage the complete condition for the correctness of an assertion made by means even of a sentence not containing a logical constant as being graspable without reference to the rules of inference. And what of that?

The point is a controversial one. Many would hold that there is nothing objectionable in supposing such a situation to obtain. And, if they are right, then it is true that *any* consistent set of rules of inference may be coherently adopted by the speakers of a language. The adoption of those rules of inference may not by itself determine the meanings of the logical constants; but there will be some range of meanings for them consonant with those rules, and determined jointly by the rules themselves and by those conditions, not characterizable in terms of deductive arguments, recognized by the speakers as ones under which an assertion of a sentence containing a logical constant is correct. Likewise, if, for the language in question, the second requirement does not hold, then the adoption of those rules of inference will have modified the meanings even of sentences not containing the logical constants. Nevertheless, in the language as now constituted, such sentences will still have perfectly determinate meanings; it is just that the condition for the correctness of an assertion made by the utterance of such a sentence will comprise circumstances under which there exists a deductive argument with it as conclusion, and therefore cannot be fully grasped without reference to the rules of inference. On such a view, there is still no room for criticism of the acceptance of any rules of inference, provided at least that they are formally consistent.

The view just outlined is a form of linguistic holism:

no one sentence of the language can be fully understood unless the entire language is understood. The understanding of a sentence comprises a readiness to recognize each possible means by which it might be deductively derived from true sentences; and, because we can place no restrictions upon which sentences *might* occur in the course of such a derivation, there is no proper fragment of the language of which we can say that, once it has been mastered, then a complete understanding of that sentence has been attained. Language, on such a view, is a game with an immensely complicated system of rules, and, in order to grasp the significance of any one move in the game, you must know *all* the rules.

Intuitionism agrees with platonism in rejecting such a holistic view of language. A holist will almost certainly be unwilling to consider the language of mathematics in isolation from the rest of the language - from the language of physical sciences, for example; but, if he were, then he would be perfectly willing to countenance an account of the significance of mathematical sentences in terms of their provability by classical reasoning; the forms of argument employed in such reasoning would not require justification in terms of anything else, but would simply determine the meanings of the mathematical expressions. Holism thus becomes, in effect, indistinguishable from formalism, which is the doctrine that mathematical formulas are not genuine sentences at all, and thus do not carry a content of the kind required for an assertion, although, of course, the game played by mathematicians with these pseudo-sentences has its own rules, which mathematicians are entitled to lay down as they please, without any responsibility to anything else: there is, therefore, no notion of truth for mathematical sentences, distinct from that of their derivability by means of the accepted rules, and with respect to which we may require that the rules of proof be truth-preserving. Formalism of this kind, which denies to mathematical formulas a genuine sentential meaning, necessarily also puts accepted mathematical practice beyond the reach of criticism, and must be rejected by intuitionists and platonists alike, to whom it is common doctrine that mathematical sentences have a meaning comparable to that of other sentences, and that mathematics is therefore what it appears to be, one sector in the quest for truth. The effective collapse of holism into formalism is not obviated by taking mathematical language as only a part of the wider language, as a holist will naturally do; for then a mathematical theory ceases to have any independent significance, and becomes merely a complex of paths for deriving consequences within some empirical theory; and, since the empirical theory stands or falls only as a whole, no question can arise over whether such derivations are justified in themselves. Such a view is, in practice, indistinguishable from that variety of formalism which lays stress on the applications of mathematical theories, such applications being seen as supplying empirical interpretations of previously uninterpreted calculi. A mathematical theory needs, on this view, no further justification than that it 'works'; and it 'works' just in case it can be incorporated into some successful empirical theory; but we cannot distil out the contribution made to the composite theory by its mathematical component, since it has substance only as a component of the whole.

From an intuitionistic, as from a platonistic, standpoint, such a conception of language is inadequate. Our grasp of the content of a sentence must be capable of being represented in isolation, as it were, from the rest of the language; otherwise we should have no command over what it was that the sentence said, since the multiplicity of ways in which the condition for the correctness of an assertion made by it could in no way be surveyed. That is not to claim that an understanding of any sentence could exist on its own, without a knowledge of any of the rest of the language: every sentence is composed of words or signs which could not be understood unless it were known how to use them in at least some other sentences. The understanding of any given sentence will depend upon the mastery of some fragment of the language, more or less extensive according to the complexity or depth of the sentence. But it is essential to this view that sentences can be ranked in a hierarchy, according to their complexity, and that such a ranking constitutes at least a quasi-ordering. We must not only have some general model for that in which the understanding of a sentence consists, but the understanding, as represented in this model, of each particular sentence must be derivable from the understanding of its component words, again construed in terms of that model.

There is here no objection to supposing that the understanding of some given word depends upon the understanding of some simpler expressions of the language. The most obvious kind of example is any in which to know the meaning of some word depends upon an explicit knowledge of some verbal explanation of it; but it would, equally well, be perfectly consistent with this view to hold that the understanding of arithmetical statements depended upon a prior grasp of the use of number-words (words for natural numbers) to give the cardinal number of objects of a given sort satisfying some predicate. What would render the functioning of language unintelligible would be to suppose that the relation of (immediate or remote) dependence of the meaning of one word on that of others might not be asymmetrical, that, in tracing out what is required for an understanding of a given sentence, and, therefore, of the words in it, we should be led in a circle. It is for these two interconnected reasons -- the derivability of the meaning of a sentence from the meanings of its constituents, and the asymmetry of meaning-dependence -- that we cannot be content with the general explanation that a grasp of the meaning of each mathematical sentence consists in an apprehension of the condition for its provability in accordance with some arbitrary specified modes of reasoning. If, for example, we consider a disjunctive sentence, then (if if is indeed the notion of provability that is to be taken as central in our representation of the meanings of mathematical sentences) our understanding of the specific condition for its provability must result from our grasp of its structure, and, in the first instance, of its formation, by means of the connective "or", from two subordinate sentences. There must therefore be some general conception of the way in which the provability of a disjunctive sentence depends upon the conditions for the provability of its two constituent sub-sentences. Now, of course, if we admit classical reasoning, no such general condition for the provability of a disjunctive sentence can be stated. This is not sufficient by itself to show classical reasoning to be at fault, since we cannot simply *assume* that provability is the correct notion in terms of which to represent our understanding of mathematical sentences. What it does show is that we cannot give a general explanation of a grasp of the significance of mathematical sentences in terms of a knowledge of the conditions for their provability by means of specified modes of reasoning, as a holist might do if he were prepared to consider the language of mathematics on its own, unless those modes of reasoning satisfy a fairly stringent requirement: namely that they permit us to explain how we can determine the specific condition for the provability of each individual sentence from its internal composition, where our grasp of the meanings of its component words is again explained in terms of this model, that is to say, in terms of their contribution to the provability-conditions of sentences in which they occur.

It is this anti-holistic conception of language, common, as already remarked, to the intuitionistic and platonistic philosophies of mathematics, that imposes a requirement that a rule of inference be capable of justification in terms of some semantic notion of logical consequence. In attempting to find a semantics with respect to which a given logical system is both sound and complete, a logician is not merely seeking an algebraic rather than a proof-theoretic characterization of the deducibility-relation of that system: a semantical theory proper (such as, for instance, the Beth trees purport to provide) is to be distinguished from a merely algebraic valuation system (such as the topological interpretation of intuitionistic logic with respect, say, to the real line). If any distinction between a semantical theory and a purely algebraic valuation system can be drawn at all, the ground of it must lie in the fact that the semantical theory is connected with the way in which both logical and non-logical expressions are given meaning, while a purely algebraic one is not. On such a view, a valuation system constitutes a genuine semantics just in case it can be extended to a plausible theory of meaning for the language. This means that a proposed semantical theory is itself subject to judgement, according as a plausible theory of meaning can or cannot be erected upon it as foundation. Just as a formalization of some part of logic is to be judged in accordance with whether it can be shown to be sound, and, if possible, complete, relative to some semantic theory, so the semantic theory itself is to be judged by criteria that do not belong to logic, properly so called, but to the philosophy of language. On such a view, linguistic practice in general, and the acceptance of modes of inference in particular, are not self-justifying. Linguistic practice is coherent only if we can find some workable theory of meaning, some model for what the understanding of a sentence consists in and how that understanding is derived from the understanding of its component words, on which that practice can be justified: if we cannot, then that practice demands revision, no matter how well established it may be.

The forms of reasoning employed in classical mathematics, namely those embodied in classical logic, can be justified only by the two-valued semantics, or, at least, by a semantical valuation system whose elements form a Boolean algebra. The generalization to an arbitrary Boolean algebra is needed for the case in which a mathematical theory is taken, not as having a single intended model, but as comprising the truths which hold in each of some range of equally admissible models: each element of the valuation system will then be the set of models in which some given sentence of the theory comes out true, and the sole designated element will be the set of all the admissible models. On an extreme platonistic view, for example, we have an inchoate intuition of the mathematical structure which, in doing set theory, we are trying to describe, an intuition which determines the intended model of that theory up to isomorphism: every set-theoretical statement is, therefore, determinately either true or false, and the independence of the continuum hypothesis and other statements merely reflects the fact that we have not succeeded in rendering our intuition sufficiently explicit to embody all its features within our system of axioms. But an alternative view, which would involve no interference with classical logic, would attribute the independence results to an indeterminacy in our conception of the mathematical structures which our theory describes; on this view, we do not have any even inchoate or implicit ground for preferring one model to another, and the structures that we are describing must be taken to be all those which are models of our existing axiom system, or all those of a certain kind (e.g. all which yield a standard model of the natural numbers); hence there will be set-theoretic statements which are neither absolutely true nor absolutely false (but are true in some models and false in others). The distinction between these two views is irrelevant, from an intuitionistic standpoint. For both, the fundamental semantic notion is that of truth in a model, and they differ only as regards the number of admissible models of a given theory; for both, the logic which is justified by the semantics resulting from a correct theory of meaning for the language of mathematics is classical logic.

The claim made earlier, that a grasp of the meaning of a sentence is to be identified with an apprehension of the condition under which an assertion made by means of it is correct was an intentionally vague one. It is evident that it is fundamental to the notion of an assertion that it be capable of being either correct or incorrect; and therefore, in so far as assertion is taken to be the primary mode of employment of sentences, it is fundamental to our whole understanding of language that sentences are capable of being true or false, where a sentence is true if an assertion could be correctly made by uttering it, and false if such an assertion would be incorrect. But, within this general framework, many different conceptions of what it means to say of an assertion that it is correct, and therefore of the appropriate notion of truth for our sentences, are possible. It is an essential feature of any theory of meaning that will yield a semantics validating classical logic that each sentence is conceived of as possessing a determinate truth-value, independently of whether or not we know it or have at our disposal the means to discover it. If we are operating with the straightforward two-valued semantics, then this simply reduces to saying that each sentence is determinately true or false, independently of our knowledge or means of knowing; if we are considering a sentence as having been given meaning by relating it to a range of mathematical structures (models), then it amounts to saying that it is determinately true or false in each of the admissible models.

Now if, as a matter of fact, we do possess a means of determining the truth-value of each sentence of the language of some given mathematical theory, then the fact that this truth-value is in principle independent of our knowledge is of no importance; but, even for quite elementary mathematical theories, such as first-order arithmetic, it has to be acknowledged that there are sentences whose truth-value we not merely do not know but may not even have the means of knowing. There is no a priori reason to suppose that the modes of reasoning which we are capable of comprehending and of recognizing as valid are sufficiently powerful to yield a proof or disproof of every arithmetical statement, or even of each statement of a one-quantifier form (i.e. one resulting from the application of a single unbounded quantifier to a decidable predicate). Hence, on the assumption that every such statement is determinately either true or false, the notion of the truth of an arithmetical statement is not to be explained in terms of the procedures available to us for recognizing such statements as true. Since the sentential operators lead from decidable statements only to decidable statements, what, in the case of arithmetical statements, creates this situation in the first place is the use, in forming them, of unbounded quantification over the natural numbers. Since the theory of meaning underlying classical mathematics, as conceived by the platonist, requires that the understanding of a sentence consists in a knowledge of the condition for it to be true (or for it to be true in a particular model), that is, in an awareness of what has to be the case for it to be true, we must possess an understanding of quantification over an infinite domain which does not relate to our own restricted means of recognizing as true sentences formed by such quantification, but does yield a conception of truth for such sentences as something which they, determinately, either do or do not possess. The nub of the intuitionistic critique of classical mathematics is the contention that we do not, and could not, have any such conception of mathematical truth; that we suppose ourselves to have it only by an illusion based upon a false analogy.

If it is agreed that an understanding of a sentence consists in an awareness of the condition under which a correct assertion may be made by the utterance of that sentence, it becomes indisputable that such an understanding may be represented as consisting in a knowledge of the condition for the sentence to be true. But, now, let us ask what it means to ascribe such knowledge to someone. One case in which the ascription of such knowledge is quite unproblematic is that in which his knowledge consists in the capacity to *state* the condition for the truth of the sentence in some non-circular manner, that is, when the knowledge in question is explicit or verbalizable knowledge. It is, however, evident that we cannot take explicit knowledge of this kind as a model, of universal application, for the grasp of the meanings of words, expressions, and sentences of a language: it is impossible to have a non-circular system of verbal explanations for all the words of a language, and the mastery of the language could not possibly consist in the ability to give circular explanations. Hence, if the understanding of a language consists in the ability to derive, for each sentence of the language, a knowledge of the condition for its truth, such knowledge must, for many of the sentences, be merely implicit knowledge; and so we need to inquire what it means to ascribe to someone an implicit knowledge of the condition for the truth of a sentence. If we take for granted that we can determine, from a person's linguistic or other behaviour, when he manifests an acceptance of a sentence as true, then there will be no difficulty in saying what is required for someone to know the condition for a sentence to be true, provided that the condition in question is one which he is capable of recognizing as obtaining whenever it in fact obtains, namely that he should, whenever the condition obtains, accept the sentence as true. For very few sentences is it possible to make such a claim: but, when a sentence is decidable, then, although we shall not always recognize the condition for its truth as obtaining whenever it does obtain, we are able, at will, to get ourselves into a position in which we can recognize whether it obtains or not; so, in such a case, we may identify someone's knowledge of the condition for the sentence to be true as consisting in his readiness to accept it as true whenever the condition for its truth obtains and he is in a position to recognize it as obtaining, together with his practical knowledge of the procedure for arriving at such a position, as manifested by his carrying out that procedure whenever suitably prompted.

However, our question, in what a speaker's knowledge of the condition for a sentence to be true consists, has been answered only for certain quite special types of case. If the underlying assumption of a platonistic theory of meaning is correct, that we have, for all mathematical statements, a conception of truth for which the principle of bivalence holds, then there will be many sentences which are not decidable and the condition for whose truth cannot be stated without circularity. Among these will be some of which we shall know that, if the condition for their truth holds, then it is possible that we shall find ourselves in a position to recognize it; for which, moreover, we have an effective procedure that will eventually bring us into such a position, provided that the condition holds, but for which there is no bound on how far the procedure needs to be carried to obtain a positive outcome: typically, arithmetical statements with a single, existential, quantifier. For any such statement, it would be possible to identify a knowledge of its truth-condition with a capacity to recognize the statement as true, when in a position to do so, together with a knowledge of the procedure which leads to such recognition, if the statement is true. It is, however, obscure what is involved in claiming that the speaker is aware that the statement *is* determinately either true or false, or, what amounts to the same thing, in claiming that, by knowing the condition for the truth of such a statement, he thereby knows the condition for the truth of its (classical) negation (typically, an arithmetical statement with a single, universal, quantifier). For an arithmetical statement involving the universal quantifier, there will be no guarantee that, if it is true, we shall be able to recognize its truth-condition as fulfilled: not only do we not have any means to bring ourselves into a position to be able to recognize this, but, for all we know, the condition may not be one which any human being will ever be capable of recognizing as obtaining. For such statements -- among which belong, of course, all mathematical statements save the very simplest -- the notion of a knowledge of the condition for their truth has apparently lost all substance.

In short: since ex hypothesi, from the supposition that the condition for the truth of a mathematical statement, as platonistically understood, obtains, it cannot in general be inferred that it is one which a human being need be supposed to be even capable of recognizing as obtaining, we cannot give substance to the conception of our having an implicit knowledge of what that condition is, since nothing that we do can amount to a manifestation of such knowledge.

The solution is to abandon the principle of bivalence, and suppose our statements to be true just in case we have established that they are, i.e., if mathematical statements are in question, when we have proved them, or when we at least have an effective method of obtaining a proof of them. An understanding of a sentence may now be taken to consist in a knowledge of the condition under which a statement has been conclusively established to be true; and no difficulty can any longer arise over what such knowledge consists in, since the relevant condition is, necessarily, one which we are capable of recognizing; in fact, meaning is now being explained directly in terms of what we actually learn to do when we learn to use the sentences of our language.

An argument of this kind is based upon a fundamental principle, which may be stated briefly, in Wittgensteinian terms, as the principle that a grasp of the meaning of an expression must be exhaustively manifested by the *use* of that expression. That is, as already observed, the understanding of an expression cannot, in general, be taken to consist in the ability to give a verbal explanation of it, and hence must constitute implicit knowledge of its contribution to determining the condition for the truth of a sentence in which it occurs; and an ascription of implicit knowledge must always be explainable in terms of what counts as a manifestation of that knowledge, namely the possession of some practical capacity. When it is a knowledge of the meaning of a word that is in question, then the practical capacity which constitutes that knowledge must itself be a linguistic ability, an ability to use or react to sentences containing the word in some manner that can, ultimately, be specified without appeal to any semantic notions assumed as already understood.

The platonist may counter this line of argument in one of three ways. First, he may accept the underlying principle, that a grasp of meaning is (no more than) a mastery of a use, but argue that a grasp of the conditions under which mathematical sentences, as platonistically understood, are true is manifested precisely by that difference in linguistic behaviour which distinguishes the classical mathematician from the intuitionist, namely the employment of classical modes of reasoning. This answer appears thin. It is undoubtedly the case that *if* we have a grasp of some conception of truth for mathematical statements with respect to which the principle of bivalence holds, then the laws of classical logic are valid; but it is hardly plausible that the mere propensity to reason in accordance with those laws should *constitute* a grasp of such a notion of truth. If we consider any other class of statements for which it would be generally agreed that we do not possess a notion of truth subject to the principle of bivalence -- for example, counterfactual conditionals -- we can readily imagine that we had been induced, by childhood training, to apply the laws of classical logic to them, and we can recognize that, in such circumstances, we might be under a strong compulsion to suppose that we did have a notion of truth for such statements according to which each was determinately either true or false. Indeed, in the case of counterfactuals, such a temptation does in fact sometimes afflict us: it is easy to fall into wondering what would have happened if we had made some important decision in our lives otherwise than we did, in a frame of mind in which we submit to the illusion that such a question must have a definite answer, that there must be some determinate truth to the matter. Nevertheless, there seems no merit to the suggestion that, merely by undergoing a training in applying the laws of classical logic to these statements, we should thereby acquire, what we now lack, a conception of truth for them under which each must be determinately either true or false.

It seems, therefore, that the platonist is compelled to repudiate the principle that meaning is use, although he is bound to admit that it is only from a training in the use of expressions in any given range that we derive a grasp of their meanings. He can do this in either of two ways. On the one hand, he may choose to emphasize the *theoretical* character of a theory of meaning. Within such a theory, we explain a speaker's understanding of an expression or sentence by ascribing to him knowledge of some feature of it or by saying that he associates some semantic element or complex with it; but, on the present account, we do not then need to explain what it is for him to have this knowledge or make this association in terms of his linguistic behaviour. In constructing a theory of meaning, we are not, on such a view, attempting to articulate the complex of practical abilities that make up mastery of a language into its constituents, conceived of as isolable, though interconnected, practical abilities; we are merely aiming at what any theory attempts to provide, a picture which, taken as a whole, makes sense of a complex phenomenon, that is, makes it surveyable, even though there is no one-one correspondence between the details of the picture and observable features of the phenomenon. On this view, an acceptance of classical reasoning in mathematics does not *constitute* a grasp of a notion of truth for mathematical statements subject to the principle of bivalence, as it did according to the first of the possible platonist replies that we considered; rather, it *warrants* the ascription of a grasp of such a notion of truth to the individual concerned. This position does not represent a complete retreat into holism, since it still allows the necessity of finding some theory of meaning, some general form of representation of that in which the understanding of a sentence consists, even though a theory of this kind does not need to be justified piece-meal. It is the most sophisticated of the three platonist replies here considered. Whether it is acceptable or not depends on whether the conception of an explanatory theory implicit in it can or cannot be sustained, a question we shall not attempt to explore here.

Thirdly, the platonist may adopt a more naive manner of repudiating the principle that meaning is use: he may hold that, although it is only from a training in the employment of a language that we derive our apprehension of the meanings of its expressions, still what is required for such an apprehension is not a mere aptitude in observing the rules governing the use of expressions, but the formation of the right mental conception of the principles underlying those rules. Use therefore does not constitute meaning, as if we were computers being programmed in one way rather than another; it *guides* us, as rational creatures, to select the intended mental representation from among different possible candidates. We indeed learn the most primitive parts of language by connecting their use with our own actual capacities: for the simplest kinds of sentence, our knowledge of their truth-conditions does indeed consist in our capacity, when suitably placed, to recognize those conditions as obtaining; in the case of empirical statements, by observation, in the case of mathematical ones, by computation. Having mastered this lowest level of language, we proceed to higher levels by analogy. We come to understand the condition for the truth of a sentence belonging to one of these higher linguistic strata in terms of what it would be to be able effectively to recognize their truth or falsity in a direct manner, an ability which we do not ourselves possess but of which we can form a conception by analogy with those abilities we do have. For instance, we first learn the meanings of the quantifiers (or other expressions of generality) for finite and surveyable domains: in the case of arithmetical statements, we learn how to determine, by inspection of each instance in turn, the truth or falsity of a statement involving bounded quantification. We now form the conception of the condition for the truth of a statement involving unbounded quantification by analogy with this, by imagining a being who, unlike ourselves, could in a finite time check the truth-values of denumerably many instances of such a statement. In doing this, we are guided by the rules of inference we are trained to accept for such statements: precisely because it is established practice to apply the law of excluded middle to them, we apprehend that we are intended to understand them as being determinately either true or false. We therefore naturally have recourse to that conception of what would be required to determine them as true or as false which lies closest to the means we ourselves possess to determine the truth-values of sentences formed in a linguistically similar manner. On this view, a conception of what it is for a sentence to be true always consists in a picture of what it would be to be able to recognize it as true whenever it was true, a picture which may involve appeal to hypothetical faculties which we do not possess but which are thought of as extensions of those we do. It is because of the seductive character of this line of thought that platonists are prone to play down, as purely contingent, the restrictions imposed on our own powers of observation and mental operations, as witness Russell's remark that it is 'a mere medical impossibility' to carry out infinitely many tasks in a finite time. On this view, therefore, our understanding of, for example, the condition for the truth of an arithmetical statement involving unbounded quantification is not in itself *constituted* by our acceptance of classical logic for such statements, as with the first platonist reply; nor does our acceptance of that logic merely *warrant* the ascription to us of that understanding, without further explanation, as with the second reply; but it *prompts* us to acquire such an understanding, which is to be explained in terms of our analogical conception of what would be possible for a being with powers exceeding our own.

Of the three platonist replies, the first is certainly the weakest, the second the strongest debating position, and the third that which has in practice had the strongest appeal. From an intuitionistic standpoint, it is an unacceptable defence. The language that we use, when we are engaged in mathematics as in other activities, is *our* language, and its meaning must be connected with our own capacities: it cannot be derived from the hypothetical conception of capacities which we do not have, and the attempt to explain it by such means only illustrates the illusions implicit in our misunderstandings of our own language. The debate can of course be carried on from this point at great length; but this is where we shall leave it in this exposition, which is intended to do no more than bring out the issues involved.

In attempting to stage a debate between an intuitionist and a platonist, it is essential to find terms on which they can communicate. For the platonist, there being among the natural numbers at least one which satisfies a given decidable predicate 'P(x)' just *is* a condition which, determinately, either obtains or does not, and hence there can be no doubt that it is the right condition to select as necessary and sufficient for the truth of '$\exists n\, P(n)$'. He cannot, however, put this contention to the intuitionist, because, although the intuitionist attaches a meaning to the words 'There is among the natural numbers at least one which satisfies "P(x)" ', he does not construe them as expressing a condition which determinately either obtains or does not obtain: the whole issue between them is, in the first place, about the kind of meaning that such a sentence may be taken as having. The issue has therefore had to be presented as concerned with the question by what means we are to conceive of meaning as being conferred on mathematical statements. One consequence of this attempt to set the debate without making any presupposition about whether or not arithmetical sentences of the above kind state conditions which, determinately, either obtain or do not obtain is that the issue has been represented as one lying within the quite general philosophy of language, not as specific to the philosophy of mathematics. Every consideration so far adduced for or against the intuitionistic critique of classical mathematics has turned on some quite general point about the philosophy of language, that is, about the form which a theory of meaning should take, and is therefore applicable to a much wider field than the language of mathematics. If the arguments hitherto cited for the intuitionistic conception of the meaning of mathematical statements are correct, they will likewise impede a realistic conception of statements of many other kinds than mathematical ones, and, with it, the validity of classical logic as applied to them. (That is not to say that, on that hypothesis, the appropriate logic for such statements will be precisely an intuitionistic one; the fact that, unlike mathematical statements, empirical statements lack the property that, once verified, they remain verified must prevent the application to them of anything closely resembling the semantics appropriate to statements of intuitionistic mathematics.) The correctness of such arguments cannot, therefore, be judged solely in relation to mathematics; they will stand or fall according to whether or not a plausible general theory of meaning can be worked out upon principles which respect those arguments.

With the occacional exception of remarks in Brouwer's more philosophical writings, intuitionists themselves are, for the most part, chary of claiming that their ideas about meaning are applicable to a wider field than mathematics; so we need to inquire whether there is any way of presenting the intuitionistic critique of classical mathematics as dependent upon features particular to the subject-matter of the statements to which it relates. This is usually effected by representing intuitionists as repudiating the realistic view of mathematical reality adhered to by platonists: for a platonist, mathematical statements are about an objective reality, comprised of abstract objects related to one another to form a variety of abstract structures, existing independently of ourselves and of our thought about it, and thus determining those statements as true or as false independently of our knowledge, just as, on a realistic conception of the physical universe, that universe constitutes an objective reality, independent of our knowledge of it, and rendering determinately true or false the material-object statements which we make. For an intuitionist, on the other hand, what makes our mathematical statements true or false is our own mathematical activity, which is essentially mental activity: mathematical reality is, therefore, not something existing independently of ourselves, though partially apprehended by us, but simply the product of our own thought.

It cannot be contested that the difference, so described, between an intuitionistic and a platonistic conception of mathematical reality is genuine. The question is, however, whether the ontological position adopted by each is, for him, a *premiss* from which he derives his view of the way in which it is possible to give meaning to mathematical statements, and therefore the interpretation he wants to put on them. For anyone who followed the very general line of argument, previously sketched, for the intuitionistic interpretation of mathematical statements, the intuitionistic ontology would be a *consequence* of the intuitionistic theory of meaning, not a premiss for it. From such a perspective, a realistic interpretation of statements of some given class just is the supposition that we possess, for them, a conception of truth subject to the principle of bivalence, and this is a question to be settled by inquiry into the way in which meaning is in fact conferred on them, that is, into the correct theory of meaning for them. The metaphysical view -- realist or non-realist -- that we adopt is, therefore, on this way of looking at the matter, consequent upon the position we take up concerning the theory of meaning, not something to be decided in advance of our selection of that position; to affirm, or deny, the existence of an objective reality described by our statements, and rendering them determinately true or false, is to adopt a picture which accords with one or other conception of the kind of meaning which those statements have, but a picture which has in itself no substance otherwise than as a representation of the given conception of meaning.

If the metaphysical view taken by intuitionists of mathematical reality is, on the contrary, to be a premiss for the intuitionistic conception of the meaning of mathematical statements, then the rejection of the classical interpretation of the logical constants, and, in the first place, of quantification over a denumerable domain, cannot be grounded on the very general considerations we adduced above, but must depend upon the (alleged) fact that, in making mathematical statements, we are not purporting to describe an *external* reality. It would, then, for instance, be open to us -- at least, as far as any intuitionistic arguments go -- to adopt a realistic conception of physical reality; and, in that case, where quantification over the natural numbers will not yield statements that are determinately either true or false, quantification over intervals of infinite time, or regions of infinite space, or bodies disposed in such time or space, will still do so; it will be the fact that material-object statements do, but mathematical statements do not, refer to an external reality that makes the difference.

To decide whether a cogent argument can be constructed along these lines, two questions have to be settled: how, if the ontological view is to serve as a premiss for the view about meaning, it is itself in turn to be grounded; and what is the path from it to the view about meaning. At first sight, the answer to the second question is obvious: since there is no independently existing reality to render the quantified statement true or false, it can be true or false only in so far as we have been able to recognize it as such. But, on second thoughts, the matter is not so simple. Let us suppose that we are concerned with an arithmetical statement of the form '$\exists n\, P(n)$', with '$P(x)$' decidable. Someone might hold the natural numbers to be objective, independently existing abstract objects, to each of which the predicate '$P(x)$' determinately either applies or does not apply, and still hold the statement '$\exists n\, P(n)$' not to be one of which we can assume that it is determinately either true or false, because he argues, along the lines of our earlier exposition of one mode of argument for the intuitionistic position, that we do not possess a conception of truth, transcending our own capacity for recognition, with respect to which the principle of bivalence would hold for that statement. What we are at present interested in, however, is the opposite: would it be possible for someone to regard the natural numbers as mental constructions, as the products of human thought, and yet think that '$\exists n\, P(n)$' must, independently of our knowledge, be either true or false?

It is not, after all, apparent that the view that he holds about the ontological status of the natural numbers should, in itself, make any difference to his interpretation of the quantifier. Fictional characters are creations of the human imagination (which is, of course, quite a different thing from what is meant by calling mathematical objects creations of human *thought*); but, for all that, a statement to the effect that there is one among Shakespeare's characters who has one legitimate and one illegitimate son has a determinate truth-value. This comparison might be faulted in many ways. First, someone might hold that, unlike Shakespeare's characters, the natural numbers do not form a *definite* totality, i.e. that we do not so understand the Peano axioms as to determine the natural numbers up to isomorphism; but this is not an intuitionistic view. Or, again, it might, correctly, be insisted that, in quantifying over the natural numbers, we are not, as in the case of Shakespeare's characters, quantifying only over constructions that *have* been made, but also over ones which *could* be made in accordance with fixed principles of construction, over possible mental constructions as well as actual ones; this would be to make the point turn on the potential or uncompleted character of an infinite totality as intuitionistically conceived. There is, indeed, such a disanalogy between the two cases; but, so long as it is agreed that the principles of construction are fully determinate, it is not clear why the fact that not every natural number has actually been constructed should deprive the quantified statement of a definite truth-value. (Note that it would be illicit, at this point, to appeal to the fact that we cannot in practice survey the whole of an infinite totality, and so cannot effectively decide the truth-value of the quantified statement, since this would hold, equally well, for a statement involving quantification over infinite past or future time, which, on a realistic view of the physical universe, would be allowed to have a determinate truth-value.) Finally, it might be urged that the comparison tells the other way. The actual statement taken as an example was, as it happens, true; but, since the principle of bivalence does *not* hold for statements about individual fictional characters -- the statement 'Hamlet wore a moustache' is neither true nor false -- it does not hold either for an arbitrary statement involving quantification over fictional characters of however restricted a domain.

The point of this third objection would, naturally, be to repudiate the principle of bivalence even for the instances '$P(\bar{k})$' of the quantified statement, that is, for the result of applying a decidable predicate to a specific number. Plainly, if we cannot assume that each statement in the sequence '$P(0)$, '$P(1)$, '$P(2)$', ... is determinately either true or false, then we cannot assume that of the statement '$\exists n\, P(n)$' or the statement '$\forall n\, P(n)$'. Of course, it is, intuitionistically, legitimate to assert the law of excluded middle '$P(\bar{k}) \vee \neg P(\bar{k})$' for any such statement; but the justification for that is that we can, if we choose, find a proof either of '$P(\bar{k})$' or of '$\neg P(\bar{k})$', and therefore of any statement 'Q' which we can show to follow both from '$P(\bar{k})$' and from '$\neg P(\bar{k})$'; it does not follow that we must regard '$P(\bar{k})$' as already being either true or false.

We have hitherto taken the meanings of decidable statements as unproblematic, allowing the platonist the right to ascribe to them determinate truth-values, because this yielded a classical logic which was in any case correct, intuitionistically, for such statements. Here, however, this concession is challenged: though the logical laws governing such statements are not in dispute, what is being questioned is whether, even for them, we possess a notion of truth which will allow a determinate truth-value to statements obtained from them by quantification even on the platonist assumption that we may form the logical sum or product of an infinite sequence of truth-values. When there is an external reality to which our statements relate, then they may be regarded as possessing determinate truth-values independently of whether we in fact know these or not. But, just as a fictional character can have only those properties he is described as having, and an object of perception (sense-datum) can have only those properties it is perceived as having, so a mental construction can have only those properties which we understand it as having, i.e. which we have either stipulated or proved it to have; hence, even if we have an effective means of deciding whether or not a given natural number k satisfies a predicate 'P(x)', it neither satisfies nor fails to satisfy the predicate until we have so decided, and hence, until then, the statement 'P($\bar{k}$)' is not either true or false.

The question what notion of truth is admissible even for decidable statements of arithmetic thus comes to take on a critical importance. This is not surprising, since the metaphysical question, what there really is -- not so much what *objects* the universe contains, but what *facts* obtain -- is the very same question as the question which statements we can suppose to possess a determinate truth-value. On the first line of argument we considered, we thought of those statements for which, on a constructivist view, the laws of classical logic fail, that is to say, among mathematical statements, the undecidable ones, as alone being problematic in this regard; and hence we thought of the question whether there is, as the platonist supposes, an objective reality which determines each of those statements as true or as false as being settled by answering the prior question in what our understanding of those statements can be taken to consist. But, on this second approach, we have raised the question of the appropriate conception of truth for decidable mathematical statements, that is, for ones for which we need entertain no serious doubt about the account that is to be given of our understanding of them: that understanding consists, uncontroversially, in our mastery of the relevant decision procedure (and perhaps also, and, if so, equally unproblematically, in our grasp of their empirical applications).

What reason can we have, other than a prior belief in the existence of a mathematical reality which renders it so, for supposing a statement like '10^{20} is the sum of two primes' to be either true or false? The obvious answer is that we have an (in principle) effective procedure for determining its truth-value, and that that procedure, if applied, would yield one result or the other. To this it may be retorted that, where $\rightarrow$ represents the subjunctive conditional of ordinary discourse, $(P \rightarrow Q) \vee (P \rightarrow R)$ can no more be validly inferred from $P \rightarrow Q \vee R$ than it can when $\rightarrow$ represents the conditional of intuitionistic mathematics. Hence, from the truth of 'If we were to apply our decision procedure, we should establish either that 10^{20} is the sum of two primes or that it is not' we cannot infer that either 'If we were to apply our decision procedure, we should establish that 10^{20} is the sum of two primes' or 'If we were to apply our decision procedure, we should establish that 10^{20} is not the sum of two primes' is true. The cases in which we *are* prepared to allow the inference from $P \rightarrow Q \vee R$ to $(P \rightarrow Q) \vee (P \rightarrow R)$ are precisely those in which we do believe in the existence of an objective reality which determins one or other conditional as true, even in a case in which the antecedent is no longer capable of fulfilment: for instance, save on a very idealistic view of the past or of the physical world, we should allow that one or other of the statements 'If the audience at the lecture had been counted, it would have been found to amount to 50 or more' and 'If the audience at the lecture had been counted, it would have been found to fall short of 50' must have been true (even if the audience has now dispersed beyond recall).

If, however, we consider the obvious counter-examples to the rule

$$\frac{P \rightarrow Q \vee R}{(P \rightarrow Q) \vee (P \rightarrow R)},$$

with $\rightarrow$ representing the subjunctive conditional, they fall into two types. First are cases in which there is some other determining factor, neither implied nor presupposed by the antecedent: cases in which we should be prepared to assert $P \& S \rightarrow Q$ and also $P \& \neg S \rightarrow R$, so that, asked whether, if it had been the case that P, it would have been the case that Q or that R, we can only reply, 'It would have depended whether S or not'. The second type is composed of cases in which we believe that there is a genuine indeterminacy, either on quantum-mechanical grounds or because some voluntary agency is involved, so that, although $P \rightarrow Q \vee R$ holds, no addition to the antecedent which did not, in conjunction with P, logically imply one or other disjunct would allow us to strengthen the consequent either to Q or to R. The arithmetical example fits neither category. It does not fall under the first head, because the decision procedure, applied to a given statement, must yield the same outcome under whatever conditions it is applied: we should not think of it as a decision procedure if its outcome depended upon any extraneous factor. The example does not fall under the second head, because the decision procedure is of itself sufficient, if applied, to determine the truth or falsity of the statement: it is a decision procedure precisely because it will always yield a result, without the need for any supplementation or for the exercise of judgement on our part. Such a case therefore seems a paradigmatic one for claiming that one or other subjunctive conditional concerning the outcome of the procedure must hold good, independently of our actually carrying out that procedure, and that therefore the decidable mathematical statement *has* a definite truth-value, independently of our actually knowing it.

Again, we shall leave the debate at this point, without attempting to resolve it. The upshot of our review of this second approach is that the status of mathematical objects, as existing independently of us or as the products of our own thought, is irrelevant to whether a classical interpretation of the logical constants is admissible or whether they can be interpreted only in the intuitionistic sense, unless the thesis that such objects are the products of our thought is understood in the most radical manner possible, namely as entailing that even primitive predicates (and ones compounded from these by the sentential operators and quantification over a finite domain) are true of them only when we have expressly recognized them to be. To what extent such a radical anti-realism with respect to the objects of mathematics is defensible, and to what extent it is compatible with realism about the contents of the physical universe, are questions left to the reader to think through.

7.2. THE NOTION OF A PROOF

The standard explanations of the intuitionistic logical constants are those which are given by laying down, for each logical constant, the condition for a mathematical construction to be a proof of a statement of which that logical constant is thc principal operator, it being assumed known how to recognize a construction as a proof of any one of the immediate sentential constituents of that statement. These intuitive explanations do not, of themselves, make up an actual semantical theory for intuitionistic logic, since the terms in which they are formulated are not, as they stand, amenable to mathematical treatment of the kind required for a completeness proof or the like. A sustained attempt has been made by Kreisel and by Goodman to develop a mathematical theory of constructions which should serve precisely the purpose of providing a semantical theory incorporating the intuitive explanations of the logical constants; it will not be described here, since it is not only complicated but, as yet, still in an imperfect state of development. But there is no doubt that these standard intuitive explanations of the logical constants determine their intended intuitionistic meanings, so that anything which can be accepted as the correct semantics for intuitionistic logic must be shown either to incorporate them or, at least, to yield them under suitable supplementary assumptions.

In order to evaluate the claim embodied in the positive

intuitionistic thesis, it is therefore necessary, in the first place, to inquire whether these explanations of the logical constants are coherent or not, whether they confer intelligible meanings on them; if this question can be answered affirmatively, there are still many other questions to be raised concerning the notions of choice sequences and of species; but, if it has to be answered negatively, the whole conception, inherent in intuitionistic mathematics, of how mathematical statements are to be given meaning will have been shown to be defective.

The principal reason for suspecting these explanations of incoherence is their apparently highly impredicative character: if we know which constructions are proofs of the atomic statements of any first-order theory, then the explanations of the logical constants, taken together, determine which constructions are proofs of any of the statements of that theory; yet the explanations require us, in determing whether or not a construction is a proof of a conditional or of a negation, to consider its effect when applied to an arbitrary proof of the antecedent or of the negated statement, so that we must, in some sense, be able to survey or grasp some totality of constructions which will include all possible proofs of a given statement. The question is whether such a set of explanations can be acquitted of the charge of vicious circularity.

Of what sort are the constructions which are supposed to constitute proofs of mathematical statements, in the sense in which 'proof' is used in the intuitive explanations of the logical constants? We know that they must be mental constructions: hence they are not to be identified with the formal proofs of any formalized theory. Intuitionists usually say that written proofs are only the imperfect representations of the corresponding mental constructions: but, unless we are to acquiesce in a purely solipsistic interpretation of the whole conception, they must be communicable, and, if communicable, to be communicated by means of language; there is therefore no justification for holding that their linguistic representations may, in certain cases, necessarily be imperfect. The important point is not that the mental construction is, as it were, in a different medium from the written proof, but, rather, that the written proof is a proof in the required sense only in virtue of its being couched in an interpreted language: the features which make it genuinely a proof of its conclusion, and effectively recognizable as such, are neither identifiable with nor isomorphic to any of its purely formal characteristics as a complex structure of written signs, but belong to it solely in virtue of the meanings of those signs. We therefore have no reason to expect that any proof of some given statement will be recognizable as such by any means that falls short of demanding a full understanding of the language in which the proof is expressed.

Are we, then, to say that any (constructively) valid written proof, such as might appear in an article in a mathematical journal or in a textbook, is, considered relative to the intended meanings of the words and symbols employed, a proof in the sense in which this word is used in the explanations of the logical constants? It seems to follow from the character of those explanations themselves that we are not. Consider, first, the explanations of $\vee$ and of $\exists$. For a construction to be a proof of $A \vee B$, it is required to be a proof of A or of B; for it to be a proof of $\exists x\, A(x)$, it is required to be a proof of $A(c)$, where c denotes some element of the domain. In an ordinary informal proof, however, a statement $A \vee B$ might appear as a line of that proof, asserted not because a proof had been given of one or other disjunct, but because we have an effective method of obtaining such a proof, e.g. if A is a decidable statement and B is $\neg A$; likewise $\exists x\, A(x)$ might be asserted, as a step in an informal proof, because we have some effective method of finding an individual satisfying $A(x)$. In the sense that is given to 'proof' in the explanations of the logical constants, therefore, the informal proof does not actually provide us with a proof of the disjunctive or existential statement, or of any later statement inferred from it; it provides only a method, effective in principle, for finding such a proof. We thus appear to be forced to acknowledge a distinction between a proof, in the strict sense of the word, and a mere demonstration, the latter being related to the former by the fact that a demonstration supplies an effective means of constructing an actual proof. What appear in ordinary mathematical articles and textbooks are demonstrations, not proofs in the strict sense; and a demonstration provides an adequate ground for the unqualified assertion of its conclusion. But the primary notion is that of a proof in the strict sense, which we shall refer to as a *canonical* proof: the notion of a demonstration is a secondary one, definable in terms of that of a canonical proof; and it is by reference to the notion of a canonical proof that the logical constants are to be explained.

It might be thought that the need for any such distinction could be obviated by modifying the intuitive explanations of $\vee$ and of $\exists$: instead of requiring, for a construction to be a proof of $A \vee B$, that it actually be a proof of A or of B, we could require merely that it constitute what we can recognize as being an effective method of finding a proof either of A or of B; and, likewise, instead of requiring, for a construction to be a proof of $\exists x\, A(x)$, that it actually be a proof of some statement A(c), where c denotes an element of the domain, we could require merely that it constitute what we can recognize as being an effective method of finding a proof of such a statement. However, reflection on the intuitive explanations of $\forall$, of $\rightarrow$, and of $\neg$ yields a deeper reason for drawing a distinction between canonical proofs and demonstrations. One thing that is absolutely clear about the notion of an informal proof, such as may be used in everyday mathematical reasoning, is that it may validly appeal to the elimination rules governing these three constants, that is, to universal instantiation, to modus ponens, and to the ex falso quodlibet. Now, for a construction to be a proof of $\forall x\, A(x)$, of $A \rightarrow B$, or of $\neg A$, we are required to recognize it as operating on any element of the domain or on any proof of A to yield a proof of a certain statement (of A(c), of B, or of $0 = 1$). If we were to understand 'proof' here as meaning any ordinary informal proof, then this stipulation would place no restriction whatever on what we were to acknowledge as constituting a proof of a statement of any of these three types. Whatever we chose to accept as being a proof of $\forall x\ A(x)$, it would, provided that it itself conformed to the canons of ordinary informal proof, supply us with an effective means of finding, for any term t, a proof of $A(t)$, namely by simply appending to the proof of $\forall x\ A(x)$ a single application of universal instantiation. Likewise, whatever we chose to accept as being a proof of $A \rightarrow B$, it would, provided that it itself conformed to the canons of ordinary informal proof, supply us with an effective means of transforming any proof of A into a proof of B, namely by annexing to the proof of A the given proof of $A \rightarrow B$ and then appending a single application of modus ponens. And whatever we chose to accept as being a proof of $\neg A$, it would, on the same proviso, supply us with a means of transforming a proof of A into a proof of $0 = 1$, by combining the proof of A with the proof of $\neg A$ and appending a single application of negation elimination (ex falso quodlibet). The constraints on what constituted a proof of statements of these kinds would then all come from whatever intuitive prior notion of an informal proof we were appealing to; the explanations of the logical constants would not, themselves, impose any constraints whatever, but would merely lay down conditions which are automatically satisfied, given certain elementary and indisputable properties of informal proofs.

Obviously, however, this is not what is intended when these explanations of the logical constants are given; we are not appealing to an already understood notion of proof, of which notion the validity of the elimination rules is partially constitutive, but laying down what is to count as a proof in such a way that the validity of those rules follows as a consequence. In recognizing a construction as a proof of $\forall x\ A(x)$, we are supposed to see it as yielding, for each element of the domain, a proof that that element satisfies $A(x)$ without appeal to the fact that we have, in it, a general construction that will do this for every element of the domain; hence universal instantiation is valid just because, in any given case, we could prove its conclusion by applying the construction which constituted a proof of its premiss, without having to cite that premiss. In the same way, in recognizing a construction as a proof of $A \rightarrow B$ (or of $\neg A$), we are supposed to see it as transforming any proof of A into a proof of B (or of $0 = 1$) without appeal to the fact that we have, in it, a general construction that will do this for every proof of A; hence modus ponens and negation elimination are valid just because, in any given case, we could prove their conclusions by applying to the proof of the minor premiss the construction which constituted a proof of the major premiss, without having actually to cite the major premiss. It follows that, under the notion of proof which the explanations of the logical constants serve to specify (given that we know what counts as a proof of an atomic statement), no inference by means of an elimination rule for $\forall$, $\rightarrow$, or $\neg$ will ever occur in the main deduction (it could, of course, occur in a subordinate deduction): the citation of such a rule points to an effective means which we have for obtaining such a proof, but has itself no place within such a proof. (It is equally evident, on reflection, that the elimination rules for &, $\vee$, and $\exists$ likewise have no place in the main deduction of such a proof.) Since these elimination rules may perfectly well occur in the main deductions of ordinary informal proofs, as already remarked, we are forced to admit a distinction between the canonical proofs which exemplify that notion of proof specified by the explanations of the logical constants, and demonstrations, which include all ordinary informal proofs, when these are constructively valid.

If the threat that the intuitive explanations of the logical constants may be viciously circular is to be averted, this can only be because it is possible to impose on the proofs a hierarchy, according to their complexity, where the complexity of the proof matches the complexity of the statement to be proved; that is, for any given statement, if it can be proved at all, then it can be proved by a proof whose complexity does not exceed a bound depending on the structure of the statement. Thus, when we have to decide whether to accept a given construction as a proof of a conditional statement $A \rightarrow B$, we need to judge whether, when applied to an arbitrary proof of A, it will yield a proof of B. In so doing, we shall need to consider only those possible proofs of A that have a complexity below a certain bound determined by the complexity of A, and hence of lower complexity than is required for a proof of A → B itself. The stipulation of the condition for a construction to be a proof of a given statement may thus be seen as procedding by stages; hence, even though the stipulation relating to conditionals involves, in effect, a quantification over all proofs of the antecedent, we can always regard the domain of this quantification as having been fixed in advance of its being laid down what is to count as a proof of any particular given conditional. In reviewing possible proofs of some statement A, we shall never need to consider any proofs involving some statement of greater complexity than A; in specifying what is to be a proof of A → B, we implicitly quantify only over proofs of A, and require the recognition only of proofs of A and of B: and so no circularity is involved.

This must be so, indeed, if an intuitionistic theory of meaning is to observe the principle discussed in section 1 of this chapter, namely that an understanding of any sentence can depend only upon a prior understanding of a proper fragment of the language, a fragment containing no sentences of greater complexity or depth than the given one. Suppose, for instance, that we have an arithmetical statement of the form $\forall n\ (P(n) \rightarrow Q(n))$, where $P(x)$ and $Q(x)$ are both primitive recursive. In order to judge whether a given construction will, for each k, transform a proof of $P(\bar{k})$ into a proof of $Q(\bar{k})$, we shall need to be able, for an arbitrary k, to survey all possible proofs of $P(\bar{k})$ (not, of course, in the sense of inspecting each in turn, but in that of having a conception of the form which any such proof can assume). If the simplicity of a statement of the form $P(\bar{k})$ did not impose a restriction on the type of proof of it we needed to consider, so that we had to envisage the possibility of a proof invoking notions from, say, set theory or complex analysis, this would be an impossible task: we should be reduced to confessing that we could see nothing in common to all possible such proofs save their having $P(\bar{k})$ as their last line, which would be to abandon the full intuitionistic meaning of $\rightarrow$ in favour of one that required a proof of $A \rightarrow B$ to be a derivation of B from A as hypothesis. Moreover, we could not be said, in such a case, to know the meaning of $\forall n\ (P(n) \rightarrow Q(n))$ until we understood the whole of mathematics; indeed, there would be a straightforward circularity involved, since we should not know that there might not be a proof of $P(\bar{k})$ involving the very statement $\forall n\ (P(n) \rightarrow Q(n))$ whose meaning we are trying to derive from the explanations of $\forall$ and $\rightarrow$. If, on the other hand, we are entitled to confine ourselves to considering, as proofs of $P(\bar{k})$, primitive recursive computations, then the meaning of our statement will have been intelligibly given in terms of what is simpler and can be taken as known in advance; and it may be comparatively easy to demonstrate a means of transforming any primitive recursive computation establishing a statement $P(\bar{k})$ into one establishing the statement $Q(\bar{k})$, which is precisely what is to be required of a constructive proof of a statement of this form.

Our discussion of the distinction between canonical proofs and demonstrations suggests that it may indeed be possible to set up such a hierarchical ranking of canonical proofs according to their complexity, this complexity being a monotonic function of the complexity of their conclusions. To say that in a canonical proof no applications of the elimination rules occur is just to say that the lines of the proof increase in logical complexity from premisses to conclusion, that we do not need to consider, in the course of the proof, any statement of complexity greater than that of the conclusion. In fact, normalized natural deduction proofs provide an exact analogy for what is required of canonical proofs if the intuitionistic explanations of the logical constants are to form a system free of conceptual circularity. Naturally, it does not follow, from the normalization theorem for first-order logic, that a similar theorem will hold good for any specific formalized first-order theory for some part of intuitionistic mathematics. What our present considerations show is that it is both necessary and plausible that a normalization property should hold good of thos canonical intuitive proofs which constitute the fundamental type of mental constructions in terms of which any intuitionistic theory is given meaning.

It was emphasized in earlier chapters that it is not necessary, for the soundness of intuitionistic predicate logic, that we should assume that 0 = 1 (or ⊥) cannot be proved, and that therefore $-A$ can be proved only if A cannot be proved. If the structure of a statement sets no bound on the complexity of the proof that might be needed for that statement, then to say that 0 = 1 is unprovable is to make a very large claim, namely that intuitionistic mathematics as a whole is consistent. But this is to compare the intuitionistic claim that 0 = 1 is unprovable to the claim that 0 = 1 is unprovable in some formal system, and therefore to interpret in a holistic sense the manner in which statements are given meaning in intuitionistic mathematics in terms of what are to count as proofs of them, so that we do not know the meaning of any one statement until we understand, in principle, all possible methods of proof in the whole of mathematics; whereas, as we have seen, if a holistic view is required of mathematical language, then the intuitionists have no right, in the first place, to criticize the more widely accepted practices of their classically minded colleagues. From a platonistic standpoint, the claim that 0 = 1 is unprovable is a very trivial claim indeed, where 'provable' means 'provable by intuitively valid means', since 0 = 1 is patently false, and no false statement can have a valid proof. From an intuitionistic standpoint, the claim is equally trivial, since the structure of a statement must set a bound on the complexity of a construction required for a proof of it: in particular, numerical equations demand, for their proof or disproof, the simplest imaginable type of construction, and it is evident that there is not, among these, one that proves 0 = 1. To express this fact by saying that intuitionistic mathematics is self-evidently consistent is tolerable only if it is remembered that, from the standpoint that justifies it, this proposition is as trivial as is, to the platonist, the proposition that the system of all classically valid proofs is consistent: it does not guarantee that any methods of reasoning actually adopted by us are correct.

It may be objected, against the comparison between canonical proofs and normalized proofs in a natural deduction system, that it fails to tally with the intuitionists' insistence upon what Wittgenstein called the 'motley' of mathematics, that is, the unpredictable variety of the modes of argument it uses, the impossibility of circumscribing in advance the kind of reasoning that might be appealed to. It is on such grounds that intuitionists have always denied the possibility of attaining a complete formalization of any mathematical theory, perhaps even of logic itself. Of course, there is a hiatus in the argument: it is not the goal of even the most ambitious formalization to encompass every form of proof that could ever be given, and recognized as correct, for a statement belonging to the theory; a formalized theory is taken as complete if every statement of the language of the theory that can be proved at all can be proved in the formal system -- it is not necessary that every proof of it be reproducible in the system. Hence the motley of mathematics can be acknowledged without impugning the possibility of complete formalizations of particular theories. But, conversely, the fact that intuitionists have been disposed to look askance at that possibility shows how deeply they are impressed by the unsurveyable extent of mathematical proofs; and this surely tells against the idea that such proofs -- even when restricted to 'canonical' ones -- can be arranged in a tidy hierarchy.

There seems to be a crucial difference between normalized proofs in a natural deduction calculus and the intuitive proofs to which the intuitionistic explanations of the logical constants appeal: in a natural deduction system (whether for pure logic or for some actual mathematical theory) the introduction rules for $\forall$, $\rightarrow$, and $\neg$ necessarily take very specific forms, whereas, according to the intuitive explanations of the meanings of these logical constants, *any* construction, without restriction, will serve to establish a statement with one of them as its principal operator, provided only that we can recognize that construction as satisfying the relevant condition. Of certain operations, we may be able to recognize immediately that they carry any proof of A into a proof of B, and then they can by themselves serve as proofs of $A \rightarrow B$: but, for other operations, it may not be obvious that they do this, although it is possible to give a proof that they do; and, in such a case, what constitutes the proof of $A \rightarrow B$ will not be just the operation on its own, but it together with the proof that it carries any proof of A into a proof of B. Similarly for $\forall$ and $\neg$. Since no limit is placed on what may enter a proof that an operation satisfies the condition for being a proof of $A \rightarrow B$ (or of $\forall x\, A(x)$ or of $\neg A$), it is at just this point that any confines within which we seek to enclose the possible complexity of a proof of a given conclusion will be burst. As a result, the point of introducing the notion of a canonical proof evaporates; and we are once again faced with the danger that the intuitive explanations of the logical constants contain a vicious circle.

It would be a mistake to be stampeded by these considerations into a state of despair about the chances of showing the intuitionistic theory of meaning to be viable. The general point, about the impossibility of circumscribing in advance all possible methods of proof of a given statement, and the particular point, about the need for an adequate representation of the means available for proving conditional statements, are really independent, although, at first sight, the latter appears merely to indicate the point of application of the former. If we continue to conceive of canonical proofs after the model of proofs in a natural deduction system, then to insist on the uncircumscribability of the means of establishing conditional statements is, in such terms, to deny that any restrictions can be placed upon the subordinate deductions, and, therefore, that a canonical proof can, after all, properly be likened to a normalized natural deduction proof. There is, however, no adequate ground for such a denial. Our problem is to attain a notion of canonical proof which will allow appeal to be made to the full intuitionistic meaning of $\rightarrow$. In discussing Brouwer's proof of the Bar Theorem, we saw, however, that the effect of giving $\rightarrow$ its full intuitionistic meaning would be able to be achieved if, for every statement A, we were in a position to cite a sufficiently strong axiom of the form 'A $\rightarrow$ there exists a proof of A of such-and-such a kind', where a proof will be a suitable dressed spread, and the axiom will state restrictions on the form of the proof, depending upon the structure of the statement A. At present, we have no clear idea how to formulate such axioms, let alone how to find a set of general principles for generating them; but there is no reason to think such a task impossible. Once these axioms were formulated, they could be treated like additional hypotheses to any theorem in whose proof they were used; they would then make explicit everything that is required for the proofs of conditional statements. (What, in any particular theory, is specially required for the proofs of universally quantified statements will, of course, be embodied, in a familiar way, in axioms laying down the structure of the domain.) Given any statement C, let us symbolize by C* the conditional whose consequent is C and whose antecedent is the conjunction of all the axioms of the form 'A $\rightarrow$ there is a proof of A' or of the form '$\forall x_1 \ldots \forall x_r$ ($A(x_1, \ldots, x_r) \rightarrow$ there is a proof of $A(x_1, \ldots, x_r)$)' for every sentence A or open sentence $A(x_1, \ldots, x_r)$ that is a constituent of C (where, this time, we are treating the open sentence A(x), rather than the sentence A(t), for t a term, as a constituent of $\forall x\, A(x)$ and of $\exists x\, A(x)$). There is now no obstacle to regarding a canonical proof of C as analogous to a normalized proof of C* (rather than of C itself). There will therefore still be a bound on the complexity of canonical proofs of any given statement.

Needless to say, all this is highly programmatic: no one can at present give a detailed account of canonical proofs even of statements of first order arithmetic; all that has been argued here is, first, that the intuitionistic explanations of the logical constants are viable only if there is such a notion to be attained, and, secondly, that there exist grounds for optimism in this regard. But, if we achieved a detailed account of this notion, should we not have arrived at a complete axiomatization of intuitionistic mathematics? And do we not have good reason, both on the score of general considerations about the 'motley of mathematics' and on that of the Gödel incompletability results for arithmetic, for denying this to be possible? We do indeed: we have good ground, and, for most theories, conclusive ground, for holding that we shall never be able to give explicit expression to all the principles of proof that we can be brought to recognize as correct, at least if the standard demanded for explicitness is high enough to yield a formalization. This is not merely because of the psychological fact that our creative powers lag behind our power of understanding, but because the totality of methods of proof, within a given mathematical theory, is likely to be an indefinitely extensible one: certain methods of reasoning intuitively acceptable to us can be carried out only after we have achieved a formulation of some range of methods of proof not including them. This means, of course, that mathematics undergoes a continual process of evolution. This remark is in itself banal, and would be interpreted by a platonist as meaning that we constantly make advances in our knowledge of the unchanging mathematical reality which we are describing. On any constructivist view of mathematics, on which its subject-matter is our own mathematical activity, and meaning is given to our statements by reference to the methods of proof that we possess, this evolutionary process must be understood more radically, as entailing that the very meanings of our mathematical statements are always subject to shift. On the intuitionistic view, this evolution creates a special danger. If we look on the appeal to the full intuitionistic meaning of $\rightarrow$, in proving a statement of the form $A \rightarrow B$, as mediated by the invocation of a principle of the form '$A \rightarrow$ there is a proof of A', an advance in our apprehension of the available modes of proof may lead us to weaken such principles, because restrictions on the means whereby A could be proved which formerly seemed reasonable no longer appear so. When this happens, some proof, involving a conditional $A \rightarrow B$, that had formerly seemed acceptable, may be invalidated. Hence, because of the peculiarities of the intuitionistic interpretation of $\rightarrow$, provability is not a stable property: we cannot think of an addition to our stock of methods of proof as merely allowing us to prove more than we could before, while all the proofs we had already given remain intact, since such an addition may lead to a rejection of certain earlier proofs. The intuitionistic interpretation of $\rightarrow$ does, indeed, give to the notion of proof a self-reflexive or impredicative character, and to some degree weakens the conclusive and irreversible nature of mathematical results; mathematics becomes a subject whose results are fallible and liable to revision, like those of other sciences.

These considerations should not lead us to reject all that was said earlier about the necessity for a notion of canonical proof: it means merely that such a notion will never be completely stable. If the intuitionistic explanations of the logical constants and, more generally, of the meanings of mathematical statements are to be considered as constituting a coherent theory of meaning for the language of mathematics, then the notion of proof which is appealed to must be such that we can fully grasp the concept of a proof of any constituent of a given sentence in advance of grasping that of a proof of that sentence. It cannot, therefore, be identified with the notion of the sort of proof that we may, at some future time, come to consider valid, a proof which may appeal to modes of reasoning of which we have at present no conception; it must, rather, relate only to methods of proof that we currently recognize. The intuitive explanations of the logical constants are, therefore, best regarded as programmatic. In order to give them definite substance, there must, at any given time, be a more precise explanation ultimately attainable of a notion of canonical proof, in terms of which those programmatic explanations are, for the time being, to be interpreted; for, if not, they are merely circular, and yield no clear meanings at all. As mathematics progresses, so the relevant notion of a canonical proof will change, and hence the meanings of our mathematical statements are always, to some degree, subject to fluctuation. But, for all that, their meanings must, at any given stage in the development of mathematics, be specific; and the means whereby they may be exhibited as specific has to lie in an analysis of some notion of proof, adequate at that stage, such that a proof of a given statement never depends upon the understanding of a more complex statement.

7.3. ARE THE INTENDED MEANINGS OF THE LOGICAL CONSTANTS FAITHFULLY REPRESENTED ON BETH TREES?

The standard explanations of the logical constants, which fix their intended meanings, are given by stipulating what is to count as a proof of a complex statement, it being taken as known what is to be a proof of an atomic one. In relation to Beth trees, on the usual intuitive understanding of these, they are explained by reference to the notion of a state of information, and hence, apparently, in a quite different way. It needs, however, to be asked whether, by adopting a suitable way of construing the idea of a state of information, we can show the equivalence of the two accounts.

An interpretation with respect to a Beth tree involves laying down (in an effective manner) which (closed) atomic formulas are verified at which nodes. Where a is a node of a Beth tree, let us denote by a* the state of information which it intuitively represents; and, where A is any closed formula, let us denote by A* the statement which, under the given interpretation on the Beth tree, it is supposed to express: as usual, we take the domain to consist of the natural numbers, and assume that the formal language contains the numerals. Now, where Q is a closed atomic formula, we may obviously take the fact that Q is verified at a node a as representing our having, in a*, a proof of Q*. However, when A is a closed complex formula, we cannot take the fact that A is *true* at a as meaning that, in a*, we have a proof of A*, without acknowledging a divergence between the meanings of ∨ and of ∃ on the Beth trees and their intended meanings; since, on the standard explanations, we have a proof of

B v C only when we have a proof of B or a proof of C, and of ∃x B(x) only when we have a proof of B($\bar{m}$) for some m. This way of understanding the notion of being true at a node would in any case be wide of the mark, since the condition for an atomic formula to be true at a is weaker than that for it to be verified at a. The obvious solution is to invoke our distinction between a canonical proof and a demonstration, and say that an atomic formula Q is verified at a if, in a*, we have a canonical proof of Q*, while an arbitrary formula A is true at a if, in a*, we have a demonstration of A*. This leads us to ask how we could suitably extend the notion of being verified at a node to complex formulas.

The answer is as follows:

Definition. A closed formula A is *verified* at a iff one of the following conditions holds:

(i) A is atomic and A is verified at a;

(ii) A is B & C and both B and C are verified at a;

(iii) A is B v C and either B or C is verified at a;

(iv) A is B → C and, for every b ≤ a, if B is verified at b, then b is barred by {c|C is verified at c};

(v) A is ¬B and for no b ≤ a is B verified at b;

(vi) A is ∀x B(x) and, for every m, a is barred by {c|B($\bar{m}$) is verified at c};

(vii) A is ∃x B(x) and, for some m, B($\bar{m}$) is verified at a.

(Note clauses (iv) and (vi), which are not what one might, perhaps, at first glance expect.) Armed with this definition, we may establish the expected relation between a formula's being verified at a and its being true at a. We state first the easy

Lemma. If A is verified at a, and b ≤ a, then A is verified at b.

Proof. The proof is trivial, by induction on the complexity of A.

Theorem. A is true at a iff a is barred by {c|A is verified at c}.

Proof. By induction on the complexity of A.

(i) A is atomic. Then the theorem holds by the definition of 'true at a'.

(ii) A is B & C. A is true at a iff B and C are both true at a, and hence, by the induction hypothesis, iff a is barred by {c|B is verified at c} and by {c|C is verified at c}. Hence, by the lemma, A is true at a iff a is barred by {c|B is verified at c and C is verified at c}, which, by the definition, is {c|A is verified at c}.

(iii) A is B ∨ C. A is true at a iff a is barred by {b|B is true at b or C is true at b}, and hence, by the induction hypothesis, iff a is barred by {c|B is verified at c or C is verified at c} = {c|A is verified at c}.

(iv) A is B → C. Suppose A is true at a. Then, for every b ≤ a, if B is true at b, C is true at b. Suppose B is verified at b, where b ≤ a. Then by the induction hypothesis, B is true at b, and so C is true at b, whence, by the induction hypothesis again, b is barred by {c|C is verified at c}. A is therefore verified at a, and so, a fortiori, a is barred by {c|A is verified at c}. Now suppose, conversely, that a is barred by {c|A is verified at c}, and let B be true at b, b ≤ a. By the induction hypothesis, b is barred by {c|B is verified at c}, and so, by the lemma, by {c|A and B are both verified at c}. It follows from the definition that b is barred by {d|C is verified at d}, and hence, by the induction hypothesis, that C is true at b. It follows that A is true at a.

(v) A is ¬B. Suppose A is true at a. Then if b ≤ a, B is not true at b, and hence, by the induction hypothesis, B is not verified at b. Hence A is itself verified at a. Conversely, suppose that a is barred by {c|A is verified at c}. If b ≤ a, then, by the induction hypothesis, if B is true at b, b is barred by {c|B is verified at c}, and hence, by the lemma, by {d|B is verified at d and B is not verified at d}, a contradiction. Thus B cannot be true at b, and so A is true at a.

(vi) *A is ∀x B(x)*. A is true at a iff, for every m, $A(\bar{m})$ is true at a, and hence, by the induction hypothesis, iff, for every m, a is barred by $\{c \mid B(\bar{m})$ is verified at $c\}$, i.e. iff A is verified at a. If a is barred by $\{c \mid A$ is verified at $c\}$, then, by the definition, for every m, a is barred by $\{d \mid B(\bar{m})$ is verified at $d\}$, and so A is verified, and therefore true, at a.

(vii) *A is ∃x B(x)*. A is true at a iff a is barred by $\{b \mid$ for some m, $B(\bar{m})$ is true at $b\}$, and so, by the induction hypothesis, iff a is barred by $\{c \mid$ for some m, $B(\bar{m})$ is verified at $c\} = \{c \mid A$ is verified at $c\}$.

Suppose that clause (iv) of the definition were altered to:

(iv') A is B → C and, for every b ≤ a, if B is verified at b, then C is verified at b.

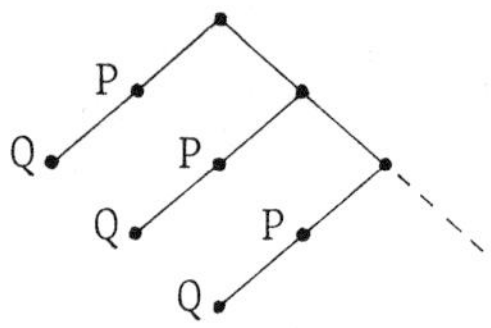

Then the Beth tree (where P and Q are atomic formulas, and are shown against those nodes on each path at which they are first *verified*) would be a counter-example to the theorem, since P → Q is true at the vertex, but not verified anywhere on the rightmost path. Similarly, if clause (vi) were altered to:

(vi') A is ∀x B(x) and, for every m, $B(\bar{m})$ is verified at a,

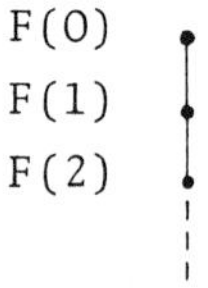

then the Beth tree (where again each formula is shown against the first node at which it is verified) would be a counterexample, since $\forall x\ F(x)$ is true at the vertex, but not verified anywhere. Under the definition as actually given, any formula of one of the forms $B \to C$, $\neg B$, and $\forall x\ B(x)$ is true at a iff it is verified at a. This is intuitively reasonable, since the standard explanations of the logical constants do not provide for any divergence between a canonical proof and a demonstration of a statement of any of these forms. We can now take the notion of a formula's being verified at a node as the fundamental one, and define a formula A to be true at a node a just in case a is barred by $\{c \mid A \text{ is verified at } c\}$.

Now if, in the conventional way, we understand the nodes of level n on a Beth tree as representing (with the terminal nodes of level $< n$) the possible states of information at the n-th temporal stage, then to say that A is true at a is to say that, in a*, we know that we shall have a canonical proof of A* in a finite time. Likewise, to say that $B \to C$ is verified at a is to say that, in a*, we know that, as soon as we obtain a canonical proof of B*, we shall in a finite time obtain a canonical proof of C*; while to say that $\forall x\ B(x)$ is verified at a is to say that, in a*, we know that, for each m, it will be only a finite time before we shall have a canonical proof of $B(\bar{m})$*. None of this accords with our concept either of a demonstration or of a canonical proof: a demonstration of a statement provides us with an effective means of obtaining a canonical proof of that statement, but with no guarantee that we ever shall obtain one, either by that means or any other: a canonical proof of $B \to C$ provides a means of transforming a canonical proof of B into a canonical proof of C, but no guarantee that we shall in fact so transform it; a canonical proof of $\forall n\ B(n)$ provides a means, for any m, of obtaining a canonical proof of $B(\bar{m})$, but, again no guarantee that we shall obtain one.

This difficulty is easily dealt with, by modifying our intuitive understanding of Beth trees. We have, in effect, to regard the temporal stages represented by the levels not as intervals, such as days, which are ineluctably replaced by their successors, so that we inescapably travel down the tree (until we reach a terminal node, if we do), but as punctuated by efforts, not necessarily successful, to obtain more information; if we never make any further such effort, we remain for ever at the same temporal stage. We already know that the Beth trees can represent the situation in which it is possible that we shall obtain more information and also possible that we never shall, namely when the node a is not terminal, and has immediately below it a node b such that the subtree T_b is isomorphic to the subtree T_a (with respect to the verification of atomic formulas as well as to the abstract tree structure), and there are also immediately below a one or more other nodes c for which this is not so: in such a case the state of information b* will coincide with a*, although a and b are distinct nodes. But, even if a is non-terminal but there is no other node b such that T_b is isomorphic to T_a, we may, on the intuitive conception now suggested, remain, if we choose, for ever in the state of information a*. If we decide to obtain further information, then the immediate outcome of our decision will be a move to the state of information b*, for some b immediately below a (which one, we cannot tell in advance). Note that, in order to allow for those trees in which b* = a*, although b stands below a, we must regard the temporal stages as punctuated, not by the actual acquisition of new information, but by the attempt to acquire it; such an attempt may fail, the failure being represented by a move from a to b, where T_b is isomorphic to T_a. Of course, we should not expect any particular Beth tree to supply a representation of every possible advance in mathematical knowledge, or even in the knowledge of propositions of (say) first-order arithmetic, starting from a given state of information; all it will show are the possible states of information concerning the truth of propositions expressible within some very limited vocabulary.

The fact that a node a is barred by a species $\underline{S}$ of nodes is now to be seen as meaning, not that, when we are in the state a*, we know that we shall in a finite time be in a state b* for some $b \in \underline{S}$, but only that we have a means, if we choose to apply it, for arriving at one of the states b*. This fits exactly the conception that what a demonstration provides is an effective means of obtaining a canonical proof. It fits equally well the conception that a canonical proof of $\forall x\ B(x)$ provides an effective means of obtaining, for any m, a canonical proof of $B(\bar{m})$, and that a canonical proof of $B \to C$ provides an effective means of obtaining, from a canonical proof of B, a canonical proof of C.

There are further problems, however. On the intuitive explanations of the logical constants, it is assumed to be decidable whether or not a given construction constitutes a (canonical) proof of a given statement, and hence, presumably, whether or not, at any given stage, we have such a proof; just as, in the theory of the creative subject, it is assumed to be decidable whether $\vdash_n A$, for any n and A. This assumption plays an important role in the intuitive explanations of the logical constants, as it does in the theory of the creative subject. In the former case, its importance lies in the fact that, in the explicatory clauses, the sentential operators are applied only to decidable statements, and the quantifiers only to decidable predicates. (If we assume that, for any construction, we can effectively determine, not merely whether or not it is a proof of any given statement, but of which statements, if any, it is a proof, then we need to take account of the use only of the universal quantifier, in the clauses relating to $\to$, $\neg$, and $\forall$.) Hence the intuitive explanations of the logical constants may be claimed as genuine *explanations*, since, in order to understand them, it is necessary to know already, not the full meanings of the logical constants to which they relate, but only their meanings in a very restricted type of context. The standard explanations of the intuitionistic logical constants are thus free of the circular character of the intuitive explanations of the classical ones; and, indeed, in the mathematical theory of constructions, these explanations appear as actual definitions. (In particular, it is obviously harmless to presuppose a knowledge of the meanings of the sentential operators as applied to decidable statements, since this is merely a matter of a practical knowledge of a decision procedure; the understanding of statements to the effect that every result of applying a given construction to an element of some decidable species has a certain decidable property is a different matter.)

On the Beth trees, on the other hand, although it is assumed to be decidable whether a given closed atomic formula is verified at a given node, this will not be the case for closed formulas in general; there is, for example, no general means of determining whether $P \to Q$ or $\forall x\, F(x)$ is verified at a given node a. Hence, if the definition of 'A is verified at a' were to be proposed as a means of explaining the meanings of the logical constants, it would display something of the same circularity as the classical explanations. Admittedly, we should not, as in the classical case, be explaining each logical constant by a condition to state which it is necessary to use the logical constant in a context of the most general kind: to explain any given logical constant, we need appeal to the meaning of that logical constant only as applied to statements of the form 'A* is verified in b*' and to quantification over possible or actual states of information subsequent to a given one. There cannot be the same presumption, however, that the meanings of the logical constants may, when they are restricted to apply to statements of this kind, be taken for granted as unproblematic, as there would be if such statements could be assumed to be decidable. In the same way, it ought, intuitively, to be decidable whether or not we have, at any given time, a demonstration of a given statement; but it is not decidable, on a Beth tree, whether a closed formula is true at a given node, and hence the definition of truth at a node, considered as intended to explain the meanings of the logical constants, would suffer from the same defect.

Worse still, the definition of truth at a node has the consequence that whatever is entailed by what is true at a node is itself true at that node. Admittedly, the same does not, in general, hold for verification. The exceptions are, however, relatively unimportant: especially is this so if the particular Beth tree is such that, whenever a closed formula $B \vee C$ is entailed by the formula verified at some node, then so is either B or C, and similarly for existential formulas. This means that we cannot, after all, really construe A's being true at a as meaning that, in a*, we recognize ourselves as having a demonstration of A* (but, at best, that we have the means to construct one, if it occurred to us how to do so).

The difficulty arises because the intuitive notion of a state of information that we are employing is a fundamentally unsatisfactory one. A state of information cannot be considered as determined just by knowing which atomic statements have been verified, nor even which are true, since the verification, or the truth, of complex formulas at a node does not depend solely upon which atomic formulas are verified or are true at that node, but on what happens at subsequent nodes. We have, therefore, to think of a state of information as comprising two things: a knowledge of which atomic statements have been verified; and an awareness of the future possibilities of verifying atomic statements, as represented by the subtree determined by the associated node. Now a Beth tree is, usually, an infinite structure, and we have therefore to think of a grasp of such a structure as consisting in a knowledge of how to construct any finite part of it, that is, essentially like the process of constructing an infinite sequence of tree-trunks. We could, indeed, think of ourselves as, at each stage, associating with each node a decidable species of formulas as being true at that node and as representing those statements of which we had, at that stage, a demonstration. To stipulate a formula as being true at a node of course imposes restrictions upon the subtree determined by that node. At a given stage, we may not, however, be aware of every feature of that subtree: that is to say, it does not follow, from the fact that we have not stipulated a formula to be true at a node, that we are free to develop the subtree having that node as vertex in such a way that the formula is not true at the node. As soon as we realize that we are not free to do this, we may add the formula in question to those laid down as being true at that node, and this corresponds to recognizing the formula to be a logical consequence of those already stipulated as true there. But this necessarily means that progress in drawing conclusions from what we have already recognized as true (e.g. at the vertex of the whole tree) is not to be represented by progress down the tree; this is because, while the construction of the tree has to begin at the vertex and proceed downwards, the determination of which formulas are verified or are true at any node proceeds upwards, depending as it does upon the verification of subformulas at lower nodes. Hence the paths of the tree cannot be viewed as representing the various possible ways in which we may come to extend our mathematical knowledge by the most usual means of doing so, namely by deductively deriving new theorems from known axioms.

This means that we cannot hope, by means of a Beth tree, to give an accurate representation of our present state of knowledge concerning, for example, some statement A of elementary number theory: we should have to stipulate as true at the vertex all those axioms of HA expressible within the vocabulary of A, together with all currently known theorems so expressible, and these would certainly be enough to determine as already true at the vertex every (closed) atomic subsentence of A or its negation (which, we should not explicitly know), and a great many more besides; we therefore do not have a sufficiently explicit knowledge of the structure of the relevant tree to be able to use it to give an informative picture of our situation in respect of the possibility of proving or disproving A and related statements. We can, on the other hand, use arithmetical statements as examples to illustrate the intuitionistic failure of classical laws of logic; but, in order to do this, we need to take account only of certain features of what we presently know of the possibility of their verification.

For instance, if we take P* to be a statement of the form $\forall n\, F^*n$, where we assume that F* is a decidable predicate, which we do not further specify, then we know that on any Beth tree constructed to give an interpretation of formulas containing the predicate-letter F we shall be entitled to take $\forall x\,(Fx \vee \neg Fx)$ as true (verified) at the vertex. Hence, for each m, the vertex will be barred by $\{b \mid F\bar{m}$ is verified at b or $\neg F\bar{m}$ is verified at b}. Further, if, for any m, $\neg F\bar{m}$ is verified at a node b, $\neg \forall x\ Fx$ will be verified at b; hence if, at a node a, $\neg\neg \forall x\ Fx$ is verified at a, then for no m will $\neg F\bar{m}$ be verified at a node below a, and so, for each m, a will be barred by $\{b \mid F\bar{m}$ is verified at b}, and $\forall x\ Fx$ will therefore be verified at a. If, now, we aim to construct a Beth tree to give an interpretation only of formulas compounded out of the sentence-letter P (which corresponds intuitively to $\forall x\ Fx$), we may take $\neg\neg P \rightarrow P$ as verified at the vertex. From what we have been given, no reason appears why we should ever come to know any statement compounded out of P* stronger than $\neg\neg P^* \rightarrow P^*$. Since every statement is equivalent either to $P^* \vee \neg P^*$, to $\neg P^*$ or to P^*, we may represent our situation by means of the Beth tree

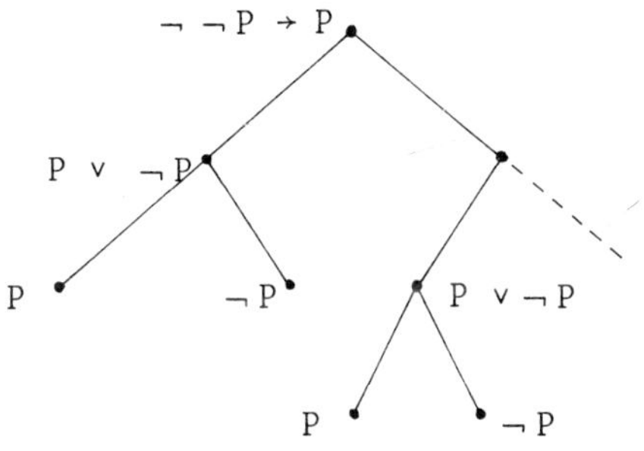

on which $\neg\neg P \rightarrow P$ is true at the vertex, but $P \vee \neg P$ is not true anywhere on the rightmost path; the tree therefore yields a counter-example to $(\neg\neg P \rightarrow P) \rightarrow P \vee \neg P$. This is a quite different thing from giving a specific statement which fails to satisfy this law. In order to give a specific example along these lines, we should have to be sure that we were not now in a position to assert $\forall n\ F^*n \vee \neg \forall n\ F^*n$, and therefore that we were not now in a position to verify $\neg F^*\bar{m}$ for any m, although we might later find ourselves in such a position; and it is difficult to see how we could construct a predicate F* satisfying these conditions, if saying that we are not now in a position to prove something means, not merely that we have not yet proved it, but that it definitely does *not* follow from those propositions which we do presently acknowledge as true.

(If, on the other hand, we choose the weaker meaning, that we have not yet in fact proved it, we cannot be sure that we can give a full representation of our situation by means of a Beth tree.)

Partly for this reason, the Beth trees cannot be seen as providing an actual semantics for an intuitionistic first-order language. Indeed, their failure to do so can be recognized from quite different considerations: in passing from a specification of what constructions serve as proofs of statements of the language to a description of the structure of the totality of possible situations in which we might obtain such proofs, we lose information that is essential to fixing the meanings of our statements. In 'Intuitionistic Logic in an intuitionistic metalanguage', de Swart claimed, on the contrary, that the Beth trees provide *the* correct semantics for an intuitionistic language. On the basis of this claim, he argued that, by giving a Beth tree, with the domain taken as $\underline{N}$, on which the formula $\neg \forall x\ (Fx \vee \neg Fx)$ is true at the vertex, he has shown that there exists a number-theoretic predicate F^* such that $\neg \forall n\ (F^*n \vee \neg F^*n)$. The Beth tree he chooses for this purpose is essentially the following:

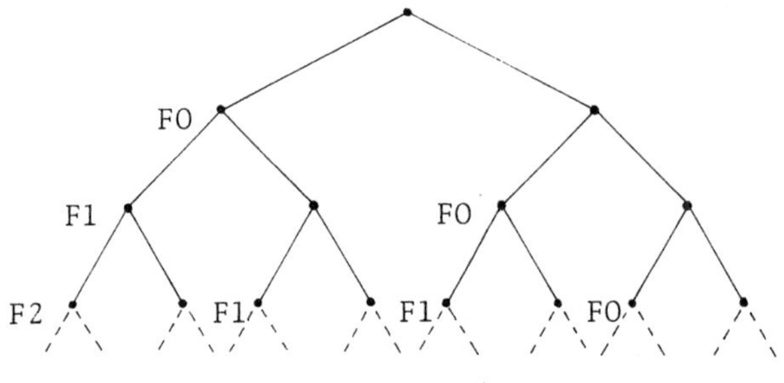

(The formulas are shown to the left of those nodes at which, on each path, they are first verified. If we represent the nodes by finite sequences of 0's and 1's, $F\bar{m}$ is verified at $\vec{u}$ iff $\vec{u}$ contains at least m + 1 0's.) But de Swart then immediately negated the claim he had just made by remarking that the problem of giving a concrete example of such a number-theoretic predicate F* remains open. If, in classical semantics, we give a model, say over the natural numbers, of a certain formula A, then no further problem arises of giving specific meanings to the predicate-letters, etc., occurring in A, such that A comes out true: by describing the model, we have indicated just such meanings. And, if the Beth trees really constituted a semantics for intuitionistic first-order languages, then any specific Beth tree, considered relative to a specific domain, would confer definite meanings upon the schematic letters. Quite evidently, however, de Swart's admission, and not his claim, is correct: by specifying such a Beth tree as the above, we do not *provide* an intuitionistic meaning for the predicate-letter F interpreted as a number-theoretic predicate; we simply indicate certain conditions the satisfaction of which by a number-theoretic predicate F* would bring out true the statement (here $\neg\forall n\ (F^*n \vee \neg F^*n)$) in which we are interested.

The situation is the same as with possible-worlds semantics for formulas of modal logic. We can give a counter-example to such a formula by describing an abstract structure of possible worlds, and laying down in which of those possible worlds each atomic formula is to be satisfied by which elements of the associated domain. But, by doing this, we do not determine, even by the standards of possible-worlds semantics, specific meanings for the predicate-letters. In order to do so, we must supplement the abstract model by a specification of which particular possible worlds are represented by the elements of the abstract structure: which represents the actual world, and in which specific ways the worlds represented by the other elements diverge from actuality. So in the case of the Beth trees. To give a definite meaning to the schematic letters, we must specify what, in particular, is to count as a verification or proof of each atomic formula (under some assignment to the free variables): how, specifically, we are to recognize in what state of information we are at any time.

This raises the question with what right we require a Beth tree to have a particular structure. Earlier, we supposed that we were considering some particular statement --

say, an arithmetical statement -- whose meaning we already knew intuitively, and trying to see how to construct a Beth tree to represent the possibilities actually open to us, or open so far as we knew, of proving it or related statements. But the traffic between intuitive meaning and semantic representation must flow both ways. If, in the classical case, we understand a first-order sentence, then we can also describe its intended interpretation within the two-valued semantics; but, as we have already noted, we can, conversely, gain an understanding of a sentence by being given the semantic values of its component predicates, individual constants, etc.; and, within the general framework of classical semantics, there are no restrictions upon the interpretation that may be given. The standard use of Beth trees is, likewise, not to provide a representation of that understanding of some sentence which we already possess, but to illustrate, by stipulation, the consistency of some formula, as in the example quoted from de Swart. But since, as we have seen, we do not succeed, simply by stipulating that a Beth tree shall have a certain structure, in conferring definite meanings on the sentences involved, the question arises whether such stipulation is subject to restrictions; that is, whether, for any Beth tree we like to describe, there actually are statements the possible verifications of which are related as on the tree. For instance, granted that, by describing the above tree, de Swart did not determine a specific number-theoretic meaning for the predicate-letter F, can we nevertheless be sure that there will be *some* number-theoretic predicate F* for which that Beth tree will provide an accurate diagram?

In that example, there are several restrictions on the possibility of verifying statements involving F; in particular:

(i) for $m > n$, $F\bar{m}$ can be verified only after $F\bar{n}$ has been verified;

(ii) no statement of the form $\neg F\bar{m}$ is ever verified;

(iii) $\forall x\, Fx$ is never verified.

Some restrictions we can make reasonable by suitable stipulations as to the internal structure of Fx. If, for instance,

Fx actually has the form $\forall y_{y \leq x}$ Gy, this would explain why we cannot verify F2 without having first verified F0 and F1; and if, in turn, Gy has the form Hy $\vee$ $\neg$Hy, this would justify ruling out the possibility of a verification of $\neg F\bar{m}$. But on what ground do we exclude the occurrence of a verification of $\forall x$ Fx? Under our partial interpretation, this is equivalent to $\forall x\ \forall y_{y \leq x}$ (Hy $\vee$ $\neg$Hy), which is in turn equivalent to $\forall x$ (Hx $\vee$ $\neg$Hx); so, to render the restriction reasonable, we have to find an interpretation, or partial interpretation, of Hx under which $\neg\forall x$ (Hx $\vee$ $\neg$Hx) is true. But this makes the entire enterprise a question-begging one, since the original point of it was, precisely, to find an example in which $\neg\forall x$ (Fx $\vee$ $\neg$Fx) held good.

We can have no assurance that an arbitrarily specified Beth tree represents any actual case, short of supplementing the abstract description of the tree and of the verification-relation between atomic formulas and nodes with what is required to give a genuine intuitionistic semantic interpretation, namely a specification of what a verification of each atomic statement is to consist in. If we were supplied with such a supplementation, then we might be able to recognize that there is no possibility of verifying such-and-such a compound statement, or that a given statement could be verified only after some other statement had been; but, in advance of such a supplementation, the Beth tree is only a way of expressing desired conditions, of which we are as yet uncertain whether they can be fulfilled.

It is thus plain that the Beth trees cannot be claimed, as some have claimed them, to supply an adequate semantics for an intuitionistic language. Indeed, the considerations we have just been reviewing had nothing to do with the intention that the language be understood intuitionistically: they showed that the Beth trees do not provide a complete semantic theory of any kind, intuitionistic or otherwise, since a Beth tree cannot be regarded, from any standpoint, as determining specific meanings for the predicate-letters whose interpretation is given relative to that tree. In this respect, a

Beth tree is precisely like an abstract structure of possible worlds: it gives, at best, the framework of a semantic theory, not a complete semantics.

In order to test the validity of laws of modal logic, only a semantical framework, not a complete semantics, is necessary. In the same way, tne Beth trees remain useful for the investigation of intuitionistic *logic*, since their deficiencies do not relate to the way the logical constants are understood in relation to them. They are not, indeed, to be thought of as giving a full picture of the way in which the intuitionistic logical constants are given meaning: that can only be done directly in terms of the notion of a construction and of a construction's being recognized as a proof of a statement. But, as we have seen, we can construe the notion of verification at a node in terms of that of our possessing a canonical proof, even though not of a canonical proof in any actual mathematical theory; and, if we do so, the interpretations of the logical constants on the Beth trees turn out to be faithful to their intended meanings. As a result, the Beth trees prove to be a legitimate tool for the study of intuitionistic logic, although not for the construction of an actual intuitionistic semantics.

7.4. THE NOTION OF A CHOICE SEQUENCE

A choice sequence is an infinite sequence of natural numbers whose terms are generated in succession; in the process of generating them, free choices may play a part. At one extreme, the selection of each term may be wholly determined in advance by some effective rule: a sequence generated by such a rule is a *lawlike* sequence. At the other extreme, we have a sequence the selection of each term of which is wholly unrestricted: these are the *lawless* sequences. In between are those choice sequences the selection of whose terms is partially restricted in advance, but not completely determined.

Lawless sequences were mentioned in sections 5.6 and 5.7, and in section 6.1 we described an axiomatization of them as consisting of

(1) $\forall\alpha\ \forall\beta\ (\alpha \equiv \beta \vee \alpha \not\equiv \beta)$

($\underline{A}$-2) $\forall\vec{u}\ \exists\alpha\ \alpha \in \vec{u}$

and the principle of open data

($\underline{A}$-3) $\mathrm{Ext}_{\alpha,\beta_1,\ldots,\beta_r} A(\alpha, \beta_1, \ldots, \beta_r) \rightarrow$

$$\forall\alpha\ \forall\beta_1\ldots\forall\beta_r\ [A(\alpha,\beta_1,\ldots,\beta_r)\ \&$$

$$\alpha \not\equiv \beta_1\ \&\ldots\&\ \alpha \not\equiv \beta_r \rightarrow$$

$$\exists n\ \forall\gamma_{\gamma \in \bar{\alpha}(n)}\ (\gamma \not\equiv \beta_1\ \&\ldots\&\ \gamma \not\equiv \beta_r \rightarrow$$

$$A(\gamma, \beta_1,\ldots,\beta_r))].$$

For a reason to be given below, we probably cannot make do with the special case of ($\underline{A}$-3) when r=1, namely

($\underline{A}$-3') $\mathrm{Ext}_{\alpha,\beta}\ A(\alpha,\beta) \rightarrow$

$$\forall\alpha\ \forall\beta[A(\alpha,\beta)\ \&\ \alpha \not\equiv \beta \rightarrow$$

$$\exists n\ \forall\gamma_{\gamma\in\bar{\alpha}(n)}\ (\gamma \not\equiv \beta \rightarrow A(\gamma,\beta))].$$

On the other hand, it is quite obvious how to generalize ($\underline{A}$-3') to ($\underline{A}$-3). Since a schema such as ($\underline{A}$-3) is cumbersome to write and we shall have occasion to set out a number of similar schemata, we shall adopt the convention that, if a schema (T) is related to a schema (S) as ($\underline{A}$-3) is related to ($\underline{A}$-3'), then (T) may be referred to as $(S)_r$: by this means, we shall in future need to write out only the schema corresponding to ($\underline{A}$-3'). (We are here following the convention that, in such schemata, only those choice-sequence variables actually shown or indicated by dots occur.) For the case when r = 0, ($\underline{A}$-3) of course reduces to:

($\underline{A}$-3'') $\mathrm{Ext}_{\alpha} A(\alpha) \rightarrow$

$$\forall\alpha\ (\ A(\alpha) \rightarrow \exists n\ \forall\gamma_{\gamma\in\bar{\alpha}(n)} A(\gamma)).$$

We also noted in section 6.1 that axioms ($\underline{A}$-1) to ($\underline{A}$-3) imply that, for lawless sequences, extensional equality coincides with (intensional) identity:

$(\underline{A}\text{-}4) \quad \forall\alpha\ \forall\beta(\alpha \equiv \beta \leftrightarrow \alpha = \beta)$,

since if $\alpha = \beta$ but $\alpha \not\equiv \beta$, then by ($\underline{A}$-3) we should have that, for some n, $\gamma = \beta$ for all $\gamma\in\bar{\alpha}(n)$, which is absurd by ($\underline{A}$-2). This is not meant to suggest that if α and β are intensionally distinct lawless sequences, then we know that they are sooner or later going to diverge extensionally, but only that it is absurd that we should know that they will always extensionally coincide. As a result of ($\underline{A}$-4), every predicate of lawless sequences is extensional, so that we may simplify ($\underline{A}$-3') to:

$$(\underline{A}\text{-}3''') \ \forall\alpha\ \forall\beta[A(\alpha,\beta)\ \&\ \alpha \neq \beta \rightarrow \exists n\ \forall\gamma_{\gamma\in\bar{\alpha}(n)}\ (\gamma \neq \beta \rightarrow A(\gamma,\beta))],$$

and ($\underline{A}$-3) itself to $(\underline{A}\text{-}3''')_r$.

We further noted in section 6.1 that, in ($\underline{A}$-3), the condition that α be distinct from all the β_i is indispensable, since, without it, we could take $A(\alpha,\beta)$ in ($\underline{A}$-3') to be $\alpha = \beta$, and then, by setting $\alpha \equiv \beta$, obtain the absurd result that

$$\forall\beta\ \exists n\ \forall\gamma_{\gamma\in\bar{\beta}(n)}\ \gamma = \beta.$$

The requirement that γ be distinct from the β_i in ($\underline{A}$-3) is also indispensable. For suppose that we could assert the strengthened form

$$\mathrm{Ext}_{\alpha,\beta}\ A(\alpha,\beta) \rightarrow \forall\alpha\ \forall\beta[A(\alpha,\beta)\ \&\ \alpha \not\equiv \beta \rightarrow \exists n\ \forall\gamma_{\gamma\in\bar{\alpha}(n)}\ A(\gamma,\beta)]$$

of (A-3'). Now suppose that $\alpha \neq \beta$. Then, by taking $A(\alpha,\beta)$ as $\alpha \neq \beta$, we should have that, for some n, $\gamma \neq \beta$ for all $\gamma\in\bar{\alpha}(n)$. For this n, $\beta\notin\bar{\alpha}(n)$, since otherwise $\beta \neq \beta$. Hence, for some $m < n$, $\beta(m) \neq \alpha(m)$, and we should have proved

$$\forall\alpha\ \forall\beta\ (\alpha \neq \beta \rightarrow \exists m\ \alpha(m) \neq \beta(m)),$$

which, by ($\underline{A}$-4), would yield

$$\forall\alpha\ \forall\beta\ (\alpha \not\equiv \beta \rightarrow \exists m\ \alpha(m) \neq \beta(m)),$$

which is intuitively unacceptable, expressing as it does just that interpretation of (A-4) which we repudiated above.

Obviously, the lawless sequences themselves do not comprise all the choice sequences; even if we threw in the lawlike sequences as well, we should have only the two extreme types, without any of the intermediate variety. Moreover, as Troelstra has pointed out, (A-3) in effect informs us that the identity operation is the only continuous operation under which we may assert that the lawless sequences are closed. For instance, suppose that β is a lawless sequence, and that the choice sequence α satisfies:

$$(*) \qquad \alpha(n) = \begin{cases} 1 & \text{if } \beta(n) = 0 \\ 0 & \text{if } \beta(n) = 1 \\ \beta(n) & \text{otherwise.} \end{cases}$$

Now if $\alpha = \beta$, then $\forall n\ \beta(n) > 1$, and hence by (A-3)

$$\exists m\ \forall\gamma_{\gamma\in\bar{\beta}(m)}\ \forall n\ \gamma(n) > 1,$$

which is absurd by (A-2). Hence $\alpha \neq \beta$, and so, if α is a lawless sequence, we have, by (A-3):

$$\exists m\ \forall\gamma_{\gamma\in\bar{\alpha}(m)}\ [\gamma \neq \beta \rightarrow \forall n\ ((\beta(n) = 0 \rightarrow \gamma(n) = 1)$$
$$\& \ (\beta(n) = 1 \rightarrow \gamma(n) = 0))].$$

This is again absurd by (A-2), and so α cannot be taken to be a lawless sequence. At first this may seem odd, since α and β are symmetrically related by (*). But, again, it is not a matter of there being two choice sequences, either of which we may, if we will, take as being a lawless sequence, provided that we do not so take the other, but of its being impossible that we should know (*) to hold of any two lawless sequences. If, for example, α were *defined* by means of (*), i.e. if it were generated in accordance with that rule given in terms of the lawless sequence β, then indeed α would not be a lawless sequence. If we wish to use choice sequences (together with a suitable correlation law) to yield real number generators, then we shall certainly want them to be closed under continuous operations. The fact that the lawless sequences are not so closed therefore provides an additional reason, if one were needed, for seeking a more inclusive class of choice sequences. It is because lawless sequences are not closed under continuous operations that ($\underline{A}$-3') is weaker than ($\underline{A}$-3), since we cannot code r lawless sequences $\beta_1,\ldots,\beta_r$ as a single lawless sequence β.

Guided by Brouwer's writings, considerable exploration of the concept of a choice sequence has been carried out in recent years by Kreisel, Myhill, and Troelstra, and the results are reported in detail in Troelstra's monograph *Choice Sequences* in the present series. Here we are concerned only to give an outline account of the main steps in arriving at a satisfactory explanation of the concept. We shall therefore consider in turn various notions of a choice sequence (or classes of choice sequences), just as Troelstra does, examining each to see if it satisfies the intuitive requirements for representing the general notion. The class of lawless sequences may thus serve as class $\underline{A}$ of choice sequences. For every class that we consider, axiom (1) will hold good, since (intensional) identity is always decidable: this is why it was designated by a simple numeral, and we now assume it, once for all, whatever the range of the choice-sequence variables. For each class of choice sequences, we shall need to assume an existence axiom corresponding to ($\underline{A}$-2) and an axiom schema corresponding to ($\underline{A}$-3) expressing some form of data principle. The general pattern of these will be:

(2) $\forall \underline{S}\ \exists\alpha\ \alpha\in\underline{S}$

and (3) = $(3')_r$, where (3') is:

$$(3')\quad \mathrm{Ext}_{\alpha,\beta}\, A(\alpha,\beta) \rightarrow \forall\alpha\ \forall\beta[A(\alpha,\beta)\ \&\ \alpha\not\equiv\beta \rightarrow \exists\underline{S}_{\alpha\in\underline{S}}\ \forall\gamma_{\gamma\in\underline{S}}\ (\gamma\not\equiv\beta\rightarrow A(\gamma,\beta))].$$

As they stand (2) is obviously false and (3') trivial: for each particular class of choice sequences, the species-variable in (2) and (3') will be subject to appropriate restrictions (the same for the two axioms); thus, for the class $\underline{A}$ of lawless sequences, $\underline{S}$ was required to be of the form $\{\alpha | \alpha \in \vec{u}\}$ for some $\vec{u}$. In some cases, however, the axioms corresponding to ($\underline{A}$-2) and ($\underline{A}$-3) may have to take more complicated forms, not conforming to the simple patterns (2) and (3).

The general notion of a choice sequence does not rule out the possibility that the choice of the terms of the sequence is subject to *effective* restrictions laid down in advance. The idea of a spread-law was introduced precisely to embody the conception of such restrictions. A natural first attempt at formulating the general concept of a choice sequence would therefore be to identify choice sequences with those that are lawless elements of some spread. To obtain axioms for the lawless elements of a given spread s, we need to relativize '$\forall\vec{u}$' in ($\underline{A}$-2) to those $\vec{u}$ such that $s(\vec{u}) = 0$, and to relativize all choice-sequence quantifiers in ($\underline{A}$-2) and ($\underline{A}$-3) to elements of s. Of course, to consider, as forming the species of elements of a spread s, those choice sequences that are lawless elements of *some* spread (and are elements of s) is not the same as taking them to be lawless elements of s itself: some will be lawless elements of some proper subspread of s. As a limiting case, we shall have a choice sequence that is a ('lawless') element of some one-element spread: such a choice sequence will be a lawlike one -- one that is completely determined in advance by the spread-law.

The idea with which we are here working is that a choice sequence is generated by our first imposing, by means of a spread-law, some initial restriction on the subsequent choice of terms, and then proceeding to choose the terms in succession, in accordance with that initial restriction but otherwise freely. At one extreme, the initial restriction may be completely nugatory, the spread-law being that of the universal spread; in this case, we get an (absolutely) lawless sequence. At the other extreme, the initial restriction may fully determine the terms, the spread-law being that of a one-element spread; in this case, we get a lawlike sequence. In between lie all the intermediate cases.

Under this notion of a choice sequence (yielding a class $\underline{E}$ of choice sequences), there will exist, for each choice sequence α, a spread s_α of which α is a lawless element. An imperfect axiomatization of it would be obtained by specializing (2) and (3) so as to restrict $\underline{S}$ to be of the form $\{\alpha | \alpha \in s\}$ for some spread s. This would give us:

$$(\underline{B}\text{-}2) \qquad \forall s_{\,spr(s)} \exists \alpha \; \alpha \in s$$

and

$$(\underline{B}\text{-}3') \qquad \text{Ext}_{\alpha,\beta}\, A(\alpha,\beta) \rightarrow$$

$$\forall\alpha\ \forall\beta\ [A(\alpha,\beta)\ \&\ \alpha \not\equiv \beta \rightarrow$$

$$\exists s_{\,spr(s)}\ (\alpha \in s\ \&\ \forall\gamma_{\,\gamma\in s}\ (\gamma \not\equiv \beta \rightarrow A(\gamma,\beta)))],$$

($\underline{B}$-3) of course being $(\underline{B}\text{-}3')_r$.

Let us first inquire whether we can derive from these axioms anything corresponding to ($\underline{A}$-4). Suppose that $\alpha \not\equiv \beta$ but $\alpha = \beta$. Then, taking $A(\alpha,\beta)$ in ($\underline{B}$-3') as $\alpha = \beta$, we have, for some spread s to which α belongs, that $\gamma = \beta$ for every $\gamma \in s$. This means that s is a one-element spread, so that α coincides extensionally with a lawlike sequence. We therefore obtain:

$$(\underline{B}\text{-}4) \qquad \forall\alpha\ \forall\beta(\neg\exists f\ \alpha = f \rightarrow (\alpha \equiv \beta \leftrightarrow \alpha = \beta)),$$

where f ranges over lawlike sequences (constructive functions).

We are again unable to dispense with the condition that $\alpha \not\equiv \beta$ in ($\underline{B}$-3'); although no actual contradiction would ensue from dropping it, we should, by taking $A(\alpha,\beta)$ as $\alpha = \beta$, be able to prove that every choice sequence was extensionally equal to a lawlike one. Equally, we cannot dispense with the condition that $\gamma \not\equiv \beta$; for otherwise, by reasoning similar to that appealed to in the case of lawless sequences, we should, by taking A (α,β) in ($\underline{B}$-3') as $\alpha \not\equiv \beta$, obtain:

$$\forall\alpha\ \forall\beta(\alpha \not\equiv \beta \rightarrow \exists s_{spr(s)}\ (\alpha \in s\ \&\ \beta \notin s)),$$

and so, by ($\underline{B}$-4):

$$\forall\alpha\ \forall\beta(\neg\exists f\ \alpha = f\ \&\ \alpha \not\equiv \beta \rightarrow$$
$$\exists s_{spr(s)}\ (\alpha\in s\ \&\ \beta\notin s)),$$

which is intuitively unreasonable (if, e.g., α and β are intensionally distinct absolutely lawless sequences).

For an axiomatization which embodies more of the content of this notion of a choice sequence, we must employ the notation 's_α' for the spread selected at the outset of the process by which α is generated, governed by the axiom:

(B-0)* $\quad \forall\alpha\ (spr(s_\alpha)\ \&\ \alpha\in s_\alpha)$.

In place of (B-2) we can now assert:

(B-2)* $\quad \forall s_{spr(s)}\ \forall\vec{u}_{s(\vec{u})\ =\ 0}\ \exists\alpha_{\alpha\in\vec{u}}\ s \equiv s_\alpha$

and in place of (B-3'):

(B-3')* $\quad \text{Ext}_{\alpha,\beta}\ A(\alpha,\beta) \rightarrow$
$$\forall\alpha\ \forall\beta[A(\alpha,\beta)\ \&\ \alpha \not\equiv \beta \rightarrow$$
$$\exists n\ \forall\gamma_{\gamma\in s\ ,\gamma\in\bar{\alpha}(n)}\ (\gamma \not\equiv \beta \rightarrow A(\gamma,\ \beta))].$$

As usual, we shall actually assume (B-3)* = (B-3')*$_r$. (B-3*) is based on the idea that the truth, as applied to a choice sequence α, of any extensional predicate not involving reference to α must be derivable from knowing (i) some initial segment of α and (ii) the fact that α lies in the spread s_α. It should be noted that, because the predicate is extensional, we impose on the γ of (B-3')* the condition that $\gamma\in s$, rather than the stricter condition that $s_\gamma = s_\alpha$; i.e., we make the fact that the predicate is true of α turn, not on the fact that the initial restriction on the choice of terms of α was made by imposing the spread-law s_α, but merely on the fact, by whatever means we know it, that α is an element of s_α.

We may naturally define:

lawlike $(\alpha) \leftrightarrow \exists f\ \ \alpha \equiv f$.

Suppose that, for some n, $\forall\gamma_{\gamma\in s\ ,\ \gamma\in\bar{\alpha}(n)}\ \gamma = \alpha$. Then we may define a constructive function f such that

$$f(i) = \begin{cases} \alpha(i) & \text{for } i < n \\ \text{the unique } k \text{ such that } s_\alpha(\bar{f}(i)^\frown k) = 0 & \text{for } i \geq n; \end{cases}$$

and, moreover, it is intuitively clear that $\alpha \equiv f$. To drop the condition that $\alpha \not\equiv \beta$ in (B-3')* would enable us to assert that

$$\forall\alpha\ \exists n\ \forall\gamma_{\gamma\in s_\alpha,\ \gamma\in\bar{\alpha}(n)}\ \gamma = \alpha,$$

and hence that every choice sequence was lawlike, and this would formally contradict (B-2)*. Note that, using the notation 's_α', we can define

$$\text{lawless }(\alpha) \leftrightarrow \forall\beta\ \beta\in s_\alpha\ .$$

However, the class B of choice sequences proves to be unsatisfactory, because it again fails to be closed under continuous operations. By (B-2)*, we may choose α so that $s_\alpha \equiv b$, where b is the full binary spread. Now suppose that

$$\forall n\ \beta(n) = 1 \dot{-} \alpha(n),$$

so that $\beta \not\equiv \alpha$. By (B-3')*

$$\exists m\ \forall\gamma_{\gamma\in s_\alpha,\ \gamma\in\bar{\alpha}(m)}\ (\gamma \not\equiv \beta \rightarrow \forall n\ \beta(n) = 1 \dot{-} \gamma(n)),$$

i.e. for some $m > 0$

$$\forall\gamma_{\gamma\in b,\ \gamma\in\bar{\alpha}(m)}\ \gamma = \alpha.$$

By (B-2)*, however, we may choose γ such that $s_\gamma \equiv b$, $\gamma(n) = \alpha(n)$ for $n < m$ and $\gamma(m) = \beta(m)$, which is a contradiction. Thus $\beta \notin$ B if $\alpha \notin$ B.

Since one of our objects is to ensure that our choice sequences are closed under continuous operations, we may now try generalizing our notion of choice sequences to the class C of all those obtained from (absolutely) lawless sequences by such operations. For each r = 1, 2, 3, ... , let Γ^r range over continuous r-ary functionals from choice sequences to choice sequences, and let Γ range over all such continuous functionals (with any number of arguments). Then each choice sequence α in $\underline{C}$ is generated by first choosing a continuous functional Γ_α and, where Γ_α has r arguments, setting $\alpha \equiv \Gamma_\alpha(\alpha_0,\ldots,\alpha_{r-1})$ for specific lawless sequences $\alpha_0,\ldots,\alpha_{r-1}$. In order to fix $\alpha(n)$ for any given n, we shall need to generate a sufficiently large initial segment $\bar{\alpha}_i(m_{i,n})$ of each α_i to determine $(\Gamma_\alpha(\alpha_0,\ldots,\alpha_{r-1}))(n)$. This is a genuine generalization, since for any spread s it is easy to find Γ' such that $\Gamma'(\beta)\epsilon s$ for all β, and $\Gamma'(\alpha) = \alpha$ for all $\alpha\epsilon s$.

For each particular $r \geq 1$, let us write

$$\alpha\epsilon\Gamma^r \leftrightarrow \exists\beta_0\ldots\exists\beta_{r-1}\ \alpha = \Gamma^r(\beta_0,\ldots,\beta_{r-1}).$$

Let us also use '$\alpha\epsilon\Gamma$' in a similar sense. (Since we cannot actually define '$\alpha\epsilon\Gamma$', it must be regarded as a part of primitive notation, governed by denumerably many axioms of the form:

$$\forall\Gamma\ \forall\Gamma^r\ \forall\alpha\ (\Gamma = \Gamma^r \rightarrow (\alpha\epsilon\Gamma \leftrightarrow \alpha\epsilon\Gamma^r)).)$$

Then, as a first approach to axiomatizing this notion, we may adopt the specializations of (2) and (3') obtained by restricting $\underline{S}$ to species of the form $\{\alpha|\alpha\epsilon\Gamma\}$:

($\underline{C}$-2) $\quad \forall\Gamma\ \exists\alpha\ \alpha\epsilon\Gamma$

($\underline{C}$-3') $\quad \mathrm{Ext}_{\alpha,\beta}\ A(\alpha,\beta) \rightarrow$
$\forall\alpha\ \forall\beta[A(\alpha,\beta)\ \&\ \alpha \not\equiv \beta \rightarrow$
$\exists\Gamma_{\alpha\epsilon\Gamma}\ \forall\gamma_{\gamma\epsilon\Gamma}(\gamma \not\equiv \beta \rightarrow A(\gamma,\beta))]$.

We cannot prove from these axioms that the class of choice sequences is closed under continuous operations. We could, indeed, adopt this principle as an axiom, namely by replacing ($\underline{C}$-2) by the axiom schema:

($\underline{C}$-2)$^+$ $\quad \forall\Gamma^r\ \forall\beta_0\ldots\forall\beta_{r-1}\exists\alpha\ \alpha = \Gamma^r(\beta_0,\ldots,\beta_{r-1})$

for each specific $r \geq 1$ (or simply by allowing each expression of the form '$\Gamma^r(\beta_0,\ldots,\beta_{r-1})$' to count as a functor, so that

$(\underline{C}\text{-}2)^+$ becomes a truth of logic). This would then allow us to keep $(\underline{C}\text{-}3')$ as above, since, for each r, one continuous functional Π^r has the effect of intercalating the terms of r choice sequences: if $\beta = \Pi^r(\beta_0, \ldots \beta_{r-1})$, then for each $i < r$ and each n

$$\beta(nr + i) = \beta_i(n).$$

This, however, is simply to sidestep the problem of assuring ourselves intuitively that our class $\underline{C}$ really does serve the purpose for which it was introduced, that of being closed under continuous operations.

As a means of seeing this, the axioms $(\underline{C}\text{-}2)$ and $(\underline{C}\text{-}3')$ are too weak, and fail to capture enough of the intuitive notion. One idea for improving them might be to introduce the notation 'Γ_α' governed by the axiom:

$$(\underline{C}\text{-}0)^* \qquad \forall\alpha\ \alpha\in\Gamma_\alpha,$$

and to replace $(\underline{C}\text{-}2)$ and $(\underline{C}\text{-}3')$ by the analogues of $(\underline{B}\text{-}2)^*$ and $(\underline{B}\text{-}3')^*$:

$$(\underline{C}\text{-}2)^* \qquad \forall\Gamma^r\ \forall\vec{u}\ (\exists\beta_0\ldots\exists\beta_{r-1}\Gamma^r(\beta_0,\ldots,\beta_{r-1})\in\vec{u} \rightarrow \exists\alpha_{\alpha\in\vec{u}}\ \Gamma^r \equiv \Gamma_\alpha)$$

for each r, and

$$(\underline{C}\text{-}3')^* \qquad \text{Ext}_{\alpha,\beta}\ A(\alpha,\beta) \rightarrow \forall\alpha\ \forall\beta[A(\alpha,\beta)\ \&\ \alpha \not\equiv \beta \rightarrow \exists n\ \forall\gamma_{\gamma\in\Gamma_\alpha,\ \gamma\in\bar{\alpha}(n)}\ (\gamma \not\equiv \beta \rightarrow A(\gamma,\beta))].$$

Reasonable as these axioms at first appear, they would allow us to prove, by exactly parallel arguments to those used before, that choice sequences are *not* closed under continuous operations. For let Γ' be such that, for all γ and n,

$$(\Gamma'(\gamma))(n) = \begin{cases} 0 \text{ if } \gamma(n) = 0 \\ 1 \text{ if } \gamma(n) \geq 1. \end{cases}$$

By $(\underline{C}\text{-}2)^*$ we may choose α such that $\Gamma_\alpha \equiv \Gamma'$. As before, suppose

$$\forall n\ \beta(n) = 1 \dot{-} \alpha(n).$$

$\alpha \neq \beta$, so by ($\underline{C}$-3')* for some $m > 0$

$$\forall\gamma_{\gamma\in\Gamma',\ \gamma\in\bar{\alpha}(m)}\ \forall n\ \beta(n) = 1 \dot{-} \gamma(n),$$

and so $\gamma = \alpha$ for each $\gamma\in\Gamma'$, $\gamma\in\bar{\alpha}(m)$. But, as before, we may by ($\underline{C}$-2)* take γ such that $\Gamma_\gamma \equiv \Gamma'$, $\gamma\in\bar{\alpha}(m)$, $\gamma(m) = \beta(m)$, giving a contradiction.

What has gone wrong? Suppose that $\alpha \equiv \Gamma'(\gamma)$ for some lawless sequence γ. Then, where Δ' is such that

$$(\Delta'(\delta))(n) = \begin{cases} 1 \text{ if } \delta(n) = 0 \\ 0 \text{ if } \delta(n) \geq 1 \end{cases}$$

for every δ and n, by setting $\beta \equiv \Delta'(\gamma)$ for the *same* lawless sequence γ, we ensure that $\forall n\ \beta(n) = 1 \dot{-} \alpha(n)$. On the basis of what information about α do we know this last fact? ($\underline{C}$-3')* asserts that we know it just from the fact that $\alpha\in\Gamma_\alpha \equiv \Gamma'$ and from some initial segment $\bar{\alpha}(m)$ of α. But this is not so: we need to appeal also to the fact that α is generated by applying Γ' to the particular lawless sequence γ (i.e. to the very same lawless sequence from which β is generated by the application of Δ'). It might have seemed that since γ is lawless, and therefore may be continued in any way after the initial segment $\bar{\gamma}(k)$ needed to determine $\bar{\alpha}(m)$, and since the predicate was extensional, the individuality of γ made no difference; but this will be so only when the predicate makes no reference to any other choice predicate. We may therefore assert the specialization of ($\underline{C}$-3)* in the case when there are no parameters:

($\underline{C}$-3")* $\quad \mathrm{Ext}_\alpha\ A(\alpha) \rightarrow$
$$\forall\alpha(A(\alpha) \rightarrow \exists n\ \forall\gamma_{\gamma\in\Gamma_\alpha,\gamma\in\bar{\alpha}(n)}\ A(\gamma)),$$

but not ($\underline{C}$-3')*. For this reason, ($\underline{C}$-3') also becomes dubious, although its restriction

($\underline{C}$-3") $\quad \mathrm{Ext}_\alpha\ A(\alpha) \rightarrow \exists\Gamma_{\alpha\in\Gamma}\ \forall\gamma_{\gamma\in\Gamma}\ A(\gamma),$

which is essentially the principle of analytic data, is not in doubt.

Hence, to obtain a proper axiomatization, we must employ a more complex notation. We admit a predicate 'lawless (β)', assuming as axioms ($\underline{A}$-2) and ($\underline{A}$-3) with all choice-sequence variables restricted to the domain determined by this predicate. For each r we employ the intercalation operator Π^r cited above, and employ the symbols 'k_α', 'e_α', and, for variable i, '$\alpha^{(i)}$', governed by the axiom schema:

$$(\underline{C}\text{-}0)^{**} \quad \forall\alpha[k_\alpha \geq 1 \ \& \ e_\alpha \in \underline{K} \ \& \ \forall i \ \text{lawless}\ (\alpha^{(i)}) \ \& \ \forall i_{i \geq k_\alpha} \ \alpha^{(i)} \equiv \alpha^{(k\ )} \ \& \ \forall j_{j<k_\alpha} \ \forall i_{i<j} \ \alpha^{(i)} \not\equiv \alpha^{(j)} \ \& \ (k_\alpha = r \rightarrow \alpha \equiv e_\alpha | \Pi^r(\alpha^{(0)},\ldots,\alpha^{(r-1)}))].$$

As an existence axiom schema, we shall have:

$$(\underline{C}\text{-}2)^{**} \quad \forall e_{e \in \underline{K}} \ \forall\gamma\ldots\forall\gamma_{r-1} \ [\text{lawless}\ (\gamma_0) \ \& \ \ldots \ \& \ \text{lawless}\ (\gamma_{r-1}) \ \& \ \gamma_0 \not\equiv \gamma \ \ \& \ \gamma_0 \not\equiv \gamma_2 \ \& \ \ldots \ \& \ \gamma_{r-2} \not\equiv \gamma_{r-1} \rightarrow \exists\alpha\ (k_\alpha = r \ \& \ e_\alpha \equiv e \ \& \ \alpha^{(0)} \equiv \gamma_0 \ \& \ \ldots \ \& \ \alpha^{(r-1)} \equiv \gamma_{r-1})].$$

We do not need any separate data principle - the principle of open data for lawless sequences will supply all that we have a right to assume. Where, for each $i < k_\alpha$, $j < k_\beta$, $\alpha^{(i)} \not\equiv \beta^{(j)}$, then the truth of an extensional statement $A(\alpha,\beta)$ will depend, as far as α is concerned, only on knowing e_α and suitable initial segments of the $\alpha^{(i)}$; but if $\alpha^{(i)} \equiv \beta^{(j)}$, this will no longer be so.

The class $\underline{C}$ is evidently closed under continuous operations - it was devised for just this purpose - and we now have an axiomatization, albeit cumbersome, which accords with this fact; so we may turn to another intuitive requirement on the class of choice sequences, namely that it satisfy the Continuity Principle. Troelstra devotes considerable attention to the class $\underline{C}$, and succeeds in showing that the local $\forall\alpha\ \exists!n$-Continuity Principle, in the generalized form which allows an additional parameter in the predicate, holds for it, i.e. that we have:

$$\forall n\ \mathrm{Ext}_{\alpha,\gamma}\ A(\alpha,n,\gamma) \rightarrow$$

$$\forall\gamma[\forall\alpha\exists!n\ A(\alpha,n,\gamma) \rightarrow$$

$$\forall\alpha\ \exists n\ \exists m\ \forall\beta_{\beta\in\bar{\alpha}(m)}\ A(\beta,n,\gamma)].$$

He also shows, however, that the $\forall\alpha\ \exists n$-Continuity Principle does not hold. His counter-example is obtained by taking $A(\alpha, n)$ as

$$\exists\beta\ (k_\beta = n\ \&\ \alpha = \beta).$$

$A(\alpha,n)$ is obviously extensional, and we have $\forall\alpha\ \exists n\ A(\alpha,n)$. Now suppose, as the Continuity Principle would require, that for given α and specific n and m we have:

$$\forall\gamma_{\gamma\in\bar{\alpha}(m)}\ \exists\beta(k_\beta = n\ \&\ \gamma = \beta).$$

Suppose also that $\delta_0,\ldots,\delta_n$ are lawless sequences such that $\delta_i \not\equiv \delta_j$ for $i < j \leq n$. Put $\varepsilon = \Pi^{n+1}(\delta_0,\ldots,\delta_n)$, and take $e \in \underline{K}$ such that

(i) $e(\langle k\rangle) = \alpha(k)+1$ for each $k < m$

(ii) $e(\langle m+(n+1)j+i\rangle * \bar{\varepsilon}((n+1)j+i)) = \delta_i(j) + 1$ for each $i \leq n$ and each j.

Put $\gamma = e|\varepsilon$. The purpose of (i) is to ensure that $\gamma\in\bar{\alpha}(m)$, and that of (ii) to ensure that for every $i \leq n$ and every j, some value of γ depends on $\delta_i(j)$. Since $\gamma\in\bar{\alpha}(m)$, there exists β with $k_\beta = n$ such that

$$\gamma = \beta \equiv e_\beta|\Pi^n(\beta^{(0)},\ldots,\beta^{(n-1)}).$$

Since $\delta_0,\ldots,\delta_n$ are all distinct, there is some $i \leq n$ such that δ_i is distinct from each $\beta^{(j)}$ $(j < n)$. But it now follows by an application of ($\underline{A}$-3) to δ_i that the truth of

$$e|\Pi^{n+1}(\delta_0,\ldots,\delta_n) = \beta$$

depends only on some initial segment $\overline{\delta_i}(q)$ of δ_i. Hence, where $\delta'(n) = \delta_i(n)$ for $n < q$ and $\delta'(q) = \delta_i(q) + 1$, we again have:

$$e|\Pi^{n+1}(\delta_0,\ldots,\delta_{i-1},\delta',\delta_{i+1},\ldots,\delta_n) = \beta_j$$

but, in view of (ii) above, this is absurd.

We thus need a further idea if we are to attain a satisfactory notion of a choice sequence. Such an idea, deriving originally from Brouwer himself, was first injected into recent discussion of the notion by Myhill. According to this, we should think of a choice sequence, not as determined by some initial restriction, followed by free choices of terms within the limits imposed by that restriction, but as determined by choices subject to successive restrictions which may be freely introduced at any stage. If we are regarding all restrictions on future choices as effected by means of a spread-law, this means that any choice sequence α is given by means of a sequence of free choices of ordered pairs $\langle s_n^{(\alpha)}, \alpha(\underline{n})\rangle$ subject to the general conditions:

$$\mathrm{spr}(s_n^{(\alpha)}) \ \&\ s_n^{(\alpha)} \subseteq s_{n\dot{-}1}^{(\alpha)} \ \&\ s_n^{(\alpha)}(\bar{\alpha}(n+1)) = 0.$$

but otherwise lawless. As Troelstra convincingly argues, this is precisely the notion of choice sequences on which Brouwer finally settled; let us call this class of choice sequences $\underline{D}$. It clearly justifies axiom ($\underline{B}$-2), for which we can therefore also adopt the label ($\underline{D}$-2). (In any system which admits a λ-operator forming functors for choice sequences, so weak an axiom as ($\underline{D}$-2) will be easily provable.)

At any given time, all that we can know of a choice sequence α in $\underline{D}$ is some initial segment

$$\langle s_0^{(\alpha)}, \alpha(0)\rangle, \ \ldots\ , \langle s_{n-1}^{(\alpha)}, \alpha(n-1)\rangle$$

of the ordered pairs of spreads and of terms of α, where $s_{n-1}^{(\alpha)}(\bar{\alpha}(n)) = 0$ and $s_{n-1}^{(\alpha)} \subseteq s_{n-2}^{(\alpha)} \subseteq \cdots \subseteq s_1^{(\alpha)} \subseteq s_0^{(\alpha)}$. In so far as our knowledge bears on the *extension* of α, this amounts to knowing $\bar{\alpha}(n)$ and the fact that $\alpha \in s_{n-1}^{(\alpha)}$: we therefore have a justification of axiom schema ($\underline{B}$-3), which we may accordingly label ($\underline{D}$-3) The specialization of ($\underline{D}$-3) to the case when the predicate contains no choice-sequence parameters:

($\underline{D}$-3'') $$\mathrm{Ext}_{\alpha}\, A(\alpha) \rightarrow$$
$$\forall\alpha[A(\alpha) \rightarrow \exists s_{spr(s)}\ (\alpha\in s\ \&\ \forall\gamma_{\gamma\in s}\ A(\alpha))],$$

called (SD) in section 6.1, was at one time proposed by Kreisel as an intuitively acceptable axiom schema for the general notion of a choice sequence.

It was, however, shown by Troelstra that, when taken together with the $\forall\alpha\exists f$-Continuity Principle (f ranging over constructive functions), ($\underline{D}$-3'') leads to a result intuitively questionable in itself and in conflict with Church's Thesis (for constructive functions). Moreover, and yet more damagingly, the class $\underline{D}$ again fails to be closed under continuous operations, as can be seen by exactly the same argument as we used for the class $\underline{B}$. We cannot, indeed, prove from ($\underline{D}$-2) and ($\underline{D}$-3) that choice sequences are *not* closed under continuous operations, since these axioms do not fully express the notion of a choice sequence belonging to $\underline{D}$. When spr(s) and $s(\vec{u}) = 0$, let us write $s[\vec{u}]$ for the spread s' such that

$$s'(\vec{v}) = 0 \leftrightarrow s(\vec{v}) = 0\ \&\ (\vec{u} \preccurlyeq \vec{v} \vee \vec{v} \preccurlyeq \vec{u}).$$

Then it is not merely that for each spread s there exists an element α of s in $\underline{D}$, but that, for each s, there exists in $\underline{D}$ an $\alpha\in s$ such that, for any extensional predicate A( ) not involving reference to α, the truth of $A(\alpha)$ will imply that, for some n, $s[\bar{\alpha}(n)]$ will play the role of the spread asserted to exist by ($\underline{D}$-3), i.e. that $A(\gamma)$ for all $\gamma\in s[\bar{\alpha}(n)]$. We could get this effect by explicitly introducing the notation we used informally earlier, writing $s_n^{(\alpha)}$ for the spread chosen at the n-th stage to restrict future choices of terms for α, and adopting axioms ($\underline{D}$-2)* and ($\underline{D}$-3)*, expressed in this notation, corresponding to (not, of course, identical with) ($\underline{B}$-2)* and ($\underline{B}$-3)*: we should then be able to prove that our class of choice sequences was not closed under continuous operations. Troelstra is therefore justified in remarking that Brouwer's own concept of a choice sequence did not satisfy the demands he made on it.

In order to improve on the notion of choice sequences as comprising the class D, Troelstra therefore proposes the natural step of considering a class E related to D as C was related to B. Instead of simply choosing at the outset a certain continuous operation, and generating quite freely the lawless sequences to which it is to be applied, we can now, at any stage in the process, impose a like condition upon any one or more of the auxiliary sequences which we are generating. Suppose, in order to simplify the statement of the matter, that we are concerned only with continuous operations with just one choice sequence as argument. At the n-th stage in generating a choice sequence α we do two things: we choose a finite set $R_n^{(\alpha)}$ of conditions of the form specified below, subject only to the restriction that $\underline{R}_m^{(\alpha)} \subseteq \underline{R}_n^{(\alpha)}$ for $m \leq n$; and we determine $\alpha(n)$ in accordance with $\underline{R}_n^{(\alpha)}$. Each $\underline{R}_n^{(\alpha)}$ is either empty or consists of conditions of the form

$$\alpha = \Gamma_o(\alpha_o)$$

$$\alpha_o = \Gamma_1(\alpha_1)$$

$$\ldots\ldots\ldots\ldots\ldots$$

$$\alpha_{j-1} = \Gamma_j(\alpha_j)$$

for some j, where each Γ_i is a continuous operation. (We may therefore regard each Γ_i as represented by a neighbourhood function e_i, so that $\alpha = e_o \mid \alpha_o$ and $\alpha_{i-1} = e_i \mid \alpha_i$ for each i.) Actually, as already remarked, the conditions will take a more general form, since each continuous operation may take any number of choice sequences, unmentioned in preceding conditions, as arguments. When $\underline{R}_n^{(\alpha)}$ is empty, $\alpha(n)$ may be freely chosen; otherwise it will be determined by some initial segment of α_o, which (if $\underline{R}_n^{(\alpha)}$ contains at least two conditions) will in turn be determined by an initial segment of α_1, and so on until we reach an α_j as yet undetermined by any condition, whose terms we may freely choose.

How are we to understand this? There is, first, a difficulty about the consistency of the conditions $\underline{R}_n^{(\alpha)}$ with the determination of the segment $\bar{\alpha}(n)$. Suppose, for example, that $n > 0$ and that $\underline{R}_{n-1}^{(\alpha)}$ is empty; $\alpha(0),\ldots,\alpha(n-1)$ have therefore been freely chosen. At stage n, we take $\underline{R}_n^{(\alpha)}$ to consist of the single condition $\alpha = \Gamma_o(\alpha_o)$. Given a suitable initial segment of α_o, this then determines $\alpha(n)$: but how can we be sure that $(\Gamma_o(\alpha_o))(i) = \alpha(i)$ for each $i < n$? Unless we already know a sufficiently long initial segment of α_o, we cannot be sure of this, and, in any case, it will not in general hold for an arbitrary choice of Γ_o. The best thing is to impose it as a restriction on the selection of Γ_o that, for any β, $(\Gamma_o(\beta))(i) = \alpha(i)$ for each $i < n$; e.g., where Γ_o is represented by e_o, to require that $e_o(\langle i\rangle) = \alpha(i) + 1$ for $i < n$.

But the more difficult problem concerns how we are to regard the auxiliary choice sequences $\alpha_o, \alpha_1, \ldots$. Clearly they are to be thought of as also being choice sequences in their own right; but the question remains whether the process of generating each α_i should be regarded as an ingredient in that of generating α or as extraneous to it. The former of these interpretations yields the following picture. As long as $\underline{R}_n^{(\alpha)}$ is empty, we choose $\alpha(n)$ freely. If, however, $\underline{R}_{n-1}^{(\alpha)}$ is empty, but we take $\underline{R}_n^{(\alpha)}$ as consisting of the condition that $\alpha = \Gamma_o(\alpha_o)$, then, in order to complete stage n by determining $\alpha(n)$, we have to determine a sufficiently long initial segment $\bar{\alpha}_o(m_n)$ of α_o to be able to derive the value of $\alpha(n)$. Since, at this stage, the choice of terms for α_o is as yet subject to no restriction, we shall do this by making free choices of $\alpha_o(0),\ldots,\alpha_o(m_n-1)$. In the same way, as long as $\underline{R}_r^{(\alpha)} = \underline{R}_n^{(\alpha)}$ for $r > n$, we determine $\alpha(r)$ from some initial segment $\bar{\alpha}_o(m_r)$ of α_o, if necessary making further free choices of terms of α_o in order to do so. Suppose, however, that, for $r > n$, $\underline{R}_r^{(\alpha)}$ consists of the conditions $\alpha = \Gamma_o(\alpha_o)$, $\alpha_o = \Gamma_1(\alpha_1)$, while $\underline{R}_{r-1}^{(\alpha)} = \underline{R}_n^{(\alpha)}$. The requirement for the consistency of the new condition is that

$$(\Gamma_1(\alpha_1))(i) = \alpha_o(i) \text{ for each } i < m,$$

where $m = \max(m_n, m_{n+1}, \ldots, m_{r-1})$, so we again suppose the selection of Γ_1 subject to the restriction that $(\Gamma_1(\beta))(i) = \alpha_o(i)$ for any β and for $i < m$. In order to complete stage r by fixing the value of $\alpha(r)$, we have again to find a sufficiently long segment $\overline{\alpha}_o(m_r)$ of α_o; but, where $m_r > m$, we may no longer choose the terms $\alpha_o(m),\ldots,\alpha_o(m_r-1)$ freely, but must determine them from some initial segment $\overline{\alpha}_1(q)$ of α_1. Since α_1 is as yet subject to no restrictions, we freely choose terms $\alpha_1(0),\ldots,\alpha_1(q-1)$ for this purpose.

This is a very natural, and readily intelligible, picture. The alternative is to think of the process of generating the α_i as not itself forming part of that of generating α, but as taking place independently: the conditions which refer to the α_i relate to choice sequences whose identity is not fixed by the conditions themselves, but has, as it were, been antecedently established. In this case, whenever, at some stage n, we need to know new terms of α_i in order to determine terms of α_{i-1}, and so terms of $\alpha_{i-2},\ldots$, and so, ultimately, $\alpha(n)$, we are not in a position to choose the new terms of α_i as we wish, but must simply wait for them to come in from outside before we can complete stage n. In effect, this was the picture which was forced on us, in considering the class $\underline{C}$; in order to construe *it* as closed under continuous operations, we had to allow that the same lawless sequence could enter into the process of generating different choice sequences, and could not therefore be regarded as purely internal to either process. However, as applied to our present class $\underline{E}$, this second picture does not appear to be really comprehensible, since α_i is not necessarily a lawless sequence. If α_i is being generated by some process quite external to that by which α is generated, then we are not at liberty, in the course of generating α, to impose a condition $\alpha_i = \Gamma_{i+1}(\alpha_{i+1})$: whether or not such a condition holds is determined by those means, whatever they may be, that are being used to generate α_i.

Yet it seems to be to this second picture that Troelstra appeals when he claims that the class $\underline{E}$ is closed under continuous operations. He argues as follows. Suppose that $\beta = \Gamma(\alpha)$, where α is generated by means of the sequence $\langle \underline{R}_0^{(\alpha)}, \alpha(0)\rangle, \langle \underline{R}_1^{(\alpha)}, \alpha(1)\rangle, \ldots$. Then β can be regarded as generated by means of the sequence

$$\langle \underline{R}_0^{(\beta)}, \beta(0)\rangle, \langle \underline{R}_1^{(\beta)}, \beta(1)\rangle, \ldots ,$$

where each $\underline{R}_0^{(\beta)}$ consists of $\beta = \Gamma(\alpha)$ together with the conditions in $\underline{R}_n^{(\alpha)}$. For this argument to work, it has to be possible, in generating the choice sequence β, to impose a condition $\beta = \Gamma(\alpha)$ relating to a choice sequence α whose identity is already fixed by reference to a process of generation already going on. In what sense, then, are we free to choose the terms of the sequence $\langle \underline{R}_0^{(\beta)}, \beta(0)\rangle, \langle \underline{R}_1^{(\beta)}, \beta(1)\rangle, \ldots$? What would happen if we did not include each $\underline{R}_n^{(\alpha)}$ in $\underline{R}_n^{(\beta)}$? Suppose, e.g., that $\underline{R}_0^{(\alpha)}$ consists of $\alpha = \Gamma_0(\alpha_0)$ and $\underline{R}_1^{(\alpha)}$ of it and $\alpha_0 = \Gamma_1(\alpha_1)$, but that, while we take $\underline{R}_0^{(\beta)}$ to consist of $\beta = \Gamma(\alpha)$ and $\alpha = \Gamma_0(\alpha_0)$, we take $\underline{R}_1^{(\beta)}$ as consisting of those together with, say, $\alpha_0 = \Delta(\gamma)$ (where $\gamma \neq \alpha_1$, $\Delta \neq \Gamma_1$). The question may be rejected as absurd; but, if so, its absurdity reflects only the incoherence of this conception of how a choice sequence may be generated. If the question has an answer, it can only be that, in such a case, we should get an inconsistency. To avoid such a possibility, we should need to impose restrictions on our choice of conditions $\underline{R}_n^{(\alpha)}$ which would depend on knowing which stage we have reached in the generation of any choice sequence when we are at any given stage in the generation of another. This would in effect be to acknowledge only one all-embracing process of generation, whereby all choice sequences whatever would be simultaneously generated. This would be to abandon the conception under which we may, in the course of generating one choice sequence, refer to another generated by a disjoint process, in favour of an extreme version of the opposed conception.

No problem arises, of course, over the justification of

$$(\underline{E}\text{-}2') \qquad \forall e_{e \in \underline{K}}\ \exists \alpha\ \alpha \in e,$$

where '$\alpha \in e$' abbreviates '$\exists \beta\ \alpha = e | \beta$'. (Nor, where we write '$(e \circ \Pi^r)(\beta_0, \ldots, \beta_{r-1})$, for '$e | \Pi^r(\beta_0, \ldots, \beta_{r-1})$,' would one arise for the axiom schema

$$(\underline{E}\text{-}2) \qquad \forall e_{e \in \underline{K}}\ \exists \alpha\ \alpha \in (e \circ \Pi^r),$$

which has exactly the same content as ($\underline{C}$-2).) Nor can there be any problem over the principle of analytic data:

$$(\underline{E}\text{-}3'') \qquad \mathrm{Ext}_\alpha\ A(\alpha) \rightarrow$$

$$\forall \alpha [A(\alpha) \rightarrow \exists e_{e \in \underline{K}}\ \forall \gamma_{\gamma \in e}\ A(\gamma)].$$

Recalling the doubt that was raised over ($\underline{C}$-3), however, we ought not to assert the generalization of ($\underline{E}$-3'') to the form ($\underline{E}$-3) with parameters in the predicate. But the same problem as arose for closure under continuous operations arises also for Troelstra's argument in favour of the $\forall \alpha\ \exists n$-Continuity Principle. The argument runs as follows. Suppose $A(\alpha, n)$ is extensional and $\forall \alpha\ \exists n\ A(\alpha, n)$. Then by the Axiom of Choice there exists an operation Φ such that $\forall \alpha\ A(\alpha, \Phi(\alpha))$. Φ may depend on intensional features of α, i.e. upon the way in which α is generated. On the other hand, since the process of generation can never be completed, $\Phi(\alpha)$ can depend only upon an initial segment of this process (though not necessarily only on the *outcome* of this segment): that is, for any given α, $\Phi(\alpha)$ will, for some m, be determined by

$$\langle \underline{R}_0^{(\alpha)}, \alpha(0) \rangle, \langle \underline{R}_1^{(\alpha)}, \alpha(1) \rangle, \ldots, \langle \underline{R}_{m-1}^{(\alpha)}, \alpha(m-1) \rangle \ .$$

In particular, for those β such that $\underline{R}_n^{(\beta)}$ is empty for all n, Φ must act on them like a continuous operation Γ, representable by a neighbourhood function e: for all such β, $\Phi(\beta) = e(\beta)$.

So far, the argument is irreproachable. Troelstra now claims that for *any* α we have $A(\alpha, e(\alpha))$. (Since we do not have a unique n such that $A(\alpha, n)$, it is *not* claimed that $\Phi(\alpha) = e(\alpha)$ for every α.) The argument proceeds as follows.

Consider any particular α, and suppose that $e(\bar{\alpha}(k)) > 0$. This means that for any $\gamma \in \bar{\alpha}(k)$ such that $\underline{R}_i^{(\gamma)}$ is empty for each $i < k$, $\Phi(\gamma)$ is determined as $e(\alpha) = e(\bar{\alpha}(k)) - 1$. There will then exist $m > k$ such that $\Phi(\alpha)$ is determined by the first m pairs $\langle \underline{R}_i^{(\alpha)}, \alpha(i) \rangle$, as above. If $\underline{R}_{m-1}^{(\alpha)}$ is empty, then $\Phi(\alpha) = e(\alpha)$. If $\underline{R}_{m-1}^{(\alpha)}$ consists of $\alpha = \Gamma_o(\alpha_o)$, $\alpha_o = \Gamma_1(\alpha_1)$, ... , $\alpha_{j-1} = \Gamma_j(\alpha_j)$, then consider $\gamma \in \bar{\alpha}(k + 1)$ such that $\underline{R}_i^{(\alpha)}$ is empty for $i < k$ and $\underline{R}_k^{(\gamma)}$ consists of $\gamma = \Gamma_o(\alpha_o)$, $\alpha_o = \Gamma_1(\alpha_1)$, ... , $\alpha_{j-1} = \Gamma_j(\alpha_j)$. Then $\Phi(\gamma)$ is $e(\alpha)$, and we have $A(\gamma, e(\alpha))$. But $\underline{R}_k^{(\gamma)}$ and $\underline{R}_{m-1}^{(\alpha)}$ require that $\alpha_{j-1} = \Gamma_j(\alpha_j)$, $\alpha_{j-2} = \Gamma_{j-1}(\alpha_{j-1})$, ... , $\alpha_o = \Gamma_1(\alpha_1)$, $\alpha = \Gamma_o(\alpha_o)$ and $\gamma = \Gamma_o(\alpha_o)$. Hence $\alpha = \gamma$, and, since $A(\alpha, n)$ is extensional, it follows that $A(\alpha, e(\alpha))$.

The last part of this argument depends upon assuming that the choice sequence α_j referred to in the condition $\underline{R}_k^{(\gamma)}$ is the same as the choice sequence α_j referred to in the condition $\underline{R}_{m-1}^{(\alpha)}$; that is, that not only can we, when imposing conditions in the course of generating a choice sequence, refer to specific other choice sequences (whether lawless or not), but that the same specific choice sequence may be referred to in the course of generating distinct choice sequences α and γ. If this is so, then it is not apparent why it was not sufficient to take $\underline{R}_k^{(\gamma)}$ as consisting simply of $\gamma = \Gamma_o(\alpha_o)$: the other conditions determining the generation of α_o would, as it were, be provided for extraneously. Suppose that, by the time we selected $\langle \underline{R}_k^{(\gamma)}, \gamma(k) \rangle$, $\underline{R}_m^{(\alpha)}$ had already been chosen, and contained the condition $\alpha_j = \Gamma_{j+1}(\alpha_{j+1})$: we should not then be free to select $\underline{R}_{k+1}^{(\gamma)}$ in any way we pleased. It is essential for the argument that the conditions imposed in generating distinct choice sequences should not be independent of one another; but, for that, we need some elaborate machinery to guarantee their consistency with one another, a machinery which is entirely lacking.

The resolution of this difficulty is not hard, however, and allows us to retain, in essence, Troelstra's notion of choice sequences and to validate his arguments for their closure under continuous operations and for the $\forall\alpha\,\exists n$-Continuity Principle; we have to do no more than combine the two interpretations of the notion. We assume that various disjoint processes of generation ϕ, χ, ψ, ... take place. Each such process ϕ is concerned primarily to determine a *principal* choice sequence ϕ_o: but, in the course of generating ϕ_o, other, *subordinate*, choice sequences ϕ_1, ϕ_2, ... may be generated. Each choice sequence is generated by just *one* of the various disjoint generating processes: it is principal if it is determined only by the process taken as a whole, subordinate if it is determined by some proper part of the process. As before, for each n, we have, at stage n of any generating process ϕ, first to impose a finite set of conditions $\underline{R}_n^{(\phi)}$ and then to determine $\phi_o(n)$. Each subordinate choice sequence ϕ_{i+1} will be *introduced* by some condition in $\underline{R}_r^{(\phi)}$ for some r, namely the first condition in whose right-hand side it is mentioned; it is *determined* by any condition on whose left-hand side it appears. In order to determine $\phi_o(n)$, we may need to determine certain terms $\phi_{i+1}(j)$ of those subordinate choice sequences ϕ_{i+1} which have been introduced by some $\underline{R}_m^{(\phi)}$ $(m \leq n)$: for any ϕ_{i+1} as yet undetermined by any condition, we choose the required terms $\phi_{i+1}(j)$ freely. Likewise, if ϕ_o itself is as yet undetermined (i.e. if $\underline{R}_n^{(\phi)}$ is empty), we choose $\phi_o(n)$ freely. As before, the set $\underline{R}_n^{(\phi)}$ must include $\underline{R}_m^{(\phi)}$ for $m < n$. Each condition determines either ϕ_o itself or a subordinate choice sequence ϕ_{i+1} as the result of applying a continuous operation to other choice sequences: these other choice sequences may either be newly introduced subordinate choice sequences or be choice sequences, whether principal or subordinate, generated not by ϕ but by some disjoint process χ. Such extraneous choice sequences, mentioned in some condition in some $\underline{R}_r^{(\phi)}$, will be called *auxiliary* choice sequences (with respect to the process ϕ). No choice sequence ϕ_i can be determined more than one condition; but no auxiliary choice sequence can be determined by any condition occurring in any $\underline{R}_r^{(\phi)}$. The general form of a condition is, therefore:

$$\phi_i = \Gamma_i(\phi_{i_0}, \ldots, \phi_{i_{r-1}}, \beta_0, \ldots, \beta_{q-1}),$$

where either r or q (but not both) may be 0, and each β_j is an auxiliary choice sequence determined by some disjoint generating process. If the condition occurs for the first time in $\underline{R}_n^{(\phi)}$, then either i = 0 or ϕ_i was previously introduced, i.e. was mentioned in some condition in $\underline{R}_m^{(\phi)}$ for some $m < n$. For each $j < r$, ϕ_{i_j} may be newly introduced (occur in no condition belonging to $\underline{R}_m^{(\phi)}$ for $m < n$) or may have been previously introduced but not yet determined (strictly speaking, we ought, for this purpose, to consider the $R_n^{(\phi)}$ as ordered sets): we cannot determine ϕ_i as the result of applying Γ_i to a choice sequence already determined. (This proviso is to avoid circularity.) Γ_i must satisfy the constraint that, for each term $\phi_i(j)$ of ϕ_i that has already been determined, $(\Gamma_i(\gamma_0,\ldots,\gamma_{r+q-1}))(j) = \phi_i(j)$ for all $\gamma_0,\ldots,\gamma_{r+q-1}$. When, in order to determine $\phi_0(n)$, we need to know some terms of one or more of the auxiliary choice sequences, the determination of these is no part of the process ϕ: we have to wait to be supplied with them from outside until we can pass on to stage n + 1. We may or may not reach a stage when ϕ_0 itself and every subordinate choice sequence ϕ_{i+1} that has been introduced has also been determined, that is to say, ultimately in terms of the auxiliary choice sequences: if we do, then we are no longer at liberty to introduce any new conditions. Conditions determining the auxiliary choice sequences cannot play any part in the process ϕ, and thus cannot belong to any $\underline{R}_n^{(\phi)}$.

Viewed in this way, we have a coherent notion of choice sequences, essentially that intended by Troelstra, for which his arguments for closure under continuous operations and for the $\forall\alpha\ \exists n$-Continuity Principle will go through. If β is any choice sequence, generated by a process χ, and $\alpha = \Gamma(\beta)$, then α will be generated as the principal choice sequence ϕ_0 by a process ϕ in which $\underline{R}_0^{(\phi)}$ is taken as consisting of $\phi_0 = \Gamma(\beta)$: after this choice, we have no further liberty in determining the process ϕ, so that $\underline{R}_n^{(\phi)} = \underline{R}_0^{(\phi)}$ for each n.

In a similar way, the argument for the Continuity Principle goes through as intended. If $\forall\alpha\, \exists n\, A(\alpha, n)$, then, as before, for some Φ, $\forall\alpha\, A(\alpha, \Phi(\alpha))$. If $A(\alpha, n)$ is extensional, there will be a continuous operation Γ such that, for every β determined by a process χ for which $\underline{R}_n^{(\chi)}$ is empty for every n, $\Phi(\beta) = \Gamma(\beta)$. Let Γ be represented by the neighbourhood function $e \in \underline{K}$, and suppose that $\alpha \equiv \phi_o$ (i.e. that α is determined by the process ϕ) and that $e(\bar{\alpha}(k)) > 0$. We then have that, for every process ψ such that $\psi_o \in \bar{\alpha}(k)$ and $\underline{R}_i^{(\psi)}$ is empty for each $i < k$, $\Phi(\psi_o) = e(\alpha)$. If $\underline{R}_{k-1}^{(\phi)}$ is empty, then ϕ is such a ψ, and so $\Phi(\alpha) = e(\alpha)$ and $A(\alpha, e(\alpha))$. If $\underline{R}_{k-1}^{(\Phi)}$ contains $\phi_o = \Gamma_o(\phi_1)$ or $\phi_o = \Gamma_o(\beta)$ (where β is auxiliary to ϕ), we consider a process ψ such that, for $i < k$, $\underline{R}_i^{(\psi)}$ is empty and $\psi_o(i) = \alpha(i)$, while $\underline{R}_k^{(\psi)}$ consists of $\psi_o = \Gamma_o(\phi_1)$ or of $\psi_o = \Gamma_o(\beta)$ respectively (ϕ_1 or β will be auxiliary to ψ, which is completely specified by these stipulations). Then $\Phi(\psi_o) = e(\alpha)$ and $A(\psi_o, e(\alpha))$. But since $A(\alpha, n)$ is extensional and $\psi_o = \phi_o \equiv \alpha$ (in the one case $\psi_o = \Gamma_o(\phi_1) = \phi_o$, in the other $\psi_o = \Gamma_o(\beta) = \phi_o$), $A(\alpha, e(\alpha))$.

It is worth while to analyse this argument in a little more detail. Some of its premisses are independent of the particular notion of a choice sequence being used, save that it comprehends lawless sequences, among others; without making any further assumptions about how choice sequences are generated, we may assert:

(A) $\forall\alpha\, \forall k\, \exists\beta_{\text{lawless}(\beta)}\; \beta \in \bar{\alpha}(k)$;

(B) if $\forall\alpha\, \exists n\, A(\alpha, n)$, then, for some Φ, $\forall\alpha\, A(\alpha, \Phi(\alpha))$;

(C) if $\forall n\, \text{Ext}_\alpha\, A(\alpha, n)$, and $\forall\alpha\, A(\alpha, \Phi(\alpha))$, then $\exists e_{e \in \underline{K}}\, \forall\beta_{\text{lawless}(\beta)}\; \Phi(\beta) = e(\beta)$.

Other premisses depend upon supposing that each choice sequence α is given by some process of generation ϕ_α, but are independent of any assumption about the particular nature of this generating process. Let us assume that ϕ_α completely determines the intensional features, and thus the identity, of α. We denote the initial segment of ϕ_α, consisting of the first k stages, by $\overline{\phi_\alpha}(k)$, and assume that $\overline{\phi_\alpha}(k)$ determines $\bar{\alpha}(k)$. We are then entitled to suppose that, for each α, $\Phi(\alpha)$ will be determined from some finite amount of information about the generating process ϕ_α, i.e. by $\overline{\phi_\alpha}(k)$ for some k, and we express this by writing '$\Phi(\alpha)$ det. by $\overline{\phi_\alpha}(k)$'. We may then assert:

(D) for any Φ, $\forall\alpha\, \exists k\, \Phi(\alpha)$ det. by $\overline{\phi_\alpha}(k)$;

(E) if $\Phi(\alpha)$ det. by $\overline{\phi_\alpha}(k)$ and $\overline{\phi_\beta}(k) = \overline{\phi_\alpha}(k)$, then $\Phi(\beta) = \Phi(\alpha)$;

(F) if $e \in \underline{K}$ and $\forall\beta_{\text{lawless}(\beta)}\ \Phi(\beta) = e(\beta)$, and $e(\bar{\alpha}(k)) > 0$, then $\forall\beta_{\text{lawless}(\beta),\beta \in \bar{\alpha}(k)}\ \Phi(\beta)$ det. by $\overline{\phi_\beta}(k)$.

If we have a suitable formulation of the notion of a possible initial segment $\vec{h}$ of a generating process, we shall also be prepared to assert:

(G) $\forall\vec{h}\, \exists\alpha\, \phi_\alpha \in \vec{h}$.

From (A) - (F) we can derive:

(H) if $\forall n\, \text{Ext}_\alpha\, A(\alpha, n)$ and $\forall\alpha\, \exists n\, A(\alpha, n)$, then there exist Φ and $e \in \underline{K}$ such that $\forall\alpha\, A(\alpha, \Phi(\alpha))$ and, if $e(\bar{\alpha}(k)) > 0$, then for all $\gamma \in \bar{\alpha}(k)$ such that for some lawless β $\overline{\phi_\gamma}(k) = \overline{\phi_\beta}(k)$, $\Phi(\gamma) = e(\alpha)$.

This is as far as we can take the argument without making any assumptions about the generating processes ϕ_α. To take it further, it is essential that we should not suppose that all restrictions on the subsequent choices of terms of α have to be made at the outset of the generating process, but recognize that whatever restrictions it is possible to impose may be imposed, successively, at any stage in the process. If we make this assumption, then, where we write $\phi_\alpha(k)$ for the k-th stage of ϕ_α, we shall be able to assert, as a special case of (G):

$$\text{(I)} \quad \forall k\ \forall\alpha\ \exists\gamma_{\gamma\in\bar{\alpha}(k)}\ \exists\beta_{\text{lawless}(\beta)}\ (\overline{\phi_\gamma}(k) = \overline{\phi_\beta}(k)\ \&\ \phi_\gamma(k) = \phi_\alpha(k-1)).$$

(I) is, however, still not enough for us to be able to complete the argument; for that we need the stronger assumption:

$$\text{(J)} \quad \forall k\ \forall\alpha\ \exists\gamma\ (\gamma = \alpha\ \&\ \exists\beta_{\text{lawless}(\beta)}\ \overline{\phi_\beta}(k) = \overline{\phi_\gamma}(k)).$$

The validity of (J) does not depend merely on the supposition that we may, in the course of the generating process ϕ_α, impose successively more severe restrictions, at different stages, on the subsequent choice of terms of α; in the argument as set out above, it was derived from (I) by quite special considerations about the formulation of those restrictions. For class $\underline{D}$, where the successive restrictions were expressed by spread-laws, we should not (at least in this way) be able to justify (J). For an α in $\underline{D}$, the process of generating α consisted in the choice of pairs $\langle s_i^{(\alpha)}, \alpha(i)\rangle$ such that $s_n^{(\alpha)} \subseteq s_m^{(\alpha)}$ for $n > m$, and $s_n^{(\alpha)}(\bar{\alpha}(n+1)) = 0$ for every n. Where e is as in (F), to say that $e(\bar{\alpha}(k)) > 0$ is, on this conception, to say that $\Phi(\gamma) = e(\alpha)$ for every $\gamma\in\bar{\alpha}(k)$ such that $s_i^{(\gamma)}$ is the universal spread for each $i < k$. If $e(\bar{\alpha}(k)) > 0$, then, for any $m > k$, there exists $\gamma\in\bar{\alpha}(m)$ such that $s_i^{(\gamma)}$ is the universal spread for $i < k$ and $s_i^{(\gamma)} = s_i^{(\alpha)}$ for $k \leq i < m$; but, unless $s_{m-1}^{(\alpha)}$ happens to be such that, for all β in $s_{m-1}^{(\alpha)}$, if $\beta\in\bar{\alpha}(m)$, then $\beta = \alpha$, i.e. unless α is a lawlike sequence, this will not guarantee that $\gamma = \alpha$, and so we cannot infer, from the fact that $\Phi(\gamma) = e(\alpha)$, that $A(\alpha, e(\alpha))$.

The argument is, however, unnecessarily complicated. We have so far glossed over the reasoning required for the justification of ($\underline{E}$-3"). Troelstra's argument for it runs as follows. If we know that $A(\alpha)$, then we must know it from some finite number of pairs $\langle \underline{R}_0^{(\alpha)}, \alpha(0)\rangle, \ldots, \langle \underline{R}_{m-1}^{(\alpha)}, \alpha(m-1)\rangle$ in the process of generating α. Suppose that $\underline{R}_{m-1}^{(\alpha)}$ consists of

$$\alpha = \Gamma_o(\alpha_o)$$

$$\alpha_o = \Gamma_1(\alpha_1)$$

$$\ldots\ldots\ldots\ldots\ldots$$

$$\alpha_{j-1} = \Gamma_j(\alpha_j)$$

(we here revert to Troelstra's notation under which α_o is not α itself, but the first of the subordinate or auxiliary choice sequences). Let us put $\Gamma = \lambda\beta.\ \Gamma_o(\Gamma_1(\Gamma_2\ldots(\Gamma_j(\beta))\ldots))$. Suppose that, at stage m - 1, no terms of α_j have as yet been chosen. (The terms $\alpha(0), \ldots, \alpha(m-1)$ will, of course, have been either chosen or determined; and, in determining them, we may have chosen or determined some terms of $\alpha_o, \ldots, \alpha_{j-1}$. But we can ignore this fact, because the constraints which we imposed on the choice of the $\Gamma_o, \ldots, \Gamma_j$ were precisely such as to ensure that $\alpha, \alpha_o, \ldots, \alpha_{j-1}$ had those values which had already been fixed for them. In particular, the fact that $\alpha \in \Gamma$, i.e. that $\exists\beta\ \alpha = \Gamma(\beta)$, guarantees that $\alpha(0), \ldots, \alpha(m-1)$ have the required values.) The crucial step in the argument is now this: (+) *we know that* $A(\alpha)$ *just from the fact that* $\alpha = \Gamma(\alpha_j)$. If (+) is accepted, the argument concludes as follows. At this stage, α_j is completely unrestricted. Hence the generating process is open to any continuation, which could determine α_j to be any choice sequence whatever: and we know that $A(\alpha)$, i.e. $A(\Gamma(\alpha_j))$, without knowing which continuation will be adopted. It follows that $\forall\beta\ A(\Gamma(\beta))$, i.e. $\forall\gamma_{\gamma \in \Gamma} A(\gamma)$.

The case in which, at stage m - 1, we have already determined some values of α_j, say those of $\alpha_j(0), \ldots, \alpha_j(q-1)$, involves only a minor complication to the argument. Let us put:

$$(\Gamma_{j+1}(\beta))(i) = \begin{cases} \alpha_j(i) & \text{if } i < q \\ \beta(i) & \text{if } i \geq q. \end{cases}$$

Then take δ as determined by a generating process with the initial segment $\langle \underline{R}_0^{(\delta)}, \delta(0)\rangle, \ldots, \langle \underline{R}_{m-1}^{(\delta)}, \delta(m-1)\rangle$, where (i) for each $i < m$, $\delta(i) = \alpha(i)$, (ii) for each $i < m - 1$, if $\underline{R}_i^{(\alpha)}$ is

$$\begin{aligned} \alpha &= \Gamma_0(\alpha_0) \\ \alpha_0 &= \Gamma_1(\alpha_1) \\ &\ldots\ldots\ldots\ldots\ldots\ldots \\ \alpha_{j_i-1} &= \Gamma_{j_i}(\alpha_{j_i}), \end{aligned}$$

then $\underline{R}_i^{(\delta)}$ is

$$\begin{aligned} \delta &= \Gamma_0(\delta_0) \\ \delta_0 &= \Gamma_1(\delta_1) \\ &\ldots\ldots\ldots\ldots\ldots\ldots \\ \delta_{j_i-1} &= \Gamma_{j_i}(\delta_{j_i}), \end{aligned}$$

and (iii) $\underline{R}_{m-1}^{(\delta)}$ is

$$\begin{aligned} \delta &= \Gamma_0(\delta_0) \\ \delta_0 &= \Gamma_1(\delta_1) \\ &\ldots\ldots\ldots\ldots\ldots\ldots \\ \delta_{j-1} &= \Gamma_j(\delta_j) \\ \delta_j &= \Gamma_{j+1}(\delta_{j+1}). \end{aligned}$$

Take $\Gamma' = \lambda\beta \,.\, \Gamma_0(\Gamma_1(\Gamma_2 \ldots(\Gamma_{j+1}(\beta))\ldots))$. Now to claim that we know that $A(\alpha)$ from $\langle \underline{R}_0^{(\alpha)}, \alpha(0)\rangle, \ldots, \langle \underline{R}_{m-1}^{(\alpha)}, \alpha(m-1)\rangle$ is tantamount to saying that $\langle \underline{R}_0^{(\delta)}, \delta(0)\rangle, \ldots, \langle \underline{R}_{m-1}^{(\delta)}, \delta(m-1)\rangle$

entitles us to assert that $A(\delta)$; and, by the same token, *we must know that* $A(\delta)$ *just from the fact that* $\delta = \Gamma'(\delta_{j+1})$, where δ_{j+1} is as yet completely undetermined. Hence we have $\forall\beta\, A(\Gamma'(\beta))$, i.e. $\forall\gamma_{\gamma\in\Gamma'}A(\gamma)$.

This argument for ($\underline{E}$-3") needs a minor restatement to accord with our distinction between a choice sequence that is, relative to a given generating process, subordinate and one that is auxiliary. Let us say that a generating process ϕ is *autonomous* if no condition $\underline{R}_n^{(\phi)}$ contains a reference to an auxiliary choice sequence. Then if the process ϕ_α which generates α is autonomous, the argument proceeds as before. If, however, ϕ_α is not autonomous, it no longer holds that our knowledge that $A(\alpha)$ must be derived from an initial segment $\overline{\phi_\alpha}(m)$ of ϕ_α alone, since $\underline{R}_{m-1}^{(\phi_\alpha)}$ might contain the condition $(\phi_\alpha)_i = \Gamma_i(\beta)$, where β is auxiliary to ϕ_α, and then the truth of $A(\alpha)$ might in part depend on some initial segment $\overline{\phi_\beta}(n)$ of the process ϕ_β generating β. (Of course, in the general case, where we are considering continuous operations with several arguments, $\underline{R}_{m-1}^{(\phi_\alpha)}$ might refer to more than one auxiliary choice sequence.) If, in turn, ϕ_β is not autonomous, but relates to an auxiliary choice sequence γ, γ may be said to be a second-order auxiliary of ϕ_α, and the truth of $A(\alpha)$ may depend in part on some initial segment $\overline{\phi_\gamma}(p)$ of ϕ_γ; and so on. It is, however, clear that these complications make no essential difference to the argument: our knowledge that $A(\alpha)$ must derive from *some* finite amount of information, representable by some $\overline{\phi_\alpha}(m)$ together with initial segments $\overline{\phi_{\beta_i}}(n_i)$ of the generating processes ϕ_{β_i} of finitely many auxiliaries β_i of finite order; in relation to these, taken together, the argument may run exactly as before.

As remarked, the crucial step in the argument is that labelled (+). Generalized, this amounts to the following. Let us suppose that a choice sequence α is generated by a process ϕ, at each stage n of which we both determine $\alpha(n)$ and impose some restriction $\underline{R}_n^{(\phi)}$ on subsequent choices of terms of α. Suppose that $A(\alpha)$, and that, accordingly, we know that $A(\alpha)$ from some initial segment $\bar{\phi}(m)$ of ϕ, that is from $\langle \underline{R}_0^{(\phi)}, \alpha(0)\rangle, \ldots, \langle \underline{R}_{m-1}^{(\phi)}, \alpha(m-1)\rangle$ (where $\underline{R}_0^{(\phi)} \subseteq \underline{R}_1^{(\phi)} \subseteq \ldots \subseteq \underline{R}_{m-1}^{(\phi)}$). Then (++) *if* $A(\alpha)$ *is extensional, our knowledge of the truth of* $A(\alpha)$ *depends, not on the particular stages* i *at which the various conditions* $\underline{R}_i^{(\phi)}$ $(i < m)$ *were imposed, but on the mere fact that they hold* (or, equivalently, on the mere fact that $\underline{R}_{m-1}^{(\phi)}$ holds), *together with the initial segment* $\bar{\alpha}(m)$ *of* α. That is, where the condition $\underline{R}_{m-1}^{(\phi)}$ says that a certain statement $C(\alpha)$ holds of α, then we know that $\forall\gamma_{\gamma\in\bar{\alpha}(m)}\,(C(\gamma) \to A(\gamma))$.

Let us now apply these considerations directly to the argument for the Continuity Principle. Assuming that $A(\alpha, n)$ is extensional and that $\forall\alpha\, A(\alpha, \Phi(\alpha))$, we suppose $e \in \underline{K}$ such that, for all lawless β, $\Phi(\beta) = e(\beta)$. Suppose also that $e(\bar{\alpha}(k)) = n + 1$. Earlier, we argued that this meant that

$$\forall\gamma_{\gamma\in\bar{\alpha}(k)}\,(\exists\beta_{\text{lawless}(\beta)}\,\overline{\phi_\beta}(k) = \overline{\phi_\gamma}(k) \to \Phi(\gamma) = n),$$

whence

$$\forall\gamma_{\gamma\in\bar{\alpha}(k)}\,(\exists\beta_{\text{lawless}(\beta)}\,\overline{\phi_\beta}(k) = \overline{\phi_\gamma}(k) \to \forall\delta(\gamma = \delta \to A(\delta, n))).$$

Now, however, we can argue as follows. If β is lawless and $\beta \in \bar{\alpha}(k)$, then $A(\beta, n)$. We must know the truth of $A(\beta, n)$ from $\overline{\phi_\beta}(k)$. To say that β is lawless is, however, equivalent to saying that $\underline{R}_m^{(\phi_\beta)}$ is empty for every m. By the principle (++), our knowledge that $A(\beta, n)$ must be derived from the fact that $\beta \in \bar{\alpha}(k)$ together with the fact that the conditions in $\underline{R}_{k-1}^{(\phi_\beta)}$ hold. Since, however, $\underline{R}_{k-1}^{(\phi_\beta)}$ is empty, it follows just from the fact that $\beta \in \bar{\alpha}(k)$; in other words, we may assert:

$$\forall\gamma_{\gamma\in\bar{\alpha}(k)}\,A(\gamma, n)$$

from which, since $e(\alpha) = n$, $A(\alpha, e(\alpha))$ follows at once.

That this is a *different* argument for the Continuity

Principle from that given by Troelstra for it can be seen from the fact that it works equally well for the class $\underline{D}$ of choice sequences, which, as we saw, the original argument does not. For, where β is in $\underline{D}$, to say that β is lawless is equivalent to saying that $s_m^{(\beta)}$ is the universal spread for every m. As before, where $e(\bar{\alpha}(k)) = n + 1$, if β is lawless and $\beta \in \bar{\alpha}(k)$, then $A(\beta, n)$, and the truth of $A(\beta, n)$ must be derived from $\langle s_0^{(\beta)}, \beta(0)\rangle, \ldots , \langle s_{k-1}^{(\beta)}, \beta(k - 1)\rangle$. But now we can argue, on the lines of (++), that $A(\beta, n)$ must be known from the fact that $\beta \in \bar{\alpha}(k)$ and that $\beta \in s_{k-1}^{(\beta)}$. Indeed, just this was the intuitive justification we offered above for ($\underline{D}$-3).) But, in the present instance, $s_{k-1}^{(\beta)}$ is the universal spread, and so we have:

$$\forall\gamma_{\gamma\in\bar{\alpha}(k)}\ A(\gamma, n).$$

This argument proceeds without an overt appeal to the assumption (J). It may possibly be felt that the principle (++) in effect relies on an intuitive acknowledgement of (J); but it certainly gives us no direct reason to assert (J): instead , it considerably simplifies the reasoning required to establish the $\forall\alpha\ \exists n$-Continuity Principle.

Is (++) plausible? The principle of analytic data ($\underline{E}$-3") is unlike the $\forall\alpha\ \exists n$-Continuity Principle in not being something that we can demand of any reasonable notion of choice sequences. We have, indeed, to find expression for the distinctive feature of choice sequences, considered as the intuitionistic version of infinite sequences, namely that anything that can be asserted of a choice sequence can be asserted of it on the basis of a finite amount of information about it; and certainly the Continuity Principle does not embody the full content of an ascription to them of this feature. It might be held, however, that, whatever conception we may eventually adopt of the process of generation ϕ_α of a choice sequence α, the feature is fully expressed by

$$\text{(K)}\qquad A(\alpha) \rightarrow \exists k\ \forall\gamma\ (\overline{\phi_\gamma}(k) = \overline{\phi_\alpha}(k) \rightarrow A(\gamma)),$$

where $A(\alpha)$ is not required to be extensional. Against this, it could be urged that the Continuity Principle provides, by analogy, a ground for holding that, whenever $A(\alpha)$ *is* extensional, we must be able to strengthen (K) to some suitable 'data principle', saying that the truth of $A(\alpha)$ depends on a finite amount of purely *extensional* information about α; just what form this information may be required to take will depend on our conception of the generating process ϕ_α. For $A(\alpha, n)$ not necessarily extensional, our assumptions (D) and (E), taken together with the Axiom of Choice (B), provide a kind of intensional Continuity Principle. If it is not reasonable to require that, when $A(\alpha, n)$ is extensional, $\Phi(\alpha)$ shall be determined from a finite amount of *extensional* information about α, so that Φ may be taken as a continuous operation Γ, then the whole existing development of intuitionistic analysis is in error. If it *is* reasonable, then it must be equally reasonable to make the parallel demand in the case of an extensional statement $A(\alpha)$.

Troelstra's proposed notion of choice sequences, at least when subjected to a small reformulation, thus appears to meet all plausible requirements that might be made of such a notion. It is true that we cannot construct a similar argument to justify the $\forall\alpha\ \exists\beta$-Continuity Principle: but this is no defect, since there is no intuitive reason to suppose it true. We need it, indeed, if we are to carry out the 'elimination of choice sequences'; but, while it was argued earlier that it is not shocking if choice sequences are eliminable, it can certainly not be demanded that they should be. Troelstra observes that the counter-example to $\forall\alpha\ \exists\beta$-continuity arising from Kripke's schema is inconclusive, since it invokes a generating process of a quite special kind; more exactly, it assumes that there is a process ϕ_β generating a choice sequence β in which an already given choice sequence α plays the role of an auxiliary of a quite different sort from the auxiliary choice sequences appearing in our conditions $\underline{R}_n^{(\phi)}$. It hardly helps, however, to say, as he does, that the $\forall\alpha\ \exists\beta$-Continuity Principle is to be viewed 'as imposing a certain restriction on the intended interpretation of the quantifier combination $\forall\alpha\ \exists\beta$ which is not ... already implicit in the intended meaning of this quantifier combination for choice sequences' (*Choice Sequences*, pp. 154-5). The intended meanings of the intuitionistic quantifiers are not in doubt, and we have no business to be imposing additional restrictions upon them. All that can be in doubt is the correct characterization of some particular domain of quantification, and of the way in which its elements are given to us. If we find a means of characterizing the domain of choice sequences so as to yield a justification of the $\forall\alpha\ \exists\beta$-Continuity Principle under the known meanings of $\forall$ and $\exists$, then we shall be entitled to assert it: until then, we are not.

BIBLIOGRAPHY

1. Books

BETH, E. W. *The Foundations of Mathematics*, Amsterdam, 1959.

A long work, into which the author put almost everything that interested him; the material relevant to intuitionism is contained in Chapter 15, pp. 409-63, which includes a presentation of Beth trees and the method of semantic tableaux.

BISHOP, E. *Foundations of Constructive Analysis*, New York, 1967.

The author's standpoint is that of a constructivism more restrictive than that of the intuitionists, but the book should be of interest to anyone concerned with intuitionistic mathematics; some of the proofs are very compressed.

BROUWER, L. E. J. *Collected Works*, vol. I; *Philosophy and Foundations of Mathematics*, ed. A. Heyting, Amsterdam, 1975.

This contains, some in English, some in German, and a very few in French, all those of Brouwer's writings that are devoted to intuitionism, and would be indispensable to all serious students of the subject, save that its price has been set so high that few will be able to afford it.

FITTING, M. C. *Intuitionistic Logic, Model Theory and Forcing*, Amsterdam, 1969.

Treats of Kripke trees and their relation to the Cohen independence proofs for set theory.

HEYTING, A. *Intuitionism: an Introduction*, 3rd edn., Amsterdam, 1972.

An excellent introduction to the practice of intuitionistic mathematics that skates rather lightly over logical and foundational matters and contains some interesting but sketchy philosophical material.

———— *Mathematische Grundlagen forschung, Intuitionismus, Beweis theorie*, vol. 3, no. 4 of series *Ergebuisee der Mathematik und ihrer Grenzgebiete*, Berlin, 1934.

A short book now rather out of date, of historical interest as containing an early statement by Heyting of the meanings of the intuitionistic logical constants.

KINO, A., MYHILL, J., and VESLEY, R. E. (eds.). *Intuitionism and Proof Theory*, Amsterdam, 1970.

Although this is a volume of the proceedings of a conference, it contains so many valuable articles on intuitionism that it seemed worth while to list it in this section.

KLEENE, S. C. and VESLEY, R. E. *The Foundations of Intuitionistic Mathematics*, Amsterdam, 1965.

This book broke new ground with a systematic presentation of an axiomatization of intuitionistic analysis; it contains a detailed discussion of realizability. Except in rare and illuminating expository passages, it is very heavy going because of the formalized character of the proofs.

MATHIAS, A. R. D., and ROGERS, H. (eds.). *Cambridge Summer School in Mathematical Logic*, no. 337 of series *Lecture Notes in Mathematics*, Berlin, 1973.

This is another volume of conference proceedings containing many papers relating to intuitionism.

PRAWITZ, D. *Natural Deduction, a Proof-Theoretical Study*, Uppsala, 1965.

A luminously lucid account of normalization.

RASIOWA, H. and SIKORSKI, R. *The Mathematics of Metamathematics*, Warsaw, 1963.

Deals in detail with the topological interpretation of intuitionistic logic, but not from a semantic standpoint.

TROELSTRA, A. *Principles of Intuitionism*, no. 95 in series *Lecture Notes in Mathematics*, Berlin, Heidelberg, and New York, 1969.

An admirable account of basic principles, of a difficulty one grade higher than that of the present book; it treats only glancingly of logic.

——— *Choice Sequences*, in the present series, Oxford 1976.

A detailed discussion of the philosophical and mathematical analysis of the notion.

——— (ed.). *Metamathematical Investigation of Intuitionistic Arithmetic and Analysis*, no. 344 in series *Lecture Notes in Mathematics*, Berlin, Heidelberg, and New York, 1973.

An exhaustive encyclopedia of metamathematical results.

2. Articles

Abbreviations

Ann. Math.	*Annals of Mathematics*
CiM	*Constructivity in Mathematics*, ed. A. Heyting, Amsterdam, 1959.
Compositio math.	*Compositio Mathematica.*
CSS	*Cambridge Summer School in Mathematical Logic*, ed. A. R. D. Mathias and H. Rogers, Berlin, 1973.
Ergebn. math. Kolloq.	*Ergebnisse eines mathematischen Kolloquiums.*
FG	*From Frege to Gödel*, ed. J. van Heijenoort, Harvard University Press, 1967.
Indag. math	*Indagationes Mathematicae.*
IPT	*Intuitionism and Proof Theory*, ed. A. Kino, J. Myhill, and R. E. Vesley, Amsterdam, 1970.
Jber. dt. MatVerein	*Jahresbericht der Deutschen Mathematiker-Vereinigung.*
J. Symb. Log.	*Journal of Symbolic Logic.*
KNAW	*Koninklijke Nederlandse Akademie van Wetenschappen, Proceedings of the Section of Sciences.*
*LMPS*1	*Logic, Methodology and Philosophy of Science*, ed. E. Nagel, P. Suppes, and A. Tarski, Stanford, Calif., 1962.
*LMPS*3	*Logic, Methodology and Philosophy of Science*, vol. 3, ed. B. van Rootselaar and J. F. Staal, Amsterdam, 1968.
Math. Annln	*Mathematische Annalen.*
MKNAW	*Mededelingen der Koninklijke Nederlandse Akademie van Wetenschappen, Afd. Letterkunde.*
PM	*Philosophy of Mathematics: Selected Readings*, ed. P. Benacerraf and H. Putnam, Englewood Cliffs., N.J., 1964.

BETH, E. W. 'Semantical Considerations on Intuitionistic Mathematics', *KNAW*, vol. 50, 1947, pp. 1246-51 (= *Indag. math.*, vol. 9, 1947, pp. 572-7).

——— 'Semantic Entailment and Formal Derivability', *MKNAW*, n.s. vol. 18, no. 13, 1955, pp. 309-42.

——— 'Semantic Construction of Intuitionistic Logic', *MKNAW*, n.s. vol. 19, no. 11, 1956, pp. 357-88.

——— 'Remarks on Intuitionistic Logic', in *CiM*, pp. 15-25.

BROUWER, L. E. J. *Intuitionisme en Formalisme*, Amsterdam, 1912; also in *Wiskundig Tijdschrift* , vol. 9, 1913, pp. 180-211. English translation in *Bulletin of the American Mathematical Society*, vol. 20, 1913, pp. 81-96, reprinted in *PM*, 1964, pp. 66-77.

——— 'Begründung der Mengenlehre unabhängig vom logischen Satz vom ausgeschlossenen Dritten. Erster Teil: Allgemeine Mengenlehre', *KNAW*, 1e sectie 12 no. 5, 1918.

——— 'Begründung der Mengenlehre unabhängig vom logischen Satz vom ausgeschlossenen Dritten. Zweiter Teil: Theorie der Punktmengen', *KNAW*, 1e sectie 12 no. 7, 1919.

——— 'Intuitionistische Mengenlehre', *Jber. dt. MatVerein*, vol. 28, 1919, pp. 203-8.

——— 'Besitzt jede reelle Zahl eine Dezimalbruch-Entwickelung?', *Math. Annln*, vol. 83, 1921, pp. 201-10.

——— 'Begründung der Funktionenlehre unabhängig vom logischen Satz vom ausgeschlossenen Dritten. Erster Teil: Stetigkeit, Messbarkeit, Derivierbarkeit', *KNAW*, 1e sectie 13 no. 2, 1923.

——— 'Über die Bedeutung des Satzes vom ausgeschlossenen Dritten in der Mathematik, insbesondere in der Funktionentheorie', *Journal für die reine und angewandte Mathematik*, vol. 154, 1924, pp. 1-7. English translation in *FG*, 1967, pp. 334-41. Addenda & corrigenda appear in *Indag. math.*, vol. 16, 1954, pp. 104-5 and 109-11, and in English translation, in *FG*, pp. 341-5.

——— 'Intuitionistische Zerlegung mathematischer Grundbegriffe', *Jber. dt. MatVerein*, vol. 33, 1924, pp. 251-6.

——— 'Beweis, dass jede volle Funktion gleichmässig stetig ist', *KNAW*, vol. 27, 1924, pp. 189-93, with 'Bemerkungen', ibid., pp. 644-6.

——— 'Zur intuitionistischen Zerlegung mathematischer Grundbegriffe', *Jber. dt. MatVerein*, vol. 36, 1927, pp. 127-9.

——— 'Intuitionistische Ergänzung des Fundamentalsatzes der Algebra', *KNAW*, vol 27, 1924, pp. 631-4.

——— 'Zur Begründung der intuitionistischen Mathematik', *Math. Annln*, Part I, vol. 93, 1925, pp. 244-57; Part II, vol. 95, 1926, pp. 453-72; Part III, vol. 96, 1927, pp. 451-88.

——— 'Über Definitionsbereiche von Funktionen', *Math. Annln*, vol. 97, 1927, pp. 60-75. English translation of sections 1-3 in *FG*, pp. 446-63.

——— 'Intuitionistische Betrachtungen über den Formalismus', *Sitzungsberichte der Preussischen Akademie der Wissenschaften, Physikalisch-mathematische Klasse*, 1928, pp. 48-52. English translation of section 1 in *FG*, pp. 490-2.

——— 'Mathematik, Wissenschaft und Sprache', *Monatshefte für Mathematik und Physik*, vol. 36, 1929, pp. 153-64.

——— 'Consciousness, Philosophy and Mathematics', *Proceedings of the 10th International Congree of Philosophy*, Amsterdam, 1949, pp. 1235-49. Reprinted in part in *PM*, 1964, pp. 78-84.

——— 'Historical background, principles and methods of intuitionism', *South African Journal of Sciences*, vol. 49, 1952, pp. 139-46.

——— 'Points and Spaces', *Canadian Journal of Mathematics*, vol. 6, 1954, pp. 1-17.

——— 'An example of contradictority in classical theory of functions', *KNAW*, Series A. no. 57, 1954, pp. 204-5; also *Indag. math.*, vol. 16, 1954, pp. 204-6.

DALEN, D. van. 'A note on spread-cardinals', *Compositio math.*, vol. 20, 1968, pp. 21-8.

——— 'Lectures on Intuitionism', *CSS*, pp. 1-94.

——— 'A model for *HAS* - a topological interpretation of the theory of species of natural numbers', *Fundamenta Mathematicae*, vol. 82, 1974, pp. 167-74.

——— 'Choice Sequences in Beth Models', *Notes on Logic and Computer Science*, Locos 20, Oct. 1974, Univ. of Utrecht.

——— 'The Use of Kripke's Schema as a Reduction Principle', Preprint no. 11, Sept. 1975, Dept. of Maths, Univ. of Utrecht.

——— 'An Interpretation of Intuitionistic Analysis', Preprint no. 14, Oct. 1975, Dept. of Maths. Univ. of Utrecht.

——— and TROELSTRA, A. S. 'Projections of Lawless Sequences', *IPT*, pp. 163-86.

DUMMETT, M. A. E., and LEMMON, E. J. 'Modal Logics between S4 and S5', *Zeitschrift für mathematische Logik und Grundlagen der Mathematik*, vol. 5, 1959, pp. 250-64.

DYSON, V. H., and KREISEL, G. 'Analysis of Beth's Semantic Construction of Intuitionistic Logic', *Technical Report No. 3*, Applied Mathematics and Statistics Laboratories, Stanford University, Stanford, Calif. 27th Jan. 1961, Part II.

FRIEDMAN, H. 'Some Applications of Kleene's Methods for Intuitionistic Systems', in *CSS*, pp. 113-70.

GABBAY, D. 'The Decidability of the Kreisel-Putnam System', *J. Symb. Log.*, vol. 35, 1970, pp. 431-7.

GENTZEN, G. 'Untersuchungen über das logische Schliessen', *Mathematische Zeitschrift*, vol. 39, 1935, pp. 176-210, 405-31. English translation in G. Gentzen, *Collected Papers*, ed. M. E. Szabo, Amsterdam, 1969, pp. 68-131.

GLIVENKO, V. 'Sur la logique de M. Brouwer', *Acádemie Royale de Belgique, Bulletins de la classe des sciences*, Series 5, vol. 14, 1928, pp. 225-8.

——— 'Sur quelques points de la logique de M. Brouwer', *Acádemie Royale de Belgique, Bulletins de la classe des sciences*, Series 5, vol. 15, 1929, pp. 183-88.

GÖDEL, K. 'Zum intuitionistischen Aussagenkalkül', *Akademie der Wissenschaften in Wien, Mathematisch-naturwissenschaftliche Klasse*, vol. 69, 1932, pp. 65-6. Reprinted in *Ergebn. math. Kolloq.*, Heft 4 (for 1931-2), 1933, p. 40.

——— 'Zur intuitionistischen Arithmetik und Zahlentheorie', *Ergebn. math. Kolloq.*, Heft 4 (for 1931-2), 1933, pp. 34-8. English translation in *The Undecidable*, ed. M. Davis, New York, 1965, pp. 75-81; for corrections of the translation, see *J. Symb. Log.*, vol. 31, 1966, pp. 490-1.

——— 'Eine Interpretation des intuitionistischen Aussagenkalküls', *Ergebn. math. Kolloq.*, Heft 4 (for 1931-2), 1933, p. 39.

——— 'Über eine bisher noch nicht benützte Erweiterung des finiten Standpunktes', *Dialectica*, vol. 12, 1958, pp. 280-7.

GOODMAN, N. D. 'A Theory of Constructions Equivalent to Arithmetic', in *IPT*, pp. 101-20.

——— 'The Arithmetical Theory of Constructions', in *CSS*, pp. 274-98.

HARROP, R. 'On Disjunctions and Existential Statements in Intuitionistic Systems of Logic', *Math. Annln*, vol. 132, 1956, pp. 347-61.

——— 'On the Existence of Finite Models and Decision Procedures for Propositional Calculi', *Proceedings of the Cambridge Philosophical Society*, vol. 54, 1958, pp. 1-13.

——— 'Concerning Formulas of the types $A \to B \vee C$, $A \to (Ex)\, Bx$ in Intuitionistic Formal Systems', *J. Symb. Log.*, vol. 25, 1960, pp. 27-32.

HEYTING, A. 'Die formalen Regeln der intuitionistischen Logik', *Sitzungsberichte der Prenssischen Akademie der Wissenschaften, Physikalisch-Mathematische Klasse*, 1930, pp. 42-56.

——— 'Die formalen Regeln der intuitionistischen Mathematik', ibid, pp. 57-71, 158-69.

——— 'Sur la logique intuitionniste', *Academie Royale de Belgique, Bulletins de la Classe des Sciences*, 5e serie, no. 16, 1930, pp. 957-63.

——— 'Die intuitionistische Grundlegung der Mathematik', *Erkenntnis*, vol. 2, 1931, pp. 100-15. English translation in *PM*, pp. 42-9.

HOWARD, W. A. 'Functional Interpretation of Bar Induction by Bar Recursion', *Compositio math.*, vol. 20, 1968, pp. 107-24.

——— and KREISEL, G. 'Transfinite Induction and Bar Induction of Types zero and one, and the Role of Continuity in Intuitionistic Analysis', *J. Symb. Log.*, vol. 31, 1966, pp. 325-58.

JAŚKOWSKI, S. 'Recherches sur le système de la logique intuitionniste', *Actes du Congrès International de Philosophie Scientifique*, Part VI, *Philosophie des mathématiques*, Paris, 1936, pp. 58-61.

JONGH, D. H. J. de. 'Formulas of one propositional variable in intuitionistic arithmetic', Report 73-03, June 1973, Dept. of Maths., Univ. of Amsterdam.

——— 'A Characterization of the Intuitionistic Propositional Calculus', in *IPT*, pp. 211-17.

——— and SMORYNSKI, C. 'Kripke Models and the Theory of Species', Report 74-03, Apr. 1974, Dept. of Maths., Univ. of Amsterdam.

KLEENE, S. C. 'On the Interpretation of Intuitionistic Number Theory, *J. Symb. Log.*, vol. 10, 1945, pp. 109-24.

——— 'Recursive Functions and Intuitionistic Mathematics', *Proceedings of the International Congress of Mathematicians*, Cambridge, Mass., vol. 1, 1952, pp. 679-85.

——— 'Realizability', in *CiM*, pp. 285-9.

——— 'Countable Functionals', *CiM*, pp. 81-100.

——— 'Disjunction and Existence under Implication in Elementary Intuitionistic Formalisms', *J. Symb. Log.*, vol. 27, 1962, pp. 11-18, and vol. 28, 1963, pp. 154-6.

——— 'Constructive Functions in *The Foundations of Intuitionistic Mathematics*', in *LMPS*3, pp. 137-44.

——— 'Formalized Recursive Functionals and Formalized Realizability', *Memoirs of the American Mathematical Society*, no. 89, 1969.

——— 'Realizability: A Retrospective Survey', in *CSS*, pp. 95-112.

KOLMOGOROV, A. N. 'Zur Deutung der intuitionistischen Logik', Mathematische Zeitschrift, vol. 35, 1932, pp. 58-65.

KREISEL, G. 'Elementary Completeness Properties of Intuitionistic Logic with a Note on Negations of Prenex Formulae', *J. Symb. Log.*, vol. 23, 1958, pp. 317-30.

——— 'A Remark on Free Choice Sequences and the Topological Completeness Proofs', *J. Symb. Log.*, vol. 23, 1958, pp. 369-88.

——— 'Non-derivability of $\neg (x)\ A(x) \rightarrow (\exists x)\ \neg A(x)$, A(x) primitive recursive, in intuitionistic formal systems' (abstract), *J. Symb. Log.*, vol. 23, 1958, pp. 456-7.

——— 'On Weak Completeness of Intuitionistic Predicate Logic', *Technical Report No. 3*, Applied Mathematics and Statistics Laboratories, Stanford University, Stanford, Calif., 27 Jan. 1961, Part I.

——— 'Note on Completeness and Definability', *Technical Report No. 3*, Applied Mathematics and Statistics Laboratories, Stanford University, Stanford, Calif., 27 Jan. 1961, Part III.

——— 'On Weak Completeness of Intuitionistic Predicate Logic', *J. Symb. Log.*, vol. 27, 1962, pp. 139-58.

——— 'Foundations of Intuitionistic Logic', in *LMPS*1, pp. 198-210, Stanford, 1962.

——— 'Mathematical Logic', in *Lectures on Modern Mathematics*, vol. 3, ed. T. L. Saaty, New York, 1965, pp. 95-195.

——— 'Lawless Sequences of Natural Numbers', *Composito math.*, vol. 20, 1968, pp. 222-48.

——— 'Functions, Ordinals, Species', in *LMPS*3, pp. 145-59.

——— 'Church's Thesis: a kind of Reducibility Axiom for Constructive Mathematics', in *IPT*, pp. 121-50.

——— and PUTNAM, H. 'Eine Unableitbarkeitsbeweismethode für den intuitionistischen Aussagenkalkül', *Archiv für mathematische Logik und Grundlagenforschung*, vol. 3, 1957, pp. 74-8.

——— and TROELSTRA, A. S. 'Formal Systems for some Branches of Intuitionistic Analysis', *Annals of Mathematical Logic*, vol. 1, 1970, pp. 229-387.

KRIPKE, S. 'Semantical Analysis of Intuitionistic Logic I', in *Formal Systems and Recursive Functions*, ed. J. N. Crossley and M. A. E. Dummett, Amsterdam, 1965, pp. 92-130.

KURODA, S. 'Intuitionistische Untersuchungen der formalistichen Logik', *Nagoya Mathematical Journal*, vol. 2, 1951, pp. 35-47.

ŁUKASIEWICZ, J. 'On the Intuitionistic Theory of Deduction', *KNAW*, Series A, vol. 55, 1952, pp. 202-12.

McKINSEY, J. C. C., and TARSKI, A. 'The Algebra of Topology', *Ann. Math.*, vol. 45, 1944, pp. 141-91.

——— 'On Closed Elements in Closure Algebras', *Ann. Math.*, vol. 47, 1946, pp. 122-62.

——— 'Some Theorems about the Sentential Calculi of Lewis and Heyting', *J. Symb. Log.*, vol. 13, 1948, pp. 1-15.

MOSCHOVAKIS, J. R. 'Disjunction and Existence in Formalized Intuitionistic Analysis', *Sets, Models, and Recursion Theory*, ed. J. R. Crossley, Amsterdam, 1967, pp. 309-31.

——— 'A Topological Interpretation of Second-Order Intuitionistic Arithmetic', *Composito math.*, vol. 26, 1973, pp. 261-76.

MYHILL, J. 'Notes towards an Axiomatisation of Intuitionistic Analysis', *Logique et Analyse*, vol. 35, 1967, pp. 280-97.

——— 'Formal Systems of Intuitionistic Analysis I' in *LMPS*3, pp. 161-78.

——— 'Formal Systems of Intuitionistic Analysis II: the Theory of Species' in *IPT*, pp. 151-62.

NELSON, D. 'Recursive Functions and Intuitionistic Number Theory', *Transactions of the American Mathematical Society*, vol. 61, 1947, pp. 307-68.

NISHIMURA, I. 'On Formulas in one Variable in Intuitionistic Propositional Calculus', *J. Symb. Log.*, vol. 25, 1960, pp. 327-51.

ROSE, G. F. 'Propositional Calculus and Realizability', *Transactions of the American Mathematical Society*, vol. 75, 1953, pp. 1-19.

SCOTT, D. S. 'Completeness Proofs for the Intuitionistic Sentential Calculus' in *Summaries of Talks presented at the Summer Institute of Symbolic Logic*, Cornell University Press, 1957.

——— 'Extending the Topological Interpretation to Intuitionistic Analysis', *Composito math.*, vol. 20, 1968, pp. 194-210.

——— 'Extending the Topological Interpretation to Intuitionistic Analysis II' in *IPT*, Amsterdam, 1970, pp. 235-55.

SHOESMITH, D. J. and SMILEY, T. J. 'Deducibility and Many-valuedness', *J. Symb. Log.*, vol. 36, 1971, pp. 610-22.

SMORYNSKI, C. A. 'Applications of Kripke Models', in *Metamathematical Investigation of Intuitionistic Arithmetic and Analysis*, ed. A. S. Troelstra, Berlin, Heidelberg, and New York, 1973, pp. 324-91.

SWART, H. de. 'Another Intuitionistic Completeness Proof', Mathematisch Instituut, Katholieke Universiteit, Nijmegen, 1974.

——— 'Intuitionistic Logic in an intuitionistic meta-language: Chapter I - An intuitionistically plausible interpretation of intuitionistic logic', Math. Inst., Kath. Univ. Nijmegen, 1974.

——— 'Intuitionistic Logic in an intuitionistic meta-language: Chapter II - Intuitionistic Model Theory', Math. Inst., Kath. Univ., Nijmegen 1974.

TROELSTRA, A. S. 'Finite and Infinite in Intuitionistic Mathematics', *Compositio math.* vol. 18, 1967, pp. 94-116.

——— 'The Theory of Choice Sequences' in *LMPS*3, pp. 201-23.

——— 'Informal Theory of Choice Sequences', *Studia Logica*, vol. 25, 1969, pp. 31-52.

——— 'Notes on the Intuitionistic Theory of Sequences', *Indag. math:* Part I in vol. 31, 1969, pp. 430-40; Part II in vol. 32, 1970, pp. 99-109; Part III in vol. 32, 1970, pp. 245-52.

——— 'Notions of Realizability for Intuitionistic Arithmetic and Intuitionistic Arithmetic in all Finite Types', in *Proceedings of the Second Scandinavian Logic Symposium*, ed. J. E. Fenstad, Oslo, 1971, pp. 369-405.

——— 'Notes on Intuitionistic Second Order Arithmetic', in *CSS*, pp. 171-205.

VELDMAN, W. 'An Intuitionistic Completeness Theorem for Intuitionistic Predicate Logic', May 1974, Mathematisch Instituut, Nijmegen.

INDEX

Pages where a notion is explained or defined are in bold type.